10	11	12
8B	1B	2B

13	14	15	16	17	18
3A	4A	5A	6A	7A	8A

(1 – 18 is IUPAC system, A,B designation is older U.S. System)

| | | | | | 2
He
4.0026
$1s^2$ |

Main Group (*p* block)

5 **B** 10.811 $2s^22p^1$	6 **C** 12.011 $2s^22p^2$	7 **N** 14.007 $2s^22p^3$	8 **O** 15.999 $2s^22p^4$	9 **F** 18.998 $2s^22p^5$	10 **Ne** 20.180 $2s^22p^6$
13 **Al** 26.982 $3s^23p^1$	14 **Si** 28.086 $3s^23p^2$	15 **P** 30.974 $3s^23p^3$	16 **S** 32.066 $3s^23p^4$	17 **Cl** 35.453 $3s^23p^5$	18 **Ar** 39.948 $3s^23p^6$

28 **Ni** 58.693 $4s^23d^8$	29 **Cu** 63.546 $4s^13d^{10}$	30 **Zn** 65.39 $4s^23d^{10}$	31 **Ga** 69.723 $4s^24p^1$	32 **Ge** 72.61 $4s^24p^2$	33 **As** 74.922 $4s^24p^3$	34 **Se** 78.96 $4s^24p^4$	35 **Br** 79.904 $4s^24p^5$	36 **Kr** 83.80 $4s^24p^6$
46 **Pd** 106.42 $4d^{10}$	47 **Ag** 107.87 $5s^14d^{10}$	48 **Cd** 112.41 $5s^24d^{10}$	49 **In** 114.82 $5s^25p^1$	50 **Sn** 118.71 $5s^25p^2$	51 **Sb** 121.76 $5s^25p^3$	52 **Te** 127.60 $5s^25p^4$	53 **I** 126.90 $5s^25p^5$	54 **Xe** 131.29 $5s^25p^6$
78 **Pt** 195.08 $6s^15d^9$	79 **Au** 196.97 $6s^15d^{10}$	80 **Hg** 200.59 $6s^25d^{10}$	81 **Tl** 204.38 $6s^26p^1$	82 **Pb** 207.2 $6s^26p^2$	83 **Bi** 208.98 $6s^26p^3$	84 **Po** (209) $6s^26p^4$	85 **At** (210) $6s^26p^5$	86 **Rn** (222) $6s^26p^6$
110 **Ds** (271) $7s^26d^8$	111	112	114			116		

A few very unstable nuclei each of elements # 111, 112, 114, and 116 have been created using high–energy nuclear accelerators

64 **Gd** 157.25 $6s^24f^75d^1$	65 **Tb** 158.93 $6s^24f^9$	66 **Dy** 162.50 $6s^24f^{10}$	67 **Ho** 164.93 $6s^24f^{11}$	68 **Er** 167.26 $6s^24f^{12}$	69 **Tm** 168.93 $6s^24f^{13}$	70 **Yb** 173.04 $6s^24f^{14}$
96 **Cm** (247) $7s^25f^76d^1$	97 **Bk** (247) $7s^25f^9$	98 **Cf** (251) $7s^25f^{10}$	99 **Es** (252) $7s^25f^{11}$	100 **Fm** (257) $7s^25f^{12}$	101 **Md** (258) $7s^25f^{13}$	102 **No** (259) $7s^25f^{14}$

STUDY GUIDE

Debbie Finocchio
University of San Diego

CHEMISTRY

Fourth Edition

JOHN OLMSTED
California State University, Fullerton

GREG WILLIAMS
University of Oregon

WILEY

JOHN WILEY & SONS, INC.

Cover Photos: red coffee beans on plant: Fernando Bueno/The Image Bank/Getty Images
micrograph of caffeine: Michael W. Davidson at Florida State University
roasted coffee beans: Nick Gunderson/Stone/Getty Images
decaffeination processing: Maximilian Stock LTD/Phototake

To order books or for customer service, please call 1-800-CALL-WILEY (225-5945).

ISBN 0-471-69837-7

Printed in the United States of America

10 9 8 7 6 5 4 3 2 1

Printed and bound by Malloy Lithographing, Inc.

This study guide is dedicated to
the many hard-working and determined students
I have encountered who continually inspire me
to strive to be a better teacher.

Acknowledgments

I wish to thank the following people:

- o David Shinn, for checking all of the questions in the previous edition of this guide and for providing suggestions for content clarity. His excellent attention to detail improved the quality of this document.

- o My editor, Jennifer Yee, for hiring me for this project, and for keeping me motivated with her positive feedback.

- o My husband, Al, for his daily guidance on such a large project, for acting as a sounding board for my ideas, and for his constant love and support.

- o My family and friends, for their patience and support.

- o My college mentor/advisor, Patricia S. Traylor, and my high school chemistry teacher, Robert Rockwell, for sparking my interest in chemistry.

Table of Contents

At the end of each chapter, you will find a Self-Test, an answer key for the chapter's Try Its, and an answer key for the Self-Test.

Chapter 1: The Science of Chemistry

Learning Objectives

In this chapter, you will learn how to:
- Differentiate between an element, a compound, a pure substance and a mixture.
- Predict some chemical and physical properties of a given element from its placement on the periodic table.
- Convert units of measurement.
- Report numbers in scientific notation using the correct number of significant figures.
- Set up simple mathematical problems using the relationship Density = mass/volume.
- Attack word problems using the 7-step method.

Practical Aspects

This chapter provides a detailed overview of the terms used to describe matter and the ways in which chemists approach quantitative (numerical) problems. Chemistry is a subject that uses "vertical learning" – meaning that the material you learn today will be applied to later chapters. Learn the material now, so that you can use it as a tool for later chapters. Drill yourself on converting numbers between regular and scientific notation, assessing significant figures, and unit conversions.

1.1   WHAT IS CHEMISTRY?

Key Terms:
- **Science** – attempt to organize and understand our observations of nature; the systematic study of matter and its surroundings.
- **Chemistry** – branch of science that studies the properties and interactions of matter; the study of changes in matter.
- **Hypothesis** – a testable guess; a proposed general principle.
- **Theory** – unifying principle that explains a collection of facts.

EXERCISE 1: List a few things in your daily life that involve chemistry.

>   *SOLUTION:* Here are just a few:
>   **Breathing** – your body takes in oxygen, uses it, and exhales carbon dioxide.
>   **Your senses** – taste, sight and smell depend upon specific chemicals sending messages to your brain. A given chemical's aroma depends upon the chemical's structure/shape.
>   **Clothing/fabrics** – polyester, cotton, nylon, silk, wool all behave differently because they have different chemical make-ups.
>   **Soaps and detergents** – the shapes of these chemicals make them attracted to water *and* oil.
>   **Sunscreens** – certain chemicals absorb UV light and are used to block it from your skin.
>   **Fuels** – different chemicals release different amounts of energy when they combust.
>   **Medications** – most drugs work by mimicking or blocking a given chemical's normal function in your body.

EXERCISE 2: One component of the Atomic Theory of Matter states that matter cannot be created or destroyed, only rearranged. List a case where a person could think that matter might be created or destroyed. List one test that could be done to support the theory.

> *SOLUTION:* Case: mold growing on a piece of bread – the mold wasn't there originally, so where did it come from? Was it created? A test that could be done would be to put a piece of bread in a container, seal the container, and weigh it. After a lot of mold has formed, re-weigh the container and notice that the mass did not change.

1.2   ATOMS, MOLECULES, AND COMPOUNDS

Matter can be viewed with varying levels of detail:
- **Macroscopic** – water, for example, is clear and colorless and wet.
- **Microscopic** – a sample of water might contain tiny organisms that one can't see without a microscope.
- **Molecular/Atomic** – water is "H_2O". Every molecule of water contains 2 H's and 1 O. Water is a liquid at room temperature because of strong attractions that exist between water molecules.

Key Terms:
- **Physical Property** – observable property of a substance (example: color, texture, density).
- **Chemical Property** – property that has to do with a substance's potential to change its chemical make-up (example: chlorine combines with sodium in a 1:1 ratio to make table salt).
- **Atom** – fundamental unit of a chemical substance ("atomos" means uncuttable in Greek).
- **Molecule** – combination of two or more atoms held together by attractive forces.
- **Element** – substance that contains only one *type* of atom (examples: C or H on the periodic table).
- **Compound** – combination of two or more *different* atoms held together by attractive forces.
- **Chemical Formula** – notation of the type and number of each element present in one unit of the substance (example: carbon dioxide's formula is CO_2, which designates 1 carbon and 2 oxygens per molecule).

Key Concepts:
- Atoms are *tiny*.
- Each element on the periodic table has a unique one- or two- letter symbol. The first letter is always capitalized and the second letter is lower case (example: H = hydrogen, He = helium).
- Elements can be monoatomic (Ne, C), diatomic (H_2, N_2, O_2), or polyatomic (S_8).
- Chemists color-code commonly used atoms. See Figure 1-4 in the text for a list.

EXERCISE 3: Identify each as an element, atom, molecule, and/or compound. More than one definition might apply to each. a) H_2O_2;   b) Cl_2;   c) CO;   d) He;   e) Na;   f) $C_6H_{12}O_6$

> *SOLUTION:* a) molecule, compound; b) molecule, element; c) molecule, compound; d) atom, element; e) atom, element; f) molecule, compound

EXERCISE 4: Determine the chemical formula of each:
a) a molecule that contains 7 atoms of hydrogen, 1 atom of nitrogen, and 4 atoms of carbon;
b) a chemical that contains three atoms of oxygen

For questions c and d, black = carbon, white = hydrogen, dark grey = oxygen.

c) 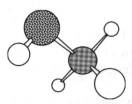   d)

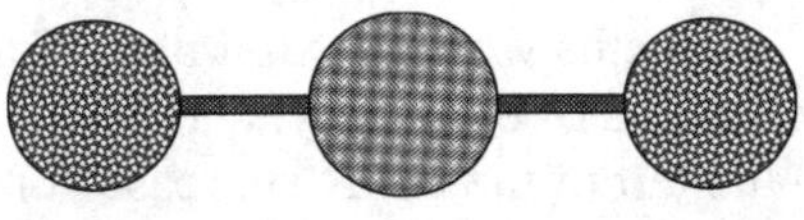

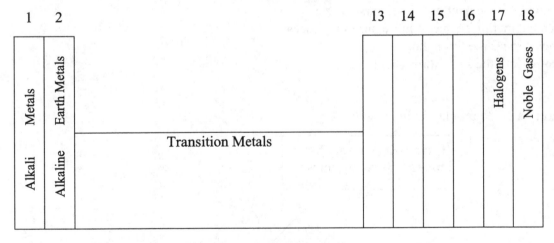

STRATEGY AND SOLUTION: Count up the number of each type of atom. (Don't worry about the order of the atoms at this point.) a) C_4H_7N b) O_3 c) CH_4O d) CO_2

1.3   THE PERIODIC TABLE OF THE ELEMENTS

Two ways to classify elements in the periodic table:

I.   Main Group vs. Transition Elements

- Main group elements are those elements in columns 1, 2, and 13-17.
- Transition elements are the "transition metals" and the "inner transition metals."

The outline of the periodic table shown here emphasizes the sections:

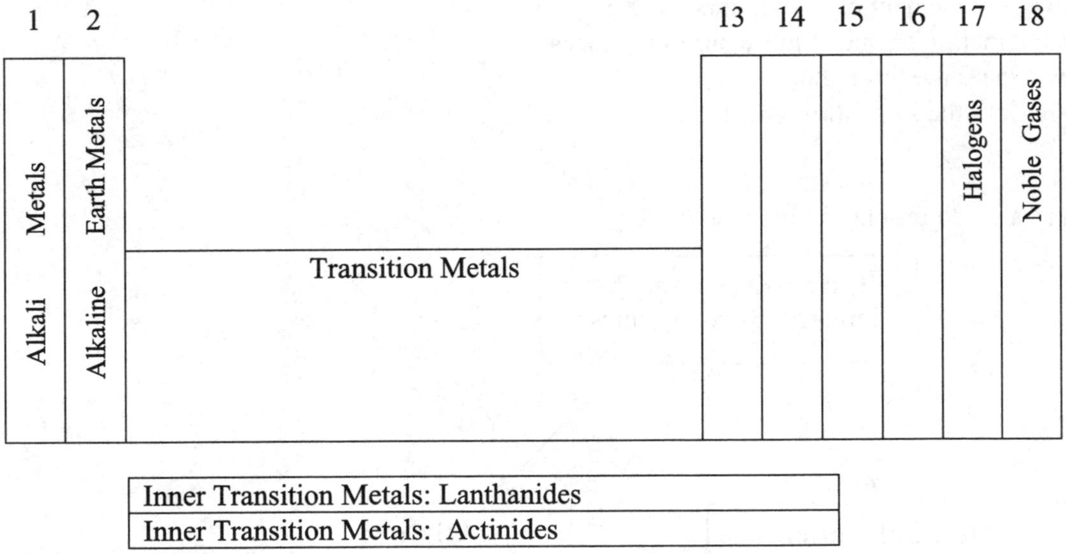

Key Concept:
- All elements within a given column have similar chemical properties and often have similar physical properties.

For example, Group I (column 1) elements, the Alkali Metals, share these features:
- Physical Properties: these elements are all soft, shiny solids at room temperature.
- Chemical Properties: these elements react violently with water; react in a 1:1 ratio with halogens; and react in a 2:1 ratio with oxygen.

II. Metals vs. Non-Metals

Key Terms: (refer to a periodic table for a visual aid)
- **Metals** – elements which exhibit metallic properties: they are usually solid at room temperature, are good conductors of electricity, and can be pounded into shapes with a hammer. These elements are found to the left of the bold "staircase" on a periodic table and include the inner transition metals. Most elements are classified as metals.
- **Non-metals** – elements which do not exhibit metallic properties. These elements are found to the right of the "staircase."
- **Metalloids** – elements which exhibit some metallic and some non-metallic characteristics. There are 6 metalloids (Si is the most abundant in nature). Note that metalloids border the "staircase."

EXERCISE 5: Give the symbol for an element which meets the requirement: a) Lithium (Li) reacts with it in a 2:1 ratio; b) it is a halogen; c) it is a metal that reacts violently with water; d) it is a noble gas; e) it exhibits both metal and non-metal behavior; it can be used to make a semiconductor.

SOLUTION: There will be more than one correct answer to each of the questions. One answer is provided here: a) O b) Br c) K d) Ar e) Si

1.4 CHARACTERISTICS OF MATTER

Key Terms:
- **Pure Substance** – an element or a compound.
- **Mixture** – a combination of one or more pure substances.
- **Homogeneous** – the same throughout.
- **Heterogeneous** – not the same throughout.

Helpful Hint
- Use this chart as a guide for classifying matter

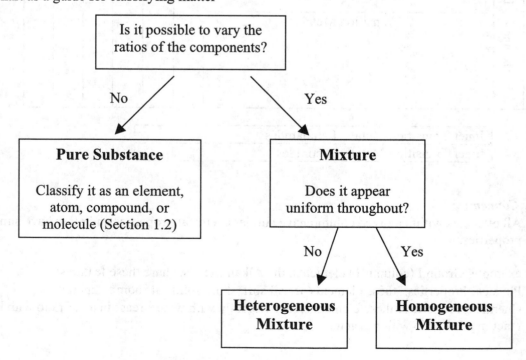

EXERCISE 6: Classify each as a pure substance, homogeneous mixture or heterogeneous mixture. a) H_2O_2; b) a chocolate chip cookie; c) a cup of tea.

STRATEGY AND SOLUTION: Use the above table as a guide to determine the proper classification. a) pure substance: it is not possible to vary the ratios of the components of this chemical; 2 H's and one O would be water, a different chemical. b) heterogeneous mixture – this is definitely a mixture because it is possible to vary the components in it (more sugar, less flour, more chocolate chips), and it is heterogeneous because one can see that it is not uniform throughout – one can detect chips versus dough. c) homogenous mixture – it appears the same throughout (assuming no tea leaves fell into it), but the proportions in it can be varied. One can make weak tea or strong tea.

Phases of Matter

Phase	Macroscopic View	Molecular/Atomic View
Solid	Definite volume and shape	Atoms/molecules are packed tightly together so that they can't move around each other
Liquid	Definite volume but no definite shape, fills container it is in from the bottom up	Atoms/molecules are packed together, but not as tightly as in solid phase. There are a few gaps here and there so molecules can roll over each other (hence "flow")
Gas	No definite shape or volume Uses up entire 3-dimensional space of container	Molecules are far apart and moving rapidly

Key Terms:
- **Physical transformation** – change in the physical properties of a substance.
- **Chemical transformation** – change in the chemical make-up of a substance; a change from one substance into another. (For example, water can be converted into hydrogen and oxygen).

Key Concept:
- If a change in chemical formula occurs, then the transformation is classified as a chemical transformation.

EXERCISE 7: Why is it that a chemical transformation, such as converting oxygen and hydrogen into water, might be misinterpreted as a physical transformation?

SOLUTION: When a chemical transformation occurs a new substance is formed, with its own unique set of chemical and physical properties. A chemical transformation therefore usually includes a change in appearance, so it is sometimes misconstrued as a physical change.

EXERCISE 8: Classify each as a chemical or physical transformation: a) iron rusting; b) rubbing alcohol evaporating on a person's skin; c) tearing a piece of paper; d) burning a piece of paper.

SOLUTION: a) chemical – iron is changed into a new chemical, iron oxide; b) physical – the chemical just changed phases; c) physical – paper still exists, just in smaller pieces; d) chemical – paper is converted into carbon dioxide, water, and ash.

EXERCISE 9: Use terms you've learned so far in this chapter to categorize the matter inside each container:

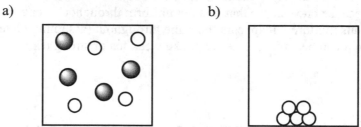

a)   b)

SOLUTION: a) a homogeneous mixture of two gases; b) a pure substance in the solid phase

Try It #1: Draw a molecular picture of a heterogeneous mixture of two liquids.

1.5   MEASUREMENTS IN CHEMISTRY

All measurements in chemistry should convey three pieces of information:
- **Magnitude** – the size of the measurement;
- **Unit** – an indicator of scale; and
- **Precision** – assessment of how carefully the number was measured.

Scientific Notation ("power of ten" notation)

<u>What</u>: Method of simplifying the writing of very large and very small quantities.
<u>When</u>: Any time you are working with potentially cumbersome numbers.
<u>How</u>: Find the number, which when multiplied by 10^x, will give you the number written in the "normal" style.

Examples:

"Normal"	Scientific Notation
0.00000534	5.34×10^{-6}
1,000,000,000	1×10^9
2,500,000	2.5×10^6
0.008767	8.767×10^{-3}

Helpful Hint
- Here's a shortcut: Notice that the exponent in the scientific notation correlates to the direction and number of places the decimal had to move to get to the "normal" number. For the number "2.5×10^6," the decimal must be moved 6 places to the right to get to 2,500,000 – similarly, 6 spaces to the right means 10^6 or 1,000,000. Practice this a few times to see the mathematical pattern.

EXERCISE 10: Convert these numbers to scientific notation: a) 4579; b) -0.05020; c) 3.8521. Convert these from scientific notation: d) 3.0033×10^4; e) 2.79×10^{-5}; f) 3×10^{11}.

    STRATEGY AND SOLUTION:

	Normal	Scientific Notation
a)	4579	4.579×10^3
b)	-0.05020	-5.020×10^{-2}
c)	3.8521	3.8521×10^0
d)	30033	3.0033×10^4
e)	0.0000279	2.79×10^{-5}
f)	300,000,000,000	3×10^{11}

EXERCISE 11: Arrange these numbers from smallest to largest:

$$11.7 \times 10^{-7} \qquad -1.48 \times 10^{-4} \qquad -2.17 \times 10^3 \qquad 3.19 \times 10^{-2}$$

    STRATEGY: Mentally arrange the numbers on a number line.

    SOLUTION: -2.17×10^3, -1.48×10^{-4}, 11.7×10^{-7}, 3.19×10^{-2}

Try It #2: Convert to scientific notation: a) 213,500,000   b) 0.0000000000450

Try It #3: Convert from scientific notation: a) 6.700×10^{-2} b) -1.20×10^7

Units

SI (Systeme International) Base Units:

Dimension	Unit	Symbol
Length	Meter	m
Mass	Kilogram	kg
Time	Second	s or sec
Temperature	Kelvin	K
Chemical Amount	Mole	mol
Electrical Current	Ampere	A
Luminous Intensity	Candela	cd

Key Term:
- **Derived Units** – units which are derived from combinations of base units.

    Examples: The dimension "volume" (length x width x height) has units of: m^3 or cm^3
                 The dimension "rate" has units of: m/sec
                 The dimension "density" has units of: g/mL
                 The dimension "molar mass" has units of: g/mol

Units and Conversions:
- $1\ cm^3 = 1\ cc = 1\ mL$

Prefixes in SI

Table 1-2 in the text lists several SI prefixes. Here are the most frequently encountered ones:

Power:	10^3	10^{-2}	10^{-3}	10^{-6}	10^{-9}
Prefix:	kilo-	centi-	milli-	micro-	nano-
Symbol:	k	c	m	μ	n

Helpful Hint
- The magnitude of a number can be simplified by using scientific notation or a prefix before the unit. For example, 756,000 g can be rewritten: 7.56×10^5 g or 756 kg.

Precision and Accuracy

Key Terms:
- **Precision** - 1) how carefully a measurement is made; or 2) how close the values within a set of measurements are to one another.
- **Accuracy** - how close a measurement is to the "true" value.

EXERCISE 12: Given that: "The true value is 5.74 m," write a data set containing 3 numbers that:
a) is accurate and precise; and b) is precise but not accurate

STRATEGY AND SOLUTION: Use the definitions of accuracy and precision to come up with an appropriate data set:
a) accurate and precise: 5.73 m, 5.74 m, 5.75 m (numbers are close to the true value *and* close to each other)
b) precise but not accurate: 6.55 m, 6.56 m, 6.55 m (numbers are close to each other but not to the true value)

Try It #4: Given that the "true" value is 5.74 m, create a data set of three numbers that are neither accurate nor precise.

Significant Figures ("sig figs" for short)

"Significant figures" is the term used to describe how many digits should be recorded in a numerical answer. The number of significant figures is dictated by how precisely (carefully) a given measurement is made. An everyday example illustrates the concept of significant figures:

If someone is asked, "what is the distance between your home and your workplace?", the person might respond, "Oh, around 30 miles." If pressed, the person may say, "I think it is 27 miles." Or the distance could be measured from the car's odometer to be 27.3 miles. In each case:

the measurement:	has this many sig figs:	and indicates:
30	1	an approximation
27	2	a more precise measurement
27.3	3	an even more precise measurement

Helpful Hint
- To determine how many sig figs are in a number, read the numerals from left to right and start counting sig figs with the first non-zero number. Zeros at the end of a number are significant only if it is obvious they were measured.

EXERCISE 13: Determine the number of sig figs in each number: a) 2060; b) 2060.; c) 3.30×10^{-5}; d) 890,000,000,000; e) 0.003040

SOLUTION:

The number	has this many sig figs:	Special note:
2060 (2.06×10^3)	3	The last zero only indicates decimal place (i.e. that the number is 2060 rather than 206).
2060. (2.060×10^3)	4	A decimal point indicates that the number was measured to the one's place.
3.30×10^{-5} (0.0000330)	3	The zero is significant; it didn't have to be written, so it must've been recorded to emphasize that the measurement was taken that carefully.
890,000,000,000 (8.9×10^{11})	2	All of the zeros indicate decimal place only.
0.003040 (3.040×10^{-3})	4	The last zero is significant.

Notice that scientific notation eliminates sig fig ambiguity.

Try It #5: How many sig figs are in these numbers: a) 420,600; b) 0.0002; c) 0.00020; d) 7.0×10^6 ?

1.6 CALCULATIONS IN CHEMISTRY

When performing calculations, the final answer for the calculation should reflect how carefully the data used within the calculation was measured.

Operations with Significant Figures

- **Multiplication/Division**: Answer can have no more sig figs than the fewest number of sig figs in the original numbers.
- **Addition/Subtraction**: Answer can have no more decimal places than the fewest places in the original numbers.

Helpful Hints
- Watch out for "exact" numbers, which are infinitely significant (Example: 1 dozen=12).
- Round off the final result to the correct number of significant figures.
- Rounding: if the number to round is ≥5, round up, if <5 then round down.

EXERCISE 14: Do the following mathematical operations and report each answer to the correct number of sig figs: a) 23.652 + 7.71; b) 55000 ÷ 5.00; c) $(6.7 \times 10^{-2})(3 \times 10^{-5})$; d) $(6.7 \times 10^{-2})(3.0 \times 10^{-5})$; e) $(6.7 \times 10^{-2})(3.00 \times 10^{-5})$; f) $(6.7 \times 10^{-2})+(3.00 \times 10^{-5})$

SOLUTION:

	Question	Numerical answer	To assign sig figs:	Answer reported to correct sig figs
a)	23.652 + 7.71	31.362	7.71: hundredth's place	31.36
b)	55000 ÷ 5.00	11000	55000 has 2 sig figs	1.1×10^{4}
c)	$(6.7 \times 10^{-2})(3 \times 10^{-5})$	2.01×10^{-6}	3×10^{-5} has 1 sig fig	2×10^{-6}
d)	$(6.7 \times 10^{-2})(3.0 \times 10^{-5})$	2.01×10^{-6}	Both #s have 2 sig figs	2.0×10^{-6}
e)	$(6.7 \times 10^{-2})(3.00 \times 10^{-5})$	2.01×10^{-6}	6.7×10^{-2} has 2 sig figs	2.0×10^{-6}
f)	$(6.7 \times 10^{-2})+(3.00 \times 10^{-5})$	0.0670300	.067: thousandth's place	0.067

Try It #6: Report each answer using the correct number of sig figs: a) 250,000 ÷ 250; b) 15956 - 72.7; c) $(7.92 \times 10^{-2})(-3.0 \times 10^{5})$

Unit Conversions

Always show all units and how they cancel out in conversions. Exercise 15 illustrates how to set up a calculation.

EXERCISE 15: An American traveler wishes to convert a $50 bill to Italian lire. The exchange rate is 280,000 lire per 1 dollar. How many lire will the traveler receive?

STRATEGY: Set up a conversion factor: 280,000 lire = 1 dollar.

$$50.00 \text{ dollars} \times \frac{280,000 \text{ lire}}{1 \text{ dollar}} = 14,000,000 \text{ lire}$$

SOLUTION: 14,000,000 lire. (2 sig figs in answer based on exchange rate data). The $50 bill and the $1 bill are exact numbers, so they are infinitely significant. They do not limit sig figs.

Notice in Exercise 15 that the given information (50 dollars) is written first and that the conversion factor is set up so that the units "dollars" will cancel and we'll be left with units of lire. If we were converting from lire to dollars, the conversion factor would be flipped:

$$14,000,000 \text{ lire} \times \frac{1 \text{ dollar}}{280,000 \text{ lire}} = 50 \text{ dollars}$$

Helpful Hints
- When converting units, start with what you're given then multiply that number by the appropriate conversion factor to get the units to cancel out.
- Always show units when setting up conversion calculations so that you can track what units you have.

EXERCISE 16: A customer places an Internet order for 3 shirts, 2 shorts, and 1 pair of shoes. The total weight of the order is 2650 g. Shipping charges are $5.25 for the first five pounds, and $1 per pound (or any portion thereof) for every pound after that. How much does the customer have to pay in shipping?

STRATEGY: Determine the shipment's mass in pounds.

$$2650 \text{g} \times \frac{1 \text{kg}}{1000 \text{g}} \times \frac{2.205 \text{ lb}}{1 \text{ kg}} = 5.84 \text{ lb}$$

SOLUTION: The shopper will have to pay $5.25 + $1 = $6.25 in shipping charges.

Key Terms:
- **Intensive properties** – properties which do not depend upon the quantity of material being observed (examples: temperature, density, color).
- **Extensive properties** – properties which *do* depend upon the quantity of material being observed (examples: mass, volume).
- **Density** – quantitative description of how closely molecules are packed in a given material. density=mass/volume. Units for density are usually "g/mL" for condensed phases.

Temperature

The SI unit for temperature is Kelvin, but most people are used to working in either Fahrenheit or Celsius degrees.

Useful Relationships:
- $T_{Kelvin} = T_{Celsius} + 273.15$
- $T_{Fahrenheit} = (9/5)T_{Celsius} + 32$

EXERCISE 17: Convert 70.°F to °C and to the Kelvin scale.

SOLUTION:: Use the appropriate temperature conversion relationships.
$$T_{Fahrenheit} = (9/5)T_{Celsius} + 32, \text{ so } T_{Celsius} = 5/9(T_{Fahrenheit} - 32) = 5/9(70. - 32) = 21°C$$
$$T_{Kelvin} = T_{Celsius} + 273.15 = 294 \text{ K}$$

Density

Useful Relationship:
- density = mass / Volume. ($\rho = m/V$).

EXERCISE 18: What is the density in g/mL of a 0.1 kg solid block that takes up 50 mL of space?

STRATEGY AND SOLUTION:

$$\rho = \frac{m}{V} = \frac{0.1 \text{kg}}{50 \text{ mL}} \times \frac{1000 \text{ g}}{1 \text{kg}} = 2 \text{ g/mL (1 sig fig)}$$

EXERCISE 19: Which occupies a greater volume: 5.00 grams of lead with a density of 11.34 g/mL OR 5.00 grams of silver with a density of 10.50 kg/L?

STRATEGY: ρ=m/V so V=m/ρ. Solve each for volume and compare the results.

$$\text{Lead: } V = \frac{m}{\rho} = \frac{5.00\,g}{11.34\,g/mL} = 0.441\,mL\ (3\text{ sig figs})$$

$$\text{Silver: } V = \frac{m}{\rho} = 5.00\,g \times \frac{L}{10.50\,kg} \times \frac{1\,kg}{1000\,g} \times \frac{1000\,mL}{1\,L} = 0.476\,mL\ (3\text{ sig figs})$$

SOLUTION: The silver sample occupies a larger volume.

1.7   CHEMICAL PROBLEM SOLVING

Any time you encounter a problem that you can't figure out how to attack, break it down using the 7-step method, as outlined in the text. Exercise 20 illustrates the 7-step approach.

EXERCISE 20: The radius of one atom of carbon is 77 pm. (1 pm = 1×10^{-12} m). How many carbon atoms lined up next to each other would it take to span 1.0 inch?
   1. **Determine what is asked for.** The number of carbon atoms that can line up in one inch.
   2. **Visualize the problem.** The carbon atoms can be visualized as little marbles that you need to line up in a row. If the atoms are lined up, then the distance that one atom occupies on the line is the atom's diameter: 2(77 pm) = 154 pm.
   3. **Organize the data.** Given data: 154 pm/atom, how many atoms/inch?, 1 pm = 1×10^{-12} m. Find a conversion factor between metric and English length units: 1 inch = 2.54 cm.
   4. **Identify the process to solve the problem.** The real question is how many times does 154 pm fit into one inch? We need to convert to similar units before comparing.
   5. **Manipulate the equations.** Set up a relationship between the data to arrive at units of "atoms of C/inch."
   6. **Substitute and calculate.**

$$\frac{\#\text{ of C atoms}}{1.0\,inch} = \frac{1\,atom\ of\ C}{154\,pm} \times \frac{1\times10^{12}\,pm}{1\,m} \times \frac{1\,m}{100\,cm} \times \frac{2.54\,cm}{1\,inch} = 1.7\times10^{8}\text{ atoms of C/inch}$$

Let's walk through how this calculation was established. First, set up the conversion to start with one of the units needed in the final result. For example, we'll start with "1 atom/154 pm" because we need "atoms of C" in the numerator in the final result. The remaining conversions become obvious after that. To get rid of "pm" in the denominator, multiply by a conversion with pm. That leads to the next conversion, because now we have to get rid of "m" in the denominator. We're constantly working towards units of "inches" in the denominator. Once the result has been determined, round off to the correct number of significant figures. The final numerical result should be reported to two significant figures because the original data was "1.0 inch" – two sig figs.
   7. **Does the result make sense?** Atoms are incredibly tiny, so this enormous number seems reasonable. 170,000,000 atoms of carbon would fit in one inch.

Chapter 1 Self-Test

You may use a periodic table, a calculator, and a list of reference conversion factors for this Self-Test. Report all numerical answers to the correct number of sig figs.

1. Propane gas is commonly used as a fuel for gas BBQ grills. Each propane molecule contains 3 atoms of carbon and 8 atoms of hydrogen.

    a) Write the chemical formula for propane.

    b) Classify each as a physical property or chemical property of propane: i) it is colorless; ii) it is highly flammable; iii) it has a low boiling point.

    c) Does the burning of propane gas constitute a physical or chemical transformation?

    d) Is propane an element, a compound or a mixture?

    e) Is the amount of heat produced when burning propane considered an intensive or extensive property?

2. Which element has chemical properties most similar to Selenium (Se): As, Br, Sb, Te, or I?

3. A cube is 10.0 cm high x 10.00 cm wide x 10.000 cm deep. It is filled with a liquid – the mass of liquid is 2.0×10^{-1} kg.
    a) What is the density of the liquid?
    b) Which measurement of the cube is the most precise: length, width, or depth?

4   A jet on its way to Paris flew over Nebraska at a speed of 548 miles per hour at an elevation of 33,000 feet. The outside temperature was –59°F. Recalculate all the data in SI units.

5. An object weighed 1000 lb (the measurement was precise to within 10 lb). Rewrite the object's weight in scientific notation.

6. a) List one physical property of the noble gases.

    b) Where are the halogens located on the periodic table?

    c) Do halogens exhibit metallic or non-metallic characteristics?

Answers to Try Its

1.

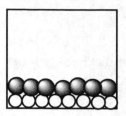

2. a) 2.135×10^8;   b) 4.50×10^{-11}
3. a) 0.06700;   b) –12,000,000
4. 6.92, 2.34, 8.99 (numbers are far from true value and far from each other)
5. a) 4;   b) 1;   c) 2;   d) 2
6. a) 1.0×10^3;   b) 15883;   c) -2.4×10^4

Answers to Self-Test

1. a) C_3H_8;   b) i. physical, ii. chemical, iii. physical;   c) chemical;   d) compound;   e) extensive
2. Te
3. a) 0.20 g/mL;   b) depth
4. 245 m/sec, 1.0×10^4 m, 223 K
5. 1.00×10^3 lb
6. a) they are gases at room temperature;   b) in column 17;   c) non-metallic characteristics

Chapter 2: The Atomic Nature of Matter

Learning Objectives

In this chapter, you will learn how to:

- Visualize chemistry at the molecular, atomic, and sub-atomic level.
- Obey the Law of Conservation of Mass by understanding how to track matter for a given physical or chemical change.
- Determine the number of protons, neutrons, and electrons in an atom or ion.
- Predict the charge on the ion that will form from an element (certain elements only).
- Determine the formula for an ionic compound.
- Interpret a simple spectrum from a mass spectrometer.
- Quantitate macroscopic to molecular-level relationships (mass-moles-molecules).

Practical Aspects

Chemists visualize matter at the molecular level. Make it your goal in this chapter to begin developing this viewpoint, as it will help tremendously in your understanding of chemistry.

2.1   ATOMIC THEORY

The Atomic Theory of Matter (John Dalton)

1. **All matter is composed of tiny particles called atoms.**
    Example: sugar is composed of atoms of carbon, hydrogen and oxygen.

2. **All atoms of a given element have identical chemical properties that are characteristic of that element.**
    Example: all atoms of chlorine have the same chemical properties.

3. **Atoms form chemical compounds by combining in whole-number ratios.**
    Example: carbon and oxygen atoms combine in a 1:2 ratio to form carbon dioxide, CO_2. If they combine in a different ratio, 1:1, they form a different compound – carbon monoxide, CO.

4. **Atoms can change how they are combined, but they are neither created nor destroyed in chemical reactions.**
    Example: When propane gas undergoes combustion, all of the carbon atoms in the propane go into making CO_2. All of the carbon atoms are accounted for or "conserved."

    This last statement is called "The Law of Conservation of Mass."

Key Term:
- **Law of Conservation of Mass** – matter cannot be created or destroyed, just rearranged.

Atoms and Molecules are Constantly in Motion

Phase	Molecular/Atomic View
Solid	• Atoms/molecules are packed tightly together. • They can't move around each other; they just vibrate in place.
Liquid	• Atoms/molecules are packed together, but not as tightly as in solid phase. • There are a few gaps here and there so molecules can roll over each other (hence, liquids "flow").
Gas	• Atoms/molecules are far apart from each other and are moving rapidly. • They are constantly colliding with each other and with the interior walls of the container. (The molecules' collisions with the walls create pressure.)

Key Term:

- **Diffusion** – the natural movement of one substance through another (without mixing).

For example: The fact that gas molecules move rapidly (and thus rapidly diffuse through each other) can be illustrated by how quickly one can smell a skunk's odor after it has sprayed – yuck!

EXERCISE 1: One function of a catalytic converter inside a car is to reduce carbon monoxide emissions by reacting any CO present with oxygen molecules to form CO_2. Draw molecular pictures to illustrate four molecules of CO undergoing this process. Be sure to conserve all atoms.

STRATEGY: This is a qualitative problem, so the 7-step method is not needed. Visualize what is occurring at the molecular level and show that in the pictures:
- oxygen molecules are diatomic,
- CO, CO_2 and O_2 are commonly known substances – all are gaseous in our atmosphere. The molecules should be separate from one another and fill the container.
- we must conserve or account for all atoms: 4 CO will make 4 CO_2, to account for the carbon atoms. Draw those molecules into the boxes. (At this point, don't worry about *how* the atoms within each CO_2 are connected. Just be sure 1 C and 2 Os are somehow connected to each other). After drawing the CO and CO_2 molecules, it will become apparent that we need 2 O_2 molecules to conserve the oxygen atoms.

SOLUTION:

Before Reaction

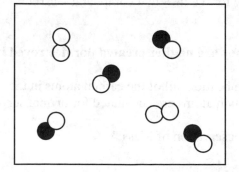

After Reaction

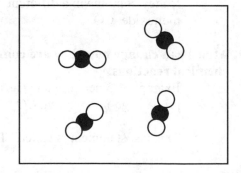

Dynamic Molecular Equilibrium

Key Term:
- **Dynamic equilibrium** – situation in which the rate of the forward process equals the rate of the reverse process.

Figure 2-9 in the text illustrates dynamic equilibrium at the molecular level.

Consider a glass of water with a lid on top. Condensation forms on the underside of the lid. Visualize the dynamic equilibrium that is established:

- **Macroscopic level** – no apparent change is occurring.
- **Molecular level** – for every liquid water molecule turning to gas, there is a gas molecule turning into liquid.
  - If you compared two photographs of the water at the molecular level, they might look identical since there is no net change. If you were to track an individual molecule, though, you'd see that it is constantly moving from gas to liquid and back to gas.

EXERCISE 2: "Dry ice" is carbon dioxide in the solid phase. It sublimes (converts directly from a solid to a gas) under normal atmospheric conditions. Using circles to depict CO_2 molecules, draw a series of molecular pictures to illustrate six CO_2 molecules: a) in the solid phase inside a closed container; b) the same container after two CO_2 molecules have sublimed; c) the same container after it has reached dynamic equilibrium, given that four CO_2 molecules will be in the gas phase; d) the same container two hours after it has established dynamic equilibrium.

SOLUTION:

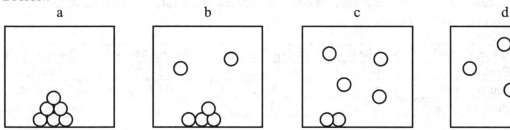

2.2   ATOMIC ARCHITECTURE: ELECTRONS AND NUCLEI

Forces

Key Concepts:
- Electrical forces exist between two charged particles: opposite charges attract, like charges repel.
- Magnetic forces exist between poles on a magnet (north pole/south pole): opposite poles attract, like poles repel.
- The path of a moving charged object will be deflected if it passes through a magnetic field (e.g., between two poles of a magnet).

Sub-Atomic Particles

Between 1895 and 1915, scientists performed several fascinating experiments to understand the structure of the atom. Some of the experiments are described in Section 2.2 of the text. Here is a summary of the conclusions made from these experiments:

Experiment	Conclusion(s)
Gas Discharge Tube	• Atoms are made up of smaller fragments that have charges • Atoms can be decomposed into negatively-charged electrons and positively-charged ions
JJ Thompson's Cathode Ray Tube Experiment	• The charge-to-mass ratio of an electron is -1.76×10^{11} C/kg
Millikan's Oil Drop Experiment	• The charge of a single electron is -1.6×10^{-19} C • The mass of a single electron is 9.1×10^{-31} kg
Rutherford's Gold Foil Experiment	The Nuclear Model of the Atom: • Most of the atom is empty space • Most of the mass and all of the positive charge in an atom resides in a tiny, concentrated area in the center of the atom called the "nucleus."

Later experiments determined that the nucleus is made up of two types of particles: protons and neutrons. Table 2-1 in the text summarizes the actual masses and charges of the three subatomic particles. For most applications it is important to know:

Sub-Atomic Particle	Symbol	Relative Charge	Relative Mass	Where it resides
Proton	p	+1	1	Nucleus
Neutron	n	0	1	Nucleus
Electron	e	-1	0	Around the nucleus

EXERCISE 3: Compare the mass of a proton (1.6726×10^{-27} kg), a neutron (1.6749×10^{-27} kg), and an electron (9.1094×10^{-31} kg) to support the information shown in the "Relative Mass" column of the above table.

STRATEGY: Take ratios of the three to compare the relative masses. Electrons have the smallest mass, so base the comparison on them.

Proton   Neutron   Electron

$$\frac{1.6726 \times 10^{-27} \text{ kg}}{9.1094 \times 10^{-31} \text{ kg}} = 1836 \qquad \frac{1.6749 \times 10^{-27} \text{ kg}}{9.1094 \times 10^{-31} \text{ kg}} = 1839 \qquad \frac{9.1094 \times 10^{-31} \text{ kg}}{9.1094 \times 10^{-31} \text{ kg}} = 1$$

SOLUTION: Protons and neutrons are about 1800 times heavier than electrons, so an electron's *relative* mass is negligible. Also, the mass of a proton is roughly equal to the mass of a neutron.

The Size of the Nucleus
- Most of an atom is empty space. To give you an idea of scale, an atom the size of a baseball stadium would have a nucleus the size of a green pea (suspended in the air somewhere above 2nd base). Furthermore, in Chapter 1, Exercise 20 of this guide, we saw that 170,000,000 atoms of carbon would fit in one inch. Imagine how small the *nucleus* is!

EXERCISE 4: The radius of a carbon atom is 77 pm, where 1 pm = 1×10^{-12} m. The radius of the nucleus of an atom is roughly 10,000 times smaller than the radius of an atom. Determine the radius of a carbon atom's nucleus in units of inches.

STRATEGY: This is a conversion factor question. Start with what is given and incorporate conversion factors to move you towards the desired goal of "inches."

$$\text{Radius of a carbon nucleus} = \frac{\text{radius of carbon atom}}{10,000}$$

$$= \frac{77\,\text{pm}}{10,000} \times \frac{1 \times 10^{-12}\,\text{m}}{1\,\text{pm}} \times \frac{100\,\text{cm}}{1\,\text{m}} \times \frac{1\,\text{inch}}{2.54\,\text{cm}} = 3.0 \times 10^{-13}\,\text{inch}$$

SOLUTION: The radius of a nucleus of a carbon atom is 3×10^{-13} inch or 0.0000000000003 inch. The answer is reported to 1 sig fig because the number "10,000" contains 1 sig fig. The answer seems reasonable because it is such a small number.

EXERCISE 5: A cathode ray tube measures the total charge of a beam of electrons to be -5.0×10^{-7} C. a) What mass (in grams) of electrons is responsible for this charge? b) How many electrons is this?

STRATEGY AND SOLUTION: We've been given a charge resulting from electrons. Use the mass to charge ratio of an electron as a conversion factor to get the mass of electrons. Then use the mass of an electron as a conversion factor to find out how many electrons this is.

$$\text{a)} \quad -5.0 \times 10^{-7}\,\text{C} \times \frac{1000\,\text{g}}{1\,\text{kg}} \times \frac{1\,\text{kg}}{-1.76 \times 10^{11}\,\text{C}} = 2.8 \times 10^{-15}\,\text{g of electrons}$$

$$\text{b)} \quad 2.8 \times 10^{-15}\,\text{g} \times \frac{1\,\text{electron}}{9.1 \times 10^{-31}\,\text{kg}} \times \frac{1\,\text{kg}}{1000\,\text{g}} = 3.1 \times 10^{12}\,\text{electrons}$$

The answers seem reasonable because there is a huge number of electrons in a very small total mass.

2.3 ATOMIC DIVERSITY: THE ELEMENTS

Key Terms:
- **Atomic number** – number of protons in an atom; defines the identity of the element.
- **Mass number** – number of protons PLUS number of neutrons in an atom.
- **Isotopes** – atoms of elements that contain the same number of protons but different numbers of neutrons. ("Isotope" comes from the Greek "iso-topos," which mean "same place" – isotopes are found in the same place on the periodic table.)

Key Concepts:

- Every square on the periodic table contains a symbol, an atomic number, and a mass number. The atomic number is the smaller, whole number. The larger number (not a whole number) is the atomic mass.
- If the number of protons equals the number of electrons, then there is no net charge.
- Two atoms of the same element may have different numbers of neutrons, thus different atomic masses and different mass numbers.
- You cannot obtain the atomic mass for an individual atom from the periodic table. The mass reported is an average of all naturally occurring isotopes.
- To designate a particular isotope of a given element (using oxygen with 10 neutrons as an example):
  - write the mass number as a superscript to the left of the element's symbol: ^{18}O
  - write the "element name-mass number": oxygen-18
- Some isotopes are unstable because they contain proton to neutron ratios that are energetically unfavorable. These isotopes are termed "radioactive", and will spontaneously attempt to stabilize themselves by ejecting a portion of the nucleus from the atom (for example, an alpha particle).

EXERCISE 6: There are three naturally occurring isotopes of carbon: carbon-12, carbon-13, and carbon-14. Determine the symbol, and the number of protons, neutrons, and electrons for these three isotopes.

STRATEGY: Find carbon on the periodic table. It is atomic number 6. All of the carbon isotopes contain 6 protons. They are electronically neutral, so they all have 6 electrons too. The number of neutrons is determined from the mass number of the isotope: mass number = # protons + # neutrons.

SOLUTION:

Name	Symbol	# of p	# of n	# of e-
Carbon-12	^{12}C	6	6	6
Carbon-13	^{13}C	6	7	6
Carbon-14	^{14}C	6	8	6

Key Concept:

- The information describing an element is highly formatted. For a given element symbol, X, information is always recorded in this manner:

$$^{\text{mass number}}_{\text{atomic number}}X^{\text{charge, if any}}_{\text{number of atoms}}$$

- The atomic number is usually not written; it can be found from the element's symbol.
- The mass number is only written to specify a particular isotope.
- A charge results when the numbers of protons and electrons are not equal.
- The number of atoms is only written if it is greater than one.

EXERCISE 7: Indicate all possible information about the subatomic particles present in: a) Cl; b) ^{35}Cl; c) Cl_2; d) $H^{37}Cl$; e) Cl^-

SOLUTION: For all of the chlorines shown, we could look on the periodic table and see that Cl's atomic number is 17, so each contains 17 protons.

a) Cl – an atom of chlorine. No mass number is shown, so we don't know which isotope this is (we can't determine number of neutrons in this atom). It has no charge, so it contains 17 electrons.

b) ^{35}Cl – an atom of chlorine-35. The mass number of this isotope is 35, so one atom of it contains 18 neutrons (35-17=18). It has no charge, so one atom contains 17 electrons.

c) Cl_2 – a molecule of chlorine that contains two atoms stuck together. Each atom contains 17 protons and 17 electrons, but we can't determine the # of neutrons without the mass number.

d) $H^{37}Cl$ – a molecule of HCl, which contains 1 atom of hydrogen and one atom of chlorine-37. The chlorine atom contains 17 electrons and 20 neutrons. The hydrogen atom contains 1 proton and 1 electron, but we don't know how many neutrons.

e) Cl^- - The mass number is not given, so we don't know the number of neutrons. It has a "-1" charge, which indicates 1 extra electron relative to protons. It has 18 electrons.

EXERCISE 8: Write the formula for: a) the elemental particle that contains 12 protons, 10 electrons, and 12 neutrons; b) a molecule of water made from oxygen-18; c) the elemental particle that contains 14 protons, 14 electrons, and an unknown number of neutrons.

STRATEGY:
a) 12 protons = Mg, 10 electrons indicates there are 2 more protons than electrons so the net charge is +2, 12 neutrons indicates that the mass number is 24.
b) The formula for water is H_2O. This water contains oxygen-18, which must be specified.
c) The number of protons = atomic number, which indicates the element's identity as Si. We can't specify which isotope because we don't have the number of neutrons. The element has no charge – the number of protons and electrons are equal.

SOLUTION: a) $^{24}Mg^{2+}$; b) $H_2{}^{18}O$; c) Si

Try It #1: Write the formula for: a) a molecule of carbon dioxide that contains carbon-14; b) the elemental particle that contains 15 protons, 18 electrons, and 16 neutrons.

Mass Spectrometry

Key Term:
- **Mass spectrometer** – instrument which determines the mass of an element or molecule by removing an electron from the substance being analyzed, and then separating charged particles with a magnetic field. (Particles with larger masses are deflected less in the magnetic field than particles with smaller masses). A chart depicting relative abundances as a function of mass is generated.

EXERCISE 9: Two isotopes of bromine exist in nature in roughly equal proportions, bromine-79 and bromine-81. If a sample of elemental bromine, Br_2, is subjected to mass spectrometry, what peaks will show up on the mass spectrum? Specify the masses and their corresponding percent abundances.

STRATEGY: We must first determine the possible combinations that exist for the bromine molecules, then determine their relative abundances, then determine the mass of each type of bromine molecule. A table can help organize all the possible combinations that exist:

Bromine Atom #1	Bromine Atom #2	Mass number of molecule composed of atom #1 and atom #2

^{79}Br	^{79}Br	158
^{79}Br	^{81}Br	160
^{81}Br	^{79}Br	160
^{81}Br	^{81}Br	162

We can see from the table that there is a 1 in 4 chance the mass number will be 158, a 2 in 4 chance it will be 160, and a 1 in 4 chance it will be 162.

SOLUTION: The mass spectrum will show a 25% abundance peak at a mass number of 158, a 50% abundance peak at 160, and a 25% abundance peak at 162.

Try It #2: a) What ions would be produced for each bromine molecule in the mass spectrometry experiment described in Exercise 9? b) Which ion would be deflected most in the magnetic field?

2.4 COUNTING ATOMS: THE MOLE

Atoms are incredibly tiny. Any sample that we would work with in a normal everyday setting would involve millions of billions of atoms. In order to relate macroscopic quantities of chemicals (i.e., 1 ring = 30 grams of silver) to their corresponding molecular quantities (i.e., how many atoms of silver are in this ring?), chemists use the mole.
A mole is a given number of things grouped together, just as a dozen is a set of twelve things grouped together.

$$1 \text{ dozen} = 12 \text{ things}$$
$$1 \text{ mole} = 6.022 \times 10^{23} \text{ things}$$

These "things" can be atoms, molecules, pennies, etc.

Key Concepts:
- one mole = 6.022×10^{23} particles.
- 6.022×10^{23} is also termed Avogadro's number and is given the symbol N_A.
- Mole is often abbreviated "mol" in calculations.
- The definition of one mole is based on the number of atoms in exactly 12 grams of ^{12}C: There are 6.022×10^{23} atoms of ^{12}C in exactly 12 g of ^{12}C.
- The molar mass of an atom is its mass shown on the periodic table. This is a weighted average of all naturally occurring isotopes.

Isotopic Abundance

Although most calculations will involve using elemental molar masses off the periodic table, it is useful to have an understanding of where these molar masses originate by understanding isotopic abundances.

EXERCISE 10: Two isotopes of boron exist in nature: ^{10}B and ^{11}B. 19.9% of naturally occurring boron is ^{10}B. What is the molar mass of boron? The isotopic masses of the two isotopes are 10.0129 g/mol and 11.0093 g/mol, respectively.

SOLUTION: If two isotopes exist then the relative abundance of ^{11}B must be:
$$100\% - 19.9\% = 80.1\%$$

A weighted average of these two isotopic masses will give the molar mass of B as it appears on the periodic table:

$$\text{MM of B} = (0.199)(10.0129 \text{ g/mol}) + (0.801)(11.0093 \text{ g/mol}) = 10.811 \text{ g/mol}$$

This number is reasonable because it is between 10 and 11 and is closer to 11.

It is also possible to determine percent abundances of different isotopes of given elements by assessing their weighted averages.

EXERCISE 11: The molar mass of chlorine is 35.453 g/mol. Chlorine has two naturally occurring isotopes, chlorine-35 with an isotopic mass of 34.97 g/mol and chlorine-37 with an isotopic mass of 36.97 g/mol. What is the percent abundance of each isotope?

SOLUTION:
Since the sum of the isotopic abundances must equal 100%, we can set up this relationship to analyze the percent abundance of each:
$$36.97(x) + 34.97(100\text{-}x) = 35.453(100)$$

$$x = \text{isotopic abundance of chlorine-37} = 24.15\%$$
$$100\text{-}x = \text{isotopic abundance of chlorine-35} = 75.85\%$$

Side note: a mass spec of a chlorine-containing compound always shows two peaks in a 3:1 ratio, 2 g/mol apart), just like this relative abundance suggests.

Mass-Mole-Atom Conversions

Many chemical calculations relate macroscopic properties to molecular properties. The mole is the pathway by which one moves from macroscopic to molecular and vice versa.

Units and Conversions:
- 6.022×10^{23} particles = 1 mol   Use this relationship to convert between atoms (or molecules) and moles of a given substance. $6.022 \times 10^{23} = N_A =$ Avogadro's number.

- Molar mass = # of grams/1 mol   Use molar mass from periodic table to convert between grams and moles of a given substance.

EXERCISE 12: Elemental sodium reacts violently with water to make sodium ions. How many atoms of sodium metal will undergo a reaction if 60.25 mg of sodium are added to water?

STRATEGY AND SOLUTION: To convert between grams and moles of Na, use molar mass of Na. Use Avogadro's number to convert between moles and atoms.

$$60.25 \text{ mg Na} \times \frac{1 \text{ g}}{1000 \text{ mg}} \times \frac{1 \text{ mol}}{22.990 \text{ g}} \times \frac{6.022 \times 10^{23} \text{ atoms}}{1 \text{ mol}} = 1.578 \times 10^{21} \text{ atoms of Na}$$

This answer seems reasonable because atoms are so small.

2.5 CHARGED ATOMS: IONS

Key Terms:
- **Ion** – charged species resulting from the gain or loss of electrons from a neutral atom or molecule.

Three terms are used to classify ions:
- **Cation** – (pronounced "cat-ion") – positively-charged ion (Examples: K^+, Mg^{2+}, Al^{3+}).
- **Anion** – (pronounced "ann-ion") – negatively-charged ion (Examples: Cl^-, S^{2-}, P^{3-}).
- **Polyatomic** – ion that contains more than one atom (Examples: SO_4^{2-} is a polyatomic anion, NH_4^+ is a polyatomic cation).

EXERCISE 13: The Law of Conservation of Mass states that matter cannot be created nor destroyed, but only rearranged. If an element loses an electron to make a cation, where does that electron go?

SOLUTION: The electron is picked up by an atom, and *that* atom becomes an anion.

Ionic Compounds

Key Term:
- **Ionic compound** – a solid that contains cations and anions in balanced whole-number ratios.

Helpful Hints
- Metal elements tend to lose electrons to become cations:
  - Group 1 metals (the alkali metals) form cations with +1 charges
  - Group 2 metals (the alkaline earth metals) form cations with +2 charges

- Non-metal elements tend to gain electrons to become anions:
  - Group 16 elements form anions with –2 charges
  - Group 17 elements (the halogens) form anions with –1 charges

EXERCISE 14: Consider these pairs of elements: Li and O, Na and I, Mg and S. When each atom forms an ion, what will its charge be? If these pairs of elements were to form ionic compounds, what would the formulas be?

STRATEGY AND SOLUTION:

Metal ion formation	Non-metal ion formation	Ratio needed	Formula for Ionic Compound
Li – alkali metal so ion will be Li^+	O – Group 16 element so ion will be O^{2-}	2 Li^+ for each O^{2-}	Li_2O
Na – alkali metal so ion will be Na^+	I – halogen so ion will be I^-	1 Na^+ for each I^-	NaI
Mg – alkaline earth metal so ion will be Mg^{2+}	S – Group 16, so ion will be S^{2-}	1 Mg^{2+} for each S^{2-}	MgS

Try It #3: Write the formula for the ionic compound that would result when calcium and chlorine combine.

Chapter 2 Self-Test

You may use a periodic table, a calculator, and a table of conversion factors for this test.

1. Draw molecular pictures to illustrate four gaseous molecules of ozone (O_3) undergoing a chemical transformation to make oxygen molecules. Be sure to obey the Law of Conservation of Mass.

2. Here is a box containing two sections that are separated by a removable partition. Redraw the box showing the left side filled with 6 atoms of He gas and the right side filled with 4 atoms of Ne gas. Now redraw the box with its partition removed. Redraw the gases in the box as they would appear after they have completely diffused and established equilibrium.

3. Complete the table:

Symbol	Name	# of protons	# of neutrons	# of electrons
$^{81}Br^-$	Bromine-81 anion			
^{65}Cu				
		34	Not known	36
		13	12	10
	Phosphorus-32			15

4. Write the formula for the ionic compound that would form between Al^{3+} and a halogen.

5. Fe_2O_3 is an ionic compound. What must be the charge of the Fe cation in this formula?

6. Can gas phase ions conduct electricity? Briefly explain your answer.

7. The mass to charge ratio for a sodium ion is 2.38×10^{-4} g/C. Given that the charge on an electron is -1.6022×10^{-19} C, determine the mass of a sodium ion.

8. A mass spectrum of an unknown element showed the following results:

Mass Number	Percent Abundance
50	4.35
52	83.79
53	9.50
54	2.36

Propose an identity for this element.

9. What is the mass of 8.9×10^{25} atoms of helium, He?

Chapter 2: The Atomic Nature of Matter

Answers to Try Its:

1. a) $^{14}CO_2$; b) $^{31}P^{3-}$
2. a) Each molecule would form a +1 ion, with a general formula of Br_2^+. The masses would vary.
   b) The ion that would be deflected most would be the lightest one, which contains two atoms of bromine-79.
3. $CaCl_2$ (Ca – alkaline earth metal so ion will be Ca^{2+}, Cl – halogen so ion will be Cl^-, 1 Ca^{2+} for 2 Cl^-)

Answers to Self-Test

1.

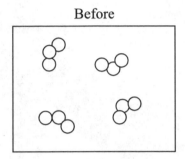

Before   After

2. One way to show it:

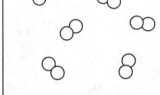

3.

Symbol	Name	# of protons	# of neutrons	# of electrons
$^{81}Br^-$	*Bromine-81 anion*	35	46	36
^{65}Cu	Copper-65	29	36	29
Se^{2-}	Selenium anion	*34*	*Not known*	36
$^{25}Al^{3+}$	Aluminum-25 cation	*13*	*12*	*10*
^{32}P	*Phosphorus-32*	15	17	*15*

4. $AlCl_3$ (or any halogen: formula will be AlX_3, where X = halogen)
5. +3 - in order to balance the charge of the three oxygens with -2 each, 2 Fe^{3+} ions are needed.
6. Yes, gas phase ions can conduct electricity – ions are charged and they can move around.
7. 3.81×10^{-23} g or 3.81×10^{-26} kg (Hint: charge on ion is equal in magnitude to charge on electron).
8. Cr – the most abundant isotope has a mass number of 52. The element on the periodic table with a mass number closest to 52 is Cr.
9. 590 grams (2 sig figs)

Chapter 3: The Composition of Molecules

Learning Objectives

In this chapter, you will learn how to:
- Name chemical compounds and write their formulas.
- Draw and interpret the many models chemists employ to illustrate structures of compounds – ball and stick, space-filling, structural formulas, and line structures.
- Calculate percent mass compositions, empirical formulas and molecular formulas.
- Prepare aqueous solutions and dilutions of specific concentrations.

Practical Aspects

One of the most important themes in chemistry is "structure determines function," meaning if we know the structure of a chemical compound, we can predict quite a bit about that chemical's physical and chemical properties. For example, in the text's chapter opener, we saw that the stimulatory behavior of caffeine is due to its similar structure to cyclic AMP.

When a chemist discovers a new chemical compound, many questions must be answered before the actual structure of the compound is determined:
- What types of elements are in the compound?
- How many of each type of element are present? (What is its chemical formula)?
- What is the molar mass of the compound?
- How are the atoms in the compound connected together? (What is its structure?)
- How should this compound be named so that other chemists could draw its structure?

It is essential to learn not only the individual skills in this chapter, but also to see how they are inter-related. By the time you're done studying this chapter, if you see a molecular formula you should think, "I know how many of each type of atom are present in this compound, I can figure out its percent composition, molar mass, the number of molecules present in a given mass of sample, how to draw a picture (structure) of the compound, and how to name that structure."

The last section of this chapter deals with making solutions and dilutions. These calculations are used in every quantitative chapter of the lecture text as well as in every chemistry laboratory class you will ever take. It will thus serve you well to thoroughly understand how to make solutions and dilutions.

3.1   REPRESENTING MOLECULES

Several types of models are used in chemistry to illustrate chemical structures. In order to be "fluent" in chemistry, one must be able to read, interpret, and draw the various models used: ball and stick, space-filling, structural formula, and line structure.

The types of models used to illustrate structures of compounds are shown in this table using 2-propenethiol as an example. (This chemical gives garlic its smell.)

Type	Example	Information emphasized in this type of model
Ball and Stick		3-D connectivities of atoms - almost like drawing a picture of a molecular model kit model. (best viewed in color – atoms are color-coded)
Space-Filling		Bulkiness of various regions within the compound. (best viewed in color – atoms are color-coded)
Structural Formula		Connectivities of atoms to include double and triple bonds.
Line Structures		Parts of the molecules that are involved in chemical reactions. *Notice that it's easier to see the double bond and the S in this structure than in the structural formula.

Line Structures and Structural Formulas

Line structures are a shorthand version of the complete structure. Chemists use line structures almost exclusively, so it is essential to know how to read the "shorthand" and how to use it to draw molecules.

Helpful Hints
- Refer to the guidelines on page 69 of the text for instructions on how to draw line structures.
- An easy way to convert from a structural formula to a line structure is to redraw the complete structure, then use thick lines to draw over the carbon backbone and any carbon-heteroatom bonds. (A heteroatom is any atom that is not carbon or hydrogen.) Finally, redraw this structure but remove the carbon atoms and any hydrogen atoms attached to carbons. Exercise 1 uses this method.

EXERCISE 1: Given the structural formula, draw the line structure.

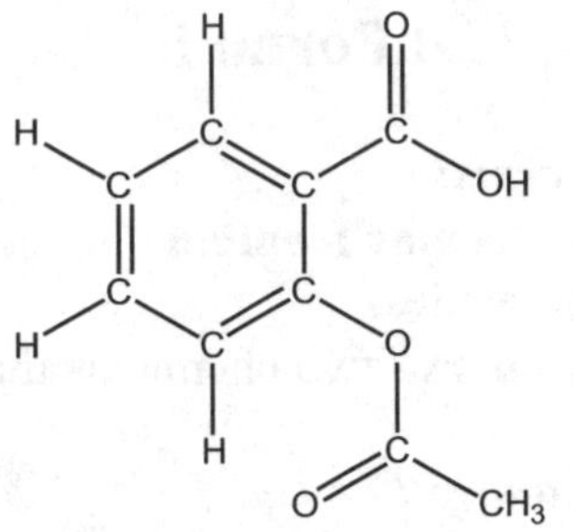

SOLUTION:

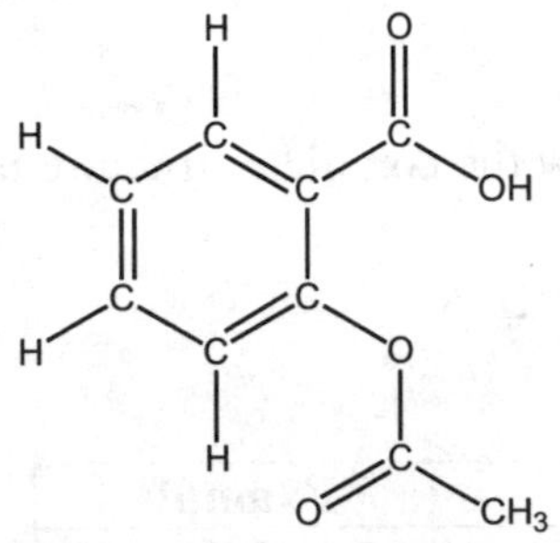

→   →

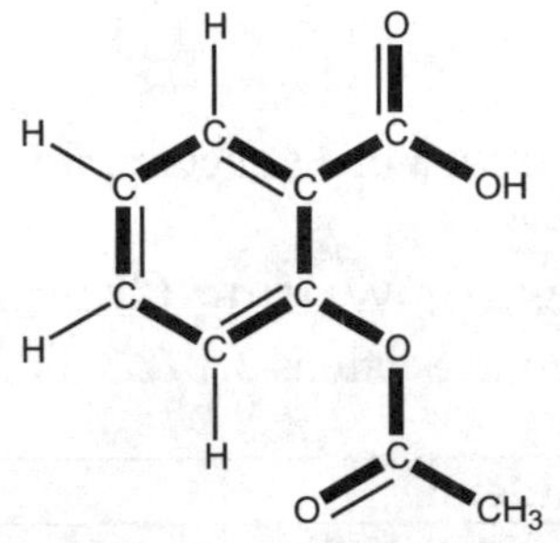

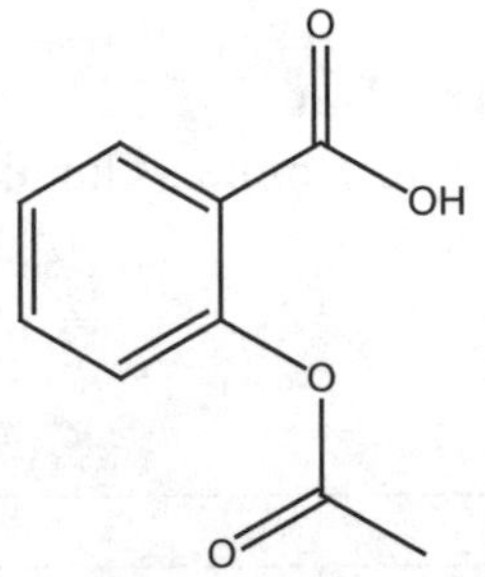

Original structure
(Aspirin)

Redrawn structure with
thick lines drawn over
carbon backbone and
bonds to heteroatoms

The final line structure
includes all multiple
bonds, heteroatoms, and H
atoms attached to
heteroatoms

Use a similar method to convert from a line structure to a structural formula. Redraw the original line structure, add carbons to any bends or ends of the line structure, and then add in all necessary hydrogens to ensure each carbon has four bonds.

EXERCISE 2: Given the line structure of acetaminophen, draw the structural formula.

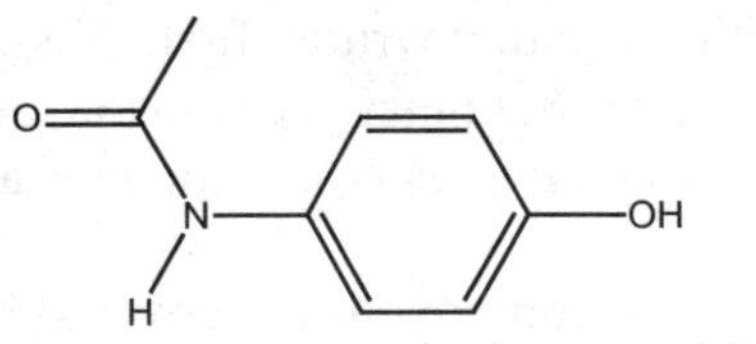

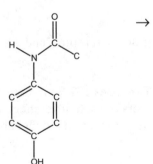

SOLUTION:

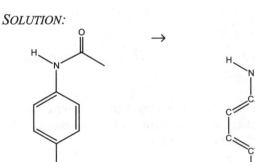   →   →

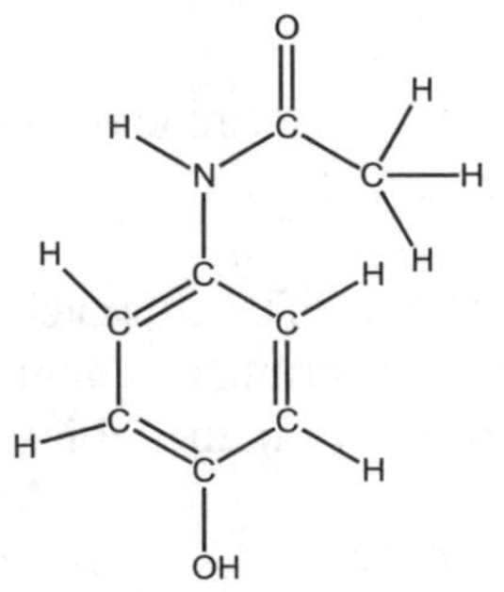

Original line structure
(rotated)

Redrawn structure with
carbon atoms shown

Final structural formula shows
hydrogen atoms attached to carbons

Chemical Formulas

Key Terms:
- **Molecular formula** – name given to a chemical formula for a substance that exists as discreet molecules.
- **Isomers** - two chemicals that have different structural formulas but the same molecular formula.

Key Concept:
- A chemical formula indicates the relative number of each type of element present in the substance.

Helpful Hint
- To determine a chemical formula from a line structure, it is best to draw the complete structure first.

Guidelines for Writing Chemical Formulas
as presented in Sections 3.1 & 3.3 of the lecture text

Guideline	Example

Binary Compounds
The element farthest left on the periodic table is written first.
 	
Exception: Hydrogen
    If H is present, it appears first ONLY if it is combined with an
    element from Group 16 or 17 on the periodic table.
 	
Note: If both elements are from the same column of the periodic
      table, the lower one is written first.
	NaCl
	
	NH_3, NaH
	HCl, H_2O
	
	SO_2

Carbon-Based Compounds (Organic Compounds) If the compound contains C & H, then it is classified as organic and is written: C then H then heteroatoms in alphabetical order.	C_2H_6 $C_{10}H_{15}NO$
Ionic Compounds The formula for an ionic compound shows the simplest ratio of cation to anion, with the cation written first. Note that if 2 different elements are present, the chemical is a binary compound, and its formula agrees with the guidelines presented in Binary Compounds above. *ionic compounds are covered in Section 3.3 of the text	Mg_3N_2 AgCl $CaSO_4$ $(NH_4)_2SO_4$

EXERCISE 3: Determine the molecular formulas of aspirin and acetaminophen (from their structures in Exercises 1 and 2).

STRATEGY: The chemicals drawn in Exercises 1 & 2 are both organic compounds. To write their chemical formulas, count up the numbers of each different element present, and record them in the order shown in the table: C then H then other atoms in alphabetical order.

SOLUTION:
Aspirin contains: 9 carbons, 8 hydrogens and 4 oxygens, so its molecular formula is: $C_9H_8O_4$.
Acetaminophen contains 8 carbons, 9 hydrogens, 2 oxygens, and 1 nitrogen, so its molecular formula is: $C_8H_9NO_2$.

EXERCISE 4: What is the chemical formula of each?

a)

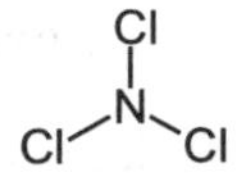

b)

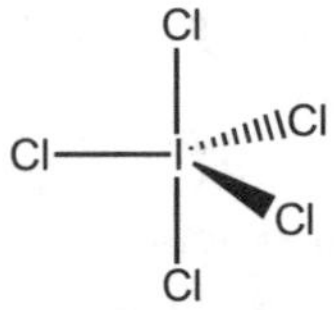

c) white = hydrogen; dark = iodine

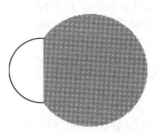

SOLUTION:

The chemical in "a" contains 1 nitrogen and 3 chlorines. It is a binary compound that doesn't contain H so its formula will be written with the left-most element on the periodic table first (N). Its formula will be: NCl_3.

The chemical in "b" contains 1 iodine and 5 chlorines. It is a binary compound with no hydrogens, but both elements are in the same column so the larger atom is written first in the formula: ICl_5.

The chemical in "c" contains 1 hydrogen and 1 iodine. Since iodine is in column 17 on the periodic table, the hydrogen will be written first: HI.

Try It #1: Here is the line structure for DEET insect repellent:
 a) Draw its structural formula.
 b) Draw the ball and stick model for DEET.

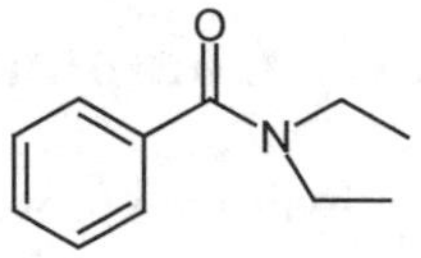

 c) Is this compound an isomer of DEET?

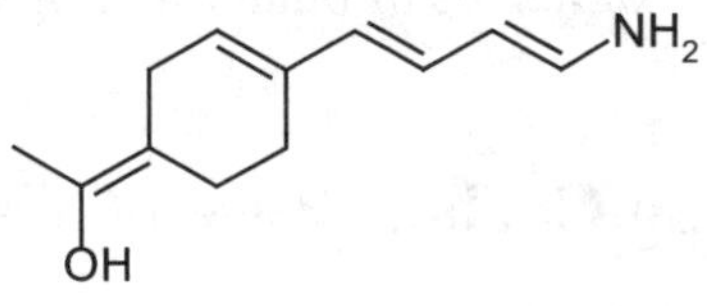

3.2   NAMING CHEMICAL COMPOUNDS

With an entire periodic table at our disposal, we can imagine there exists potential for an infinite number of different compounds, so there needs to be some system for naming them. Actually, there are several systems, so the first step is deciding which system should be used for the compound. The decision tree on page 37 of this guide will help you decide which rules to use. For now, we will focus on one set of rules at a time.

The four sets of rules you will need to master are the rules for naming:
• Non-Metallic Binary compounds (non-hydrogen)
• Binary compounds that contain hydrogen
• Carbon-based compounds (organic compounds)
• Ionic compounds (section 3.3)

Naming Non-Metallic Binary Compounds

Name the first element in the formula then the second element using these rules:
1. Drop end of second element and add -ide suffix.
2. Use prefixes to distinguish number of atoms of each element. (Note - do not use "mono" prefix on first element.)

1 = mono	2 = di	3 = tri	4 = tetra	5 = penta
6 = hexa	7 = hepta	8 = octa	9 = nona	10 = deca

EXERCISE 5: Name each compound: a) P_2O_5 b) NCl_3

SOLUTION:

a) P_2O_5:   2 phosphorus atoms = di-phosphorus
              5 oxygen atoms = penta-oxygen & oxygen is 2nd atom so it becomes ox-ide

The correct name is: <u>diphosphorus pentoxide.</u> Note: the "a" is deleted from "penta-" for ease of pronunciation.

b) NCl_3:   1 nitrogen atom = mono-nitrogen, but since it's the first atom, delete the mono and just write nitrogen
              3 chlorine atoms = tri-chlorine & chlorine is 2nd atom so it becomes chlor-ide

The correct name is: <u>nitrogen trichloride</u>

Naming Binary Compounds Containing Hydrogen

As you will learn, hydrogen provides the exception to practically every rule in chemistry. When hydrogen is in a compound, which rule to use becomes challenging. Refer to Table 3-3 in the text for a summary of names of binary hydrogen compounds.

Naming Carbon-Based Compounds (also called "organic" compounds)

To name simple organic compounds:
1. Count up the number of carbon atoms in the longest chain of carbons. This will indicate the prefix:

# of Cs:	1	2	3	4	5	6	7	8	9	10
Prefix:	Meth-	Eth-	Prop-	But-	Pent-	Hex-	Hept-	Oct-	Non-	Dec-

2. Provide a proper suffix: "-ane" = all single bonds; "-ene" = double bond; "-yne" = triple bond.

3. Account for any functional groups:
    *   Alcohols are represented by "-OH" and are given the *suffix* "-ol".
    *   Halides are represented by "-F", "-Cl", etc and are given the corresponding *prefix*: F = fluoro-, Cl = chloro-, etc.

4. Indicate locations of side chains, double bonds, triple bonds, or functional groups by specifying the carbon to which they are attached. Be sure to end up with the *lowest* number possible when counting the carbon chain.

EXERCISE 6: Name this compound:

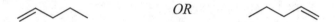

SOLUTION:
There are 6 carbons in a row, so the prefix is "hex."
All the bonds are single bonds, so "ane" suffix.
An alcohol group is present on carbon #3, so add the suffix "ol" and the prefix "3."
At this point we have: 3-hex-ane-ol. Drop the "e" on the "ane" suffix.

   The correct name is: 3-hexanol

EXERCISE 7: Draw the line structure for: 1-pentene

SOLUTION:
1-pentene: "pent" = 5 carbons in a row, "ene" indicates a double bond, "1" indicates that the double bond is on the first carbon. The line structure could be written:

OR

Both structures have a double bond originating on carbon #1.

3.3   FORMULAS AND NAMES OF IONIC COMPOUNDS

An ionic compound can usually be identified by recognizing one or more of these features within the compound:
- a Group I or II metal
- a polyatomic ion
- a binary metal oxide or sulfide

Key Terms:
- **Polyatomic ion** – an ion that contains a cluster of several atoms ("poly" = many).
- **Oxyanion** – polyatomic anion of some atom with oxygen.

Helpful Hints
- It is essential to memorize the names and corresponding formulas of the common polyatomic ions. Suggestion: make flash cards of all the polyatomic ions in Table 3-4 of the text and memorize them. Be sure to include the ion's charge when memorizing the full formula:
                Example: NH_4^+ ammonium ion

- Use the guidelines on page 78 of the text for naming oxyanions. Notice the pattern here for naming some common oxyanions:

   - The *nitrogen* oxyanions have a charge of -1: NO_3^- = nitrate, NO_2^- = nitrite
   - The *sulfur* oxyanions have a charge of -2: SO_4^{2-} = sulfate, SO_3^{2-} = sulfite
   - The *phosphorus* oxyanions have a charge of -3: PO_4^{3-} = phosphate, PO_3^{3-} = phosphite

- Use the decision tree in Figure 3-12 of the text to determine whether or not a compound is ionic.
- Guidelines for writing formulas of ionic compounds are on page 80 of the text.
- Remember, in an ionic formula the sum of the charges for the cations and anions must equal zero.

Summary of Rules for Naming an Ionic Compound:

Name cation first and anion second, following these rules:
- If the cation has more than one possible charge, indicate the charge with Roman numerals in parentheses.
- If the anion is mono-atomic, drop the ending of the element's name and add "-ide" suffix.
- If the anion is polyatomic, name it (memorize the names).
- If the compound contains H_2O, attach the appropriate Greek prefix to the word hydrate.

EXERCISE 8: Name each compound: a) Na_2O b) $CuCl_2$ c) $MgSO_4 \cdot 7H_2O$

SOLUTION:
a) Na_2O: Sodium has a +1 charge to it so its name will be sodium.
     Oxygen is the element in the anion, so its name is ox-ide.
     The correct name is: <u>sodium oxide</u>

b) $CuCl_2$: Copper has two possible charges; this one must be the +2 cation to
      counter the -1 from each Cl. Its name will be copper(II).
      Chlorine is the element in the anion, so its name is chlor-ide.
      The correct name is: <u>copper(II) chloride</u>

c) $MgSO_4 \cdot 7H_2O$:
      Magnesium has a +2 charge to it so its name will be magnesium.
      SO_4^{2-} is the polyatomic ion sulfate.
      7 waters are in one formula of this compound so hepta-hydrate.
      The correct name is: <u>magnesium sulfate heptahydrate</u>

EXERCISE 9: Write the chemical formula for: a) potassium dichromate; b) cobalt(II) chloride hexahydrate

SOLUTION:
a) potassium dichromate
Potassium has the symbol K^+. Dichromate is the polyatomic anion with the formula $Cr_2O_7^{2-}$.
Two potassium cations are needed for every dichromate anion to make the sum of the charges on all ions equal zero.
The correct formula is: $K_2Cr_2O_7$

b) cobalt(II) chloride hexahydrate
Cobalt(II) has the symbol: Co^{2+}. Chloride has the symbol: Cl^-.
Two Cl^- ions are needed for each Co^{2+}.
Hexa-hydrate indicates 6 water molecules.
The correct formula is: $CoCl_2 \cdot 6H_2O$

EXERCISE 10: The slightest difference in a chemical formula or name is significant. Notice the names and corresponding formulas of these chemicals to help you see the value of learning proper chemical nomenclature.

$Ca(NO_3)_2$ = calcium nitrate $Cu(OH)_2$ = copper(II) hydroxide
$Ca(NO_2)_2$ = calcium nitrite CuO = copper(II) oxide
Ca_3N_2 = calcium nitride Cu_2O = copper(I) oxide

Decision Tree for Naming Chemical Compounds

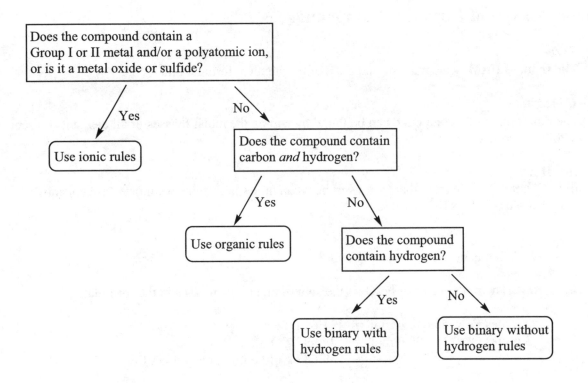

Try It #2: Name each compound: a) AgBr   b) Na_2CO_3 c) PI_3 d) CO_2

e) (black = carbon; white = hydrogen)   f)

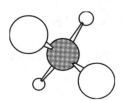

3.4   AMOUNTS OF COMPOUNDS

Helpful Hint
- You must specify what you are quantitating when using moles because the number can be used in many contexts. Using Fe_2O_3 as an example:

    One mole of Fe_2O_3 contains 2 moles of Fe atoms and 3 moles of O atoms
    One mol of Fe_2O_3 therefore contains this many atoms of each:

    2 mol Fe atoms x 6.022×10^{23} atoms/mol = 1.204×10^{24} atoms Fe
    3 mol O atoms x 6.022×10^{23} atoms/mol = 1.8067×10^{24} atoms O

Molar Masses of Chemical Compounds

Key Term:
- **Molar mass (MM)** - the mass of one mole of its chemical formula unit.

Key Concept:
- The molar mass of a compound can be found by adding the molar masses of all elements present within the compound.

Helpful Hint
Recall in Chapter 2 we learned that to convert between moles and atoms we must use Avogadro's number: 1 mole = 6.022×10^{23} particles.

EXERCISE 11: What is the molar mass of Fe_2O_3?

   SOLUTION: The molar mass of Fe_2O_3 is the sum of all atomic masses in the formula:

      2 moles Fe x 55.845 g/mol =   111.69 g Fe
      3 moles O x 15.999 g/mol =   47.997 g O
                                  =   159.687 g of Fe_2O_3 in one mole of Fe_2O_3

      The molar mass of Fe_2O_3 is 159.687 g/mol.

EXERCISE 12: How many grams of $MgCl_2$ are in 1.500 mol of magnesium chloride?

   STRATEGY: To convert between grams and moles of a substance, use its molar mass.

      Molar mass of $MgCl_2$ is 1(24.305 g/mol) + 2(35.453 g/mol) = 95.211 g/mol

   SOLUTION:

$$1.500 \text{ mol } MgCl_2 \text{ x } \frac{95.211\,\text{g}}{1\,\text{mol}} = 142.8\,\text{g of } MgCl_2$$

A similar method is employed when converting between mass of a compound and moles of ions within it as shown in the next exercise.

EXERCISE 13: How many moles of Cl⁻ are in 0.255 g of magnesium chloride?

SOLUTION:

$$0.255 \text{ g MgCl}_2 \text{ x } \frac{1 \text{ mol}}{95.211 \text{ g}} \text{ x } \frac{2 \text{ mol of Cl}^-}{1 \text{ mol MgCl}_2} = 5.36 \text{x} 10^{-3} \text{ mol of Cl}^-$$

EXERCISE 14: How many atoms of oxygen are in 75.0 g of Fe_2O_3?

STRATEGY: To convert between grams and atoms of a substance, we must use moles. Also, we see that there are 3 atoms of oxygen in each unit of Fe_2O_3, or 3 moles of oxygen atoms per each mole of Fe_2O_3. Notice that in exercise 11, we determined the molar mass of Fe_2O_3.

SOLUTION: Set up the conversions:

$$75.0 \text{ g Fe}_2\text{O}_3 \text{ x } \frac{159.678 \text{ g}}{1 \text{ mol}} \text{ x } \frac{3 \text{ mol O atoms}}{1 \text{ mole Fe}_2\text{O}_3} \text{ x } \frac{6.022 \text{x} 10^{23} \text{ atoms}}{1 \text{ mol}} = 2.16 \text{x} 10^{28} \text{ atoms of O}$$

The answer seems reasonable - atoms are quite small, so our answer should be very large.

Exercise 15 illustrates that this type of problem can be approached from more than one angle.

EXERCISE 15: How many atoms of carbon are in 1.05 g of toluene, C_7H_8?

STRATEGY AND SOLUTION: We need the molar mass of toluene to convert grams to moles.

$$1.05 \text{ g C}_7\text{H}_8 \text{ x } \frac{1 \text{ mol}}{92.141 \text{ g}} x \frac{7 \text{ mol of C}}{1 \text{ mol C}_7\text{H}_8} x \frac{6.022 x 10^{23} \text{ atoms C}}{1 \text{ mol C}} = 4.80 x 10^{22} \text{ atoms of C}$$

Alternate method:

$$1.05 \text{ g C}_7\text{H}_8 \text{ x } \frac{1 \text{ mol}}{92.141 \text{ g}} x \frac{6.022 x 10^{23} \text{ molecules C}_7\text{H}_8}{1 \text{ mol C}_7\text{H}_8} x \frac{7 \text{ atoms C}}{1 \text{ molecule C}_7\text{H}_8} = 4.80 x 10^{22} \text{ atoms of C}$$

Both methods work equally well.

EXERCISE 16: Given the structure of Fluoxetine (Prozac®), determine the mass of the compound that contains 7.689×10^{21} atoms of H.

STRATEGY AND SOLUTION: To convert between mass and atoms we will definitely need to use units of moles, so the first step is to find the molecular formula and molar mass of Fluoxetine.

Counting up all atoms yields a molecular formula of $C_{17}H_{18}F_3NO$. This compound will have a molar mass of 309.338 g/mol. Determine the number of moles of H atoms present, then how many moles of Fluoxetine contain that number of hydrogens, and then convert to grams.

$$7.689 \times 10^{21} \text{ atoms H} \times \frac{1 \text{ mol H}}{6.022 \times 10^{23} \text{ atoms H}} \times \frac{1 \text{ mol of } C_{17}H_{18}F_3NO}{18 \text{ mol H}} \times \frac{309.329 \text{ g}}{1 \text{ mol}}$$

$$= 0.02194 \text{ g of } C_{17}H_{18}F_3NO$$

Alternatively, this problem could have been done by converting atoms to molecules first.

3.5   DETERMINING CHEMICAL FORMULAS

Key Terms:
- **Mass percent composition** – the relative mass of each element present in a given compound.
- **Empirical formula (EF)** – the simplest ratio of atoms within a chemical formula.
- **Molecular formula (MF)** – the actual ratio of each element that exists within a molecule.

Chemical	Empirical Formula	Molecular Formula	Noteworthy
Propane	C_3H_8	C_3H_8	Sometimes the EF & MF are the same for a compound.
Acetylene	CH	C_2H_2	It is possible for two compounds to have the same EF but different MFs (compare acetylene to benzene).
Benzene	CH	C_6H_6	
Table Salt	NaCl	-	For an ionic compound the EF is all that is reported because (as you will see later) ionic compounds don't exist as individual molecules.

Key Concept:
- Every chemical compound has a definite chemical formula, which indicates that every chemical compound has a definite percent mass composition. If you know the chemical formula of a compound, you can calculate its percent mass composition.

$$\begin{array}{ll} \text{\% mass composition} & = \dfrac{\text{mass of element in compound}}{\text{mass of entire compound}} \text{ x 100} \\ \text{of an element in compound} & \end{array}$$

Helpful Hint
- It is often useful to base the calculation on one mole of substance because that will make the mass of compound = molar mass.

EXERCISE 17: Calculate the percent mass composition of carbon in acetaldehyde, C_2H_4O.

STRATEGY AND SOLUTION:

$$\begin{aligned} MM &= (2 \text{ mol x } 12.011\text{g/mol}) + (4 \text{ mol x } 1.0079 \text{ g/mol}) + (1 \text{ mol x } 15.999 \text{ g/mol}) \\ &= 44.053 \text{ g/mol} \end{aligned}$$

$$\%C = \frac{\text{mass of C in } C_2H_4O}{\text{mass of } C_2H_4O} \text{ x } 100 = \frac{2 \text{ mol C } (12.01115 \text{ g/mol})}{44.053 \text{ g } C_2H_4O} \text{x}100 = 54.53\% \text{ C}$$

Try It #3: Determine the % mass composition of K_2CrO_4.

Determining the Empirical Formula or Molecular Formula from % Mass Composition

Key Concept:
- If you know the percent composition of a chemical compound, you can determine its empirical formula.

Usually a chemist is trying to find the molecular formula of an unknown compound and must generate the formula from elemental analysis data. Calculations that lead to the determination of a molecular formula are:

mass % --(assume 100g) ---mass of each element

mass of each element-----(MM)---moles of each element-----mol ratios

Helpful Hints
- Apply the guidelines on page 90 in text ("Elemental Analysis") to determine empirical formulas.
- Mass percentages should add to give 100%; if they do not, assume remainder is oxygen.
- It's useful to assume 100.00 g of whole so that % can be easily translated to grams.

EXERCISE 18: Tartaric acid is produced during fermentation of wine. Solid tartaric acid sometimes deposits on corks in old wine bottles. The mass percent composition of tartaric acid is 32.01% C and 4.03% H. Determine the empirical formula of tartaric acid.

STRATEGY:
Notice that the % masses are much less than 100%. The amount missing is the %O: 63.96%.

1. Assume 100.00 g of sample. We have 32.01 g C, 4.03 g H, and 63.96 g O.
Convert the grams to moles:

$$32.01 \text{ g C } \times \text{ 1 mol}/12.011 \text{ g} = 2.665 \text{ mol C}$$
$$4.03 \text{ g H } \times \text{ 1 mol}/1.0079 \text{ g} = 3.998 \text{ mol H}$$
$$63.96 \text{ g O } \times \text{ 1 mol}/15.999 \text{ g} = 3.998 \text{ mol O}$$

2. Divide each by the smallest number of moles (in this case, moles of C):

$$\text{C}: \frac{2.665}{2.665} = 1.000 \qquad \text{H}: \frac{3.998}{2.665} = 1.500 \qquad \text{O}: \frac{3.998}{2.665} = 1.500$$

3. Multiply the numbers through by some number to obtain integer values. The hydrogens and oxygens are far off from integer values, but if we multiplied all the values through by 2, we'd have integers:

$$\text{C: 2} \qquad\qquad \text{H: 3} \qquad\qquad \text{O: 3}$$

These ratios are already integer values so no final rounding off is necessary. Place each ratio as a subscript after each element in the formula.

SOLUTION: The empirical formula of tartaric acid is: $C_2H_3O_3$.

Key Concept:
- The molecular formula indicates the actual number of each type of atom present in a compound. It can be found from the empirical formula and the molar mass.

EXERCISE 19: Tartaric acid has a molar mass of 150.09 g/mol. Determine the molecular formula for tartaric acid.

STRATEGY: The mass associated with the empirical formula is called the empirical mass, EM:

$$\text{Molar mass / empirical mass} = \text{factor by which MM and EM differ.}$$

Empirical mass for $C_2H_3O_3$
   = (2 mol C x 12.011 g/mol) + (3 mol H x 1.0079 g/mol) + (3 mol O x 15.999 g/mol) = 75.043 g

$$\text{Molar mass/empirical mass} = 150.09 / 75.04 = 2.000$$

Subscripts in the empirical formula therefore must be multiplied by 2 in order to obtain the molecular formula.

SOLUTION: The molecular formula of tartaric acid is $C_4H_6O_6$.

Elemental Analysis

Mass percents are typically obtained by some type of elemental analysis. If only one experiment is performed for the elemental analysis, then the empirical formula can be determined directly from the results of that experiment. However, if multiple analyses are performed, the data must be converted to % masses first so that they can be compared to each other. In each type of analysis, the elements within the original chemical undergo chemical reactions to get converted to other chemicals that are easily analyzed.

Elemental Analysis Method	Particularly useful for:	Example
Combustion Analysis	Carbon-based compounds	All the C in the compound gets converted to CO_2; all the H gets converted to H_2O. Measure CO_2 & H_2O quantities produced to find C & H in original sample.
Decomposition Analysis	Compounds that decompose easily into their elements	HgO becomes Hg (l) and $O_2(g)$
Precipitation Analysis*	Halide-containing compounds	All the Cl in the compound gets converted to AgCl (s)

*Precipitation reactions are covered in Section 4.5 of the text.

EXERCISE 20: 142.70 mg of a hydrocarbon were placed in a combustion chamber for analysis. The analysis showed that 477 mg of CO_2 and 111.60 mg of H_2O were produced. What is the empirical formula of the hydrocarbon?

> *STRATEGY:* Convert masses of water and CO_2 to masses of C & H. Then determine if those masses add up to the total mass of the original compound. If not, then assume the difference is oxygen. Note: we can do the calculation in units of milligrams or grams – the ratio will still be the same. We'll show the calculation in milligrams but you may wish to try it on your own in units of grams.

$$477 \text{ mg CO}_2 \text{ x} \frac{1 \text{ mol}}{44.01 \text{ g}} \text{ x} \frac{1 \text{ mol C}}{1 \text{ mol CO}_2} \text{ x} \frac{12.011 \text{ g}}{1 \text{ mol}} = 130.2 \text{ mg of C}$$

$$111.60 \text{ mg H}_2\text{O x} \frac{1 \text{ mol}}{18.015 \text{ g}} \text{ x} \frac{2 \text{ mol H}}{1 \text{ mol H}_2\text{O}} \text{ x} \frac{1.0079 \text{ g}}{1 \text{ mol}} = 12.487 \text{ mg of H}$$

The sum of these is equal to 142.69 mg, so no oxygen is present.

$$\text{Mol of C} = 130.2 \text{ mg C x} \frac{1 \text{ mol}}{12.011 \text{ g}} = 10.840 \text{ mmol C}$$

$$\text{Mol of H} = 12.487 \text{ mg H x} \frac{1 \text{ mol}}{1.0079 \text{ g}} = 12.389 \text{ mmol H}$$

Dividing by the smallest number we get: 12.389 / 10.840 = 1.143. This number is not easily rounded to an integer so we must multiply by some number to obtain an integer ratio. Multiplying by 7 seems to work well to give a ratio of 8/7.

SOLUTION: The empirical formula of the compound is therefore: C_7H_8.

3.6   AQUEOUS SOLUTIONS

Key Terms:
- **Solution** – a homogeneous mixture composed of a solvent and one or more solutes.
- **Solute** – substance present in the smaller amount in the solution.
- **Solvent** – substance present in the greater quantity in the solution. If the solvent is water, the solution is said to be "aqueous" and noted "(aq)."
- **Concentration** – amount of solute in a given volume of solution.

Units and Conversions:
- Molarity = moles of solute/volume of solution.   This is one way in which concentrations are expressed. The units for molarity are "mol/L".

Helpful Hint
- Concentration is typically shown in brackets [   ].

For example: If 1.2 moles of NaCl were dissolved in water so that the total volume of solution is 1.00 L, the solution would be 1.2 M. It would be expressed in one of these ways:

$$[NaCl] = 1.2 \text{ M} \quad \text{or} \quad 1.2 \text{ M NaCl (aq)}$$

EXERCISE 21: Convince yourself that these solutions would be 1.2 M NaCl (aq) too: a) 500.0 mL of solution that contained 0.60 moles of NaCl;   b) 3.0 L of solution that contained 3.6 moles of NaCl.

SOLUTION:
a)   500.0 mL of solution that contained 0.60 moles of NaCl

$$M = \frac{\text{Moles of solute}}{\text{Volume of solution}} = \frac{0.60 \text{ mol NaCl}}{0.5000 \text{ L solution}} = 1.2 \text{ M NaCl}$$

b)   3.0 L of solution that contained 3.6 moles of NaCl

$$M = \frac{\text{Moles of solute}}{\text{Volume of solution}} = \frac{3.60 \text{ mol NaCl}}{3.0 \text{ L solution}} = 1.2 \text{ M NaCl}$$

EXERCISE 22: Now, what volume of a 1.20 M NaCl (aq) solution contains 1.86 mol of NaCl?

STRATEGY: We can use molarity to convert between mol and volume. We're solving for volume so we'll need the volume portion of the molarity on the top.

SOLUTION: Volume of Solution $= 1.86 \text{ mol NaCl} \times \dfrac{1 \text{ L}}{1.20 \text{ mol NaCl}} = 1.55 \text{ L of solution}$

This answer makes sense numerically because there needs to be more than 1 L to get more than 1.20 moles of NaCl.

How Solute Particles Exist in Solution

Key Concepts:
- Non-ionic compounds will remain intact in solution.
- Ionic compounds – separate into ions in solution (dissociate).
- Molecular pictures help to illustrate the solute particles within solutions.

For example:

Solution	Aqueous solute will exist as:	Note:
0.1 M $C_6H_{12}O_6$ (aq)	0.1 M $C_6H_{12}O_6$	Non-ionic so solute particles stay intact
0.1 M NaCl (aq)	0.1 M Na^+ & 0.1 M Cl^-	Ionic so solute dissociates into its ions
0.1 M $MgCl_2$ (aq)	0.1 M Na^+ and 0.2 M Cl^-	Ionic so solute dissociates into its ions. [Cl^-] = 0.1 M x 2 mol Cl^-/1 mol $MgCl_2$ = 0.2 M
0.01 M $MgCl_2$ (aq)	0.01 M Na^+ & 0.02 M Cl^-	Same here. Note that this is 1/10 as concentrated as the 0.1 M solution
0.1 M $FeSO_4$ (aq)	0.1 M Fe^{2+} and 0.1 M SO_4^{2-}	Ionic so solute dissociates into ions. Note that atoms within polyatomic ions stay together
0.1 M $Fe(NO_3)_3$ (aq)	0.1 M Fe^{3+} and 0.3 M NO_3^-	Same here. Note that there are 3 NO_3^- per 1 $Fe(NO_3)_3$ so [NO_3^-] is 0.3 M

(Can you already see the value of memorizing the names and formulas of the polyatomic ions?)

EXERCISE 23: Draw a molecular picture to illustrate the solute in each:
a) 0.05 M $C_6H_{12}O_6$ (aq);   b) 0.05 M $MgCl_2$ (aq);   c) 0.01 M $MgCl_2$ (aq)

STRATEGY:
a) 0.05 M $C_6H_{12}O_6$ (aq): this solute is not ionic so the molecules will stay together. Let's use five molecules of $C_6H_{12}O_6$ in the picture.
b) 0.05 M $MgCl_2$ (aq): this solute is ionic so it will form 1 Mg^{2+} for every 2 Cl^-. The molar concentration of this solution is the same as in part "a", so we should show the same relative number of solute particles: 5 Mg^{2+} and 10 Cl^-. We'll designate Mg^{2+} as black balls and Cl^- as white balls.
c) 0.01 M $MgCl_2$ (aq): The solute is the same as in part "b", but it is 1/5 the concentration. We should show 1 Mg^{2+} and 2 Cl^-.

SOLUTION: Here are the pictures:

0.05 M $C_6H_{12}O_6$ (aq)   0.05 M $MgCl_2$ (aq)   0.01 M $MgCl_2$ (aq)

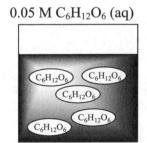 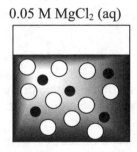 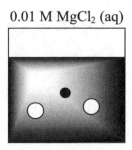

Try It #4: Draw a molecular picture to illustrate the solute in aqueous 0.1 M $FeSO_4$ and 0.1 M $Fe_2(SO_4)_3$. Show correct relative amounts of ions in each.

Preparing Solutions and Dilutions

Solutions with concentrations expressed in Molarity are prepared in volumetric flasks, which are calibrated to contain a specific total volume of liquid. It is practical in the lab to measure grams of a substance rather than moles, so molar solutions will often be prepared by weighing out a certain mass of solid and dissolving it in water in a certain size volumetric flask.

Units and Conversions:
- $1 L = 1000 mL$
- $MM = mass/moles$ Use molar mass to convert between grams and moles of a given substance.
- Molarity = moles of solute/L of solution   Use molarity as a conversion factor between volume of a solution and the number of moles of solute it contains.

EXERCISE 24: What mass of $AgNO_3$ is needed to make 250.00 mL of 0.125 M $AgNO_3$ (aq)?

STRATEGY: The volume of solution and [$AgNO_3$] are given so moles of $AgNO_3$ can be calculated. The conversion of moles of $AgNO_3$ to grams of $AgNO_3$ can be determined by using molar mass.

SOLUTION:

$$\text{Mass of } AgNO_3 = 0.25000 \text{ L solution x } \frac{0.125 \text{ mol } AgNO_3}{1 \text{ L soluttion}} \text{ x } \frac{169.874 \text{ g}}{1 \text{ mol}} = 5.31 \text{ g } AgNO_3$$

This answer seems reasonable because one should need much less than one mole of solute to prepare less than a liter of a dilute solution.

EXERCISE 25: A student weighed out 0.222 g of glucose $C_6H_{12}O_6$, transferred it to a 50.00 mL volumetric flask, and added water to the 50.00 mL line. What is [$C_6H_{12}O_6$] ?

STRATEGY: Moles of glucose can be found from grams using molar mass. We know the volume of solution, but need to account for units (mL to L).

SOLUTION:

$$[C_6H_{12}O_6] = \frac{\text{Moles of } C_6H_{12}O_6}{\text{Volume of solution}}$$

$$= 0.222 \text{ g } C_6H_{12}O_6 \text{ x } \frac{1 \text{ mol}}{180.0 \text{ g}} \text{ x } \frac{1000 \text{ mL}}{1 \text{ L}} \text{ x } \frac{1}{50.00 \text{ mL}} = 0.0247 \text{ M } C_6H_{12}O_6$$

Useful Relationship:
- $M_i V_i = M_f V_f$ Quite often chemists will use an existing solution of known concentration and dilute it to obtain a concentration that they need. The concentration of the diluted solution can be found by this relationship.
  - M_i = the concentration of the initial solution used
  - V_i = the volume of the initial solution that is transferred to the new container
  - M_f = the concentration of the final diluted solution
  - V_f = the total volume of the final diluted solution

The relationship works because the number of moles of solute being transferred is equal to the number of moles of solute in the entire container (moles$_i$ = moles$_f$). A picture can illustrate this concept:

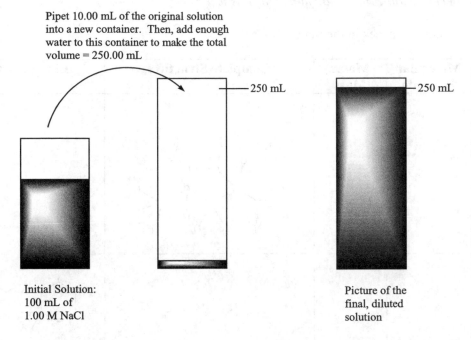

Pipet 10.00 mL of the original solution into a new container. Then, add enough water to this container to make the total volume = 250.00 mL

250 mL

250 mL

Initial Solution:
100 mL of
1.00 M NaCl

Picture of the
final, diluted
solution

The picture shows that 10.00 mL x 1.00 M = 0.010 moles of solute were transferred into the new, 250 mL container. In this case, M_i = 1.00 M, V_i = 10.00 mL, V_f = 250 mL, and M_f = the concentration of the diluted solution to be calculated.

EXERCISE 26: What is the molarity of $MgCl_2$ in a solution prepared by taking 10.00 mL of 0.750 M $MgCl_2$ (aq) and diluting it to 50.00 mL?

STRATEGY: Recognize that this is a dilution problem and identify the variables in the equation:
$$M_i V_i = M_f V_f$$

We have the original concentration of solution, the amount of it that was transferred, and the volume of the diluted solution, so we can solve the equation for M_f:

SOLUTION:
$$M_f = \frac{M_i V_i}{V_f} = \frac{(0.750\,M)(10.00\,mL)}{50.00\,mL} = 0.150\,M\,MgCl_2\,(aq)$$

Try It #5: What is the concentration of Cl⁻ in the resulting diluted solution in Exercise 28?

Helpful Hint
- Extra information could have been tossed into the above question such as: "A student prepared 500 mL of 0.750 M $MgCl_2$ (aq). She then transferred 10.00 mL of this solution to a 50.00 mL volumetric flask and diluted to the mark with water. What is the molarity of the diluted solution?" Notice that it does not matter how much solution was originally prepared. The only things that matter about the original solution are: what was its concentration? (0.750 M) and how much of it made it into the final container? (10.00 mL)

Chapter 3 Self-Test

You may use a periodic table and a calculator for this test.

1. Complete the empty spaces in the table:

	Molecular Formula	Molar Mass	Complete Structure	Line Structure
Iso-octane (gas)			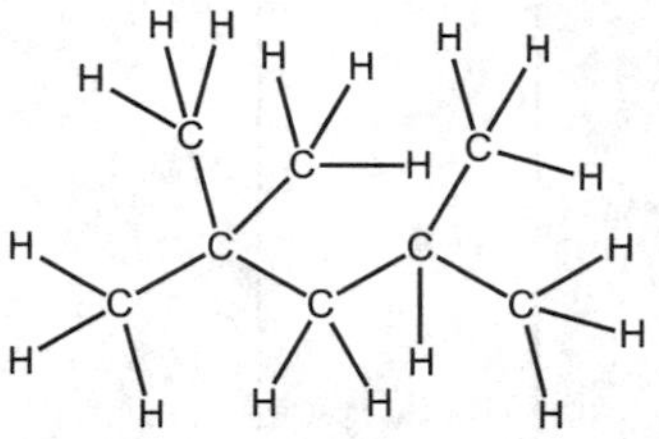	
α-ionone (aroma of violets)				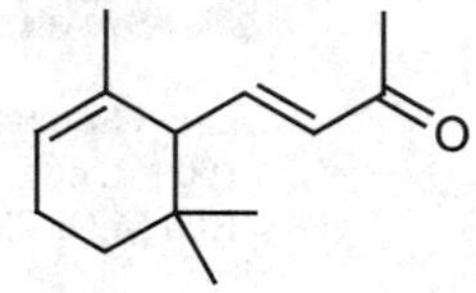

2. Draw the ball and stick model for aspirin given its line structure:

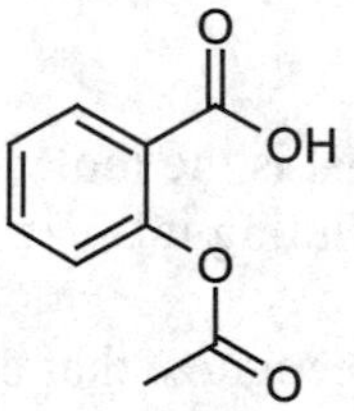

3. Draw the space-filling model for NO.

4. Provide the name or chemical formula for each:
   a) nitrogen triiodide   b) $CaCl_2 \cdot 2H_2O$ c) HCN   d) aluminum oxide

5. A sample of ammonia, NH_3, contains 4.51×10^{24} atoms of hydrogen. How many grams of ammonia are present?

6. TCDD is a highly toxic chemical that contaminated Agent Orange defoliant used in the Vietnam War. Its structure is shown below. Elemental analysis on an unknown compound – suspected to be TCDD - yielded the following data: 44.77% C, 1.25% H, 44.04% Cl, and 9.94% O. Mass spectrometry yielded a molar mass of 322 g/mol. Is it possible that this unknown is TCDD?

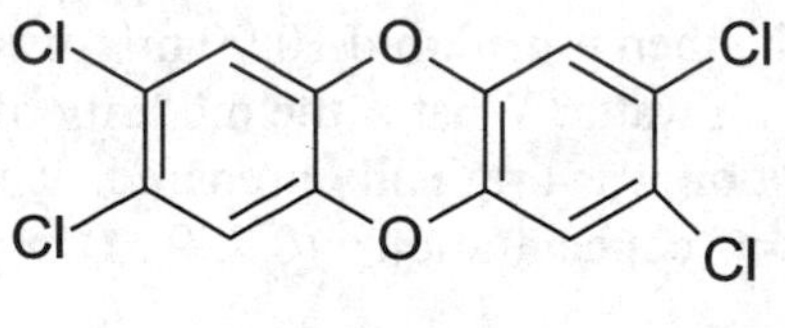

7. a) How many Cu ions are in 100.0 mL of 0.0355 M copper(II) phosphate? b) How many grams of Cu is this?

8. What volume of water would need to be added to 50.00 mL of 6.25×10^{-2} M sodium hydroxide (aq) to make a solution that is 0.00250 M in OH⁻?

9. 25.00 mL of a 1.20 M solution of $KMnO_4$ (aq) was diluted to 100.0 mL
   a) Calculate the molarity of the resulting solution.
   b) Draw a molecular picture to illustrate the original solution and the diluted solution, indicating relative quantities of solute in the picture.

Answers to Try Its

1.   a)   b)

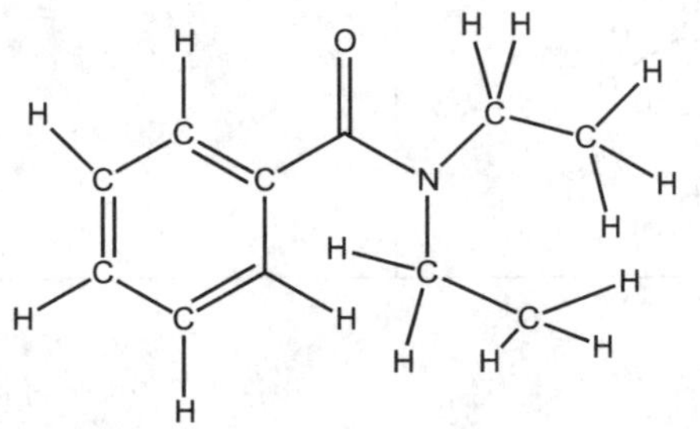

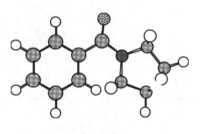

   c) No, this structure has 12 carbons and DEET has 11.

2. a) silver bromide   b) sodium carbonate   c) phosphorus triiodide   d) carbon dioxide
   e) methane   f) 1-pentyne
3. 40.27% K, 26.78% Cr, and 32.96% O
4. Fe^{2+} = black balls; SO_4^{2-} = white balls; Fe^{3+} = grey balls. What is important to show is the relative number of ions in each solution. For every one Fe^{2+} in the first solution, there are two Fe^{3+} ions in the second solution, as the formulas indicate.

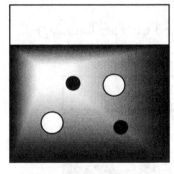   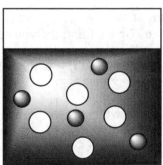

5. [Cl⁻]=0.300 M

Answers to Self-Test

1.

	Molecular Formula	Molar Mass	Complete Structure	Line Structure
Iso-octane (gas)	C_8H_{18}	114.23 g/mol	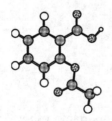	
α-ionone (aroma of violets)	$C_{13}H_{20}O$	192.30 g/mol		

2.

3.

4. a) NI_3 b) calcium chloride dihydrate   c) hydrogen cyanide   d) Al_2O_3
5. 42.5 g NH_3
6. yes, the molecular formulas are the same
7. a) 6.41×10^{21} ions of Cu   b) 0.677 g Cu (hint: did you account for 3 Cu's per formula?)
8. 1.20 L
9. a) 0.300 M   b) K^+ = black balls; MnO_4^- = white balls

Chapter 4: Chemical Reactions and Stoichiometry

Learning Objectives

In this chapter, you will learn how to:

- Write a balanced chemical equation for a given reaction.
- Predict the products for (some) chemical reactions.
- Use stoichiometry to determine the amount of chemicals formed and/or consumed in a reaction.
- Calculate the percent yield for a reaction.
- Use the concept of limiting reactant to determine the maximum amount of product that can form, as well as the amounts of any leftover unreacted starting materials.
- Categorize a given chemical reaction as either "precipitation," "acid-base," or "redox."

Practical Aspects

In the "big picture" of chemistry, this is probably the most practical chapter in the book. Chemists are interested in changes in matter. In this chapter, you'll learn how to predict what chemical change will take place when two chemicals combine AND how to track the amounts of these chemicals by using stoichiometry. In virtually every quantitative analysis that a chemist performs, stoichiometric calculations need to be done. *Mastering this chapter is crucial to your understanding of chemistry.* You will use these skills again and again throughout the rest of this text.

Every chemistry laboratory course you take will at some point use stoichiometry to solve a quantitative analysis question. For example, when you synthesize a chemical compound, you'll need to determine which chemical is the limiting reactant and the excess reactant. At the end of the synthesis, you'll need to calculate a percent yield for the reaction. This chapter therefore has a huge impact on your understanding of both the General Chemistry lecture *and* the lab.

4.1   WRITING CHEMICAL EQUATIONS

Ozone (O_3) molecules react to form oxygen molecules. We saw in Chapter 2 that the Law of Conservation of Mass indicates that all atoms must be accounted for in the process. If we start with 4 ozone molecules, we'll need to make 6 oxygen molecules, as this picture shows.

Before Reaction

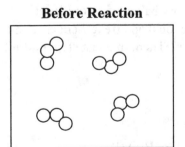

After Reaction

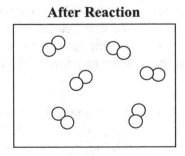

Chemists use a shorthand notation to describe chemical reactions. This shorthand indicates the recipe, or simplest whole-number ratio of chemicals involved in the reaction. The simplest whole-number ratio in the above reaction is every two ozones will make three oxygens. The chemical shorthand, or balanced chemical equation to show this process is:

$$2\ O_3\ (g) \rightarrow 3\ O_2\ (g)$$

This shorthand is read, "Two gaseous ozone molecules react to form three gaseous oxygen molecules." Notice that the arrow can be interpreted as "react to form" or "will make" or "will produce."

Key Terms:
- **Balanced chemical equation** – shorthand notation for a chemical change which shows the simplest whole-number ratio of all substances present in the reaction.
- **Reactant(s)** – starting material(s). These are shown to the left of the arrow. Different types of reactants are separated by a "+" sign.
- **Product(s)** – the chemical(s) that is/are produced in the reaction. Products are shown to the right of the arrow. Different types of products are separated by a "+" sign.
- **Coefficient** – the number in front of a chemical within a balanced equation. The coefficient indicates how many of that particular chemical will be present in the simplest whole-number ratio of substances.

Key Concepts:
- In a balanced chemical equation, both matter and charge are conserved.
- The phase of each chemical in the reaction is written in parentheses immediately following the chemical: (s) = solid, (l) = liquid, (g) = gas, (aq) = aqueous.

Try It #1: Write a balanced chemical equation for the reaction of CO with O_2 that correlates with the molecular pictures shown here:

Before Reaction	**After Reaction**
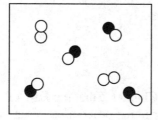	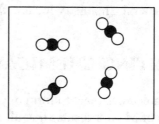

Helpful Hints
- Choose the most complicated looking molecule and begin by balancing atoms in it.
- Look for clusters of atoms that are conserved (polyatomic ions) and treat them as a unit.
- If necessary, create a fraction during the balancing process. Then, in a last step, multiply all coefficients by a factor to get rid of the fraction.

EXERCISE 1: Balance these chemical equations.
a) H_2 (g) + O_2 (g) → H_2O (l),
b) H_2SO_4 (aq) + NaOH (aq) → Na_2SO_4 (aq) + H_2O (l),
c) Fe (s) + O_2 (g) → Fe_2O_3 (s),
d) Pb^{2+} (aq) + I^- (aq) → PbI_2 (s)

STRATEGY: Follow guidelines and helpful hints to balance each equation.

a) H_2 (g) + O_2 (g) → H_2O (l). Hydrogens are balanced, but oxygens are not.

Balance Os with a coefficient of "2" in front of the water molecule: H_2 (g) + O_2 (g) → 2 H_2O (l).

Now hydrogens are no longer balanced.

Add a "2" in front of the H_2 molecule to finish accounting for the Hs: 2 H_2 (g) + O_2 (g) → 2 H_2O (l).

b) H_2SO_4 (aq) + NaOH (aq) → Na_2SO_4 (aq) + H_2O (l). Notice that "SO_4" is conserved, so treat it as a unit. There is one "SO_4" on each side of the arrow, so they are balanced.

Balance Na atoms with a "2" coefficient: H_2SO_4(aq) + 2 NaOH(aq) → Na_2SO_4(aq) + H_2O(l).

Finish balancing the equation by placing a "2" in front of the water molecules:

H_2SO_4(aq) + 2 NaOH(aq) → Na_2SO_4(aq) + 2 H_2O(l).

c) Fe (s) + O_2 (g) → Fe_2O_3 (s).

Put a "2" in front of the Fe to balance Fe atoms: 2 Fe (s) + O_2 (g) → Fe_2O_3 (s).

We need 1.5 O_2 molecules to get 3 oxygens: 2 Fe (s) + 3/2 O_2 (g) → Fe_2O_3 (s).

Multiply through by 2 to get rid of the fraction: 4 Fe (s) + $3O_2$ (g) → 2 Fe_2O_3 (s).

d) Pb^{2+} (aq) + I^- (aq) → PbI_2 (s).

Place a "2" in front of the I^-: Pb^{2+} (aq) + 2 I^- (aq) → PbI_2 (s).

This reaction contains charges. The total charge of the product is zero. Notice that the reactant charges balance with this: $[(+2) + 2(-1)] = 0$

SOLUTION: a) 2 H_2 (g) + O_2(g) → 2 H_2O(l); b) H_2SO_4(aq) + 2 NaOH(aq) → Na_2SO_4(aq) + 2 H_2O(l); c) 4 Fe (s) + $3O_2$ (g) → 2 Fe_2O_3 (s); d) Pb^{2+} (aq) + 2 I^- (aq) → PbI_2 (s)

Try It #2: Balance these chemical equations: a) NH_3 (g) + N_2O (g) → H_2O (g) + N_2 (g),
b) CS_2 (s) + O_2 (g) → CO_2 (g) + SO_2 (g), c) H_3PO_4 (aq) + KOH (aq) → H_2O (l) + K_3PO_4 (aq).

4.2   THE STOICHIOMETRY OF CHEMICAL REACTIONS

Interpretations of Balanced Equations

The reaction: 2 CO (g) + O_2 (g) → 2 CO_2 (g) can be interpreted on different levels:

Molecular: 2 molecules of CO and 1 molecule of O_2 make 2 molecules of CO_2
Molar: 2 moles of CO and 1 mole of O_2 make 2 moles of CO_2

The balanced equation acts as a recipe for the chemical reaction. If you have a specific amount of one reactant, you can figure out how much of the other reactant is needed or how much product can be made.

EXERCISE 2: Given the reaction: 2 CO (g) + O_2 (g) → 2 CO_2 (g), a) how many molecules of CO are needed to make 30 molecules of CO_2? b) how many moles of O_2 would react with 6 moles of CO?

STRATEGY: a) The recipe (balanced equation) shows that 2 molecules of CO will make 2 molecules of CO_2: they are related in a 1:1 ratio. Therefore 30 molecules of CO would be needed to make 30 molecules of CO_2. b) The recipe shows that 2 moles of CO react with 1 mole of O_2, the ratio is 2 CO:1 O_2. Therefore, 6 moles of CO would need 3 moles of O_2 to react with it.

SOLUTION: a) 30 molecules of CO; b) 3 moles of O_2

Exercise 2 was quite simple. In order to be able to track chemical amounts in all situations, chemists use a systematic accounting method called "stoichiometry." In Greek, "stoicheon" means "element," so stoichiometry means "measuring, or accounting for, the elements."

Key Terms:
- **Stoichiometry** – the study of the amounts of materials consumed and produced in a chemical reaction.
- **Stoichiometric ratio** – ratio of coefficients of chemicals in a balanced chemical equation. This ratio is set up like a conversion factor.

Key Concepts:
- All relationships between two *different* chemicals must be based on stoichiometric ratios.
- *The mole is the currency of chemistry.* Stoichiometric ratios are usually based on moles, but can be based on molecules. (Molecules and moles both count numbers of items.)

EXERCISE 3: Repeat Exercise 2, but use stoichiometric ratios to solve the problem.

STRATEGY: Use the stoichiometric ratios like conversion factors.
a) The relationship we're interested in is between what's given (molecules of CO_2) and what we're trying to find (molecules of CO). The stoichiometric ratio of these substances is: 2 CO_2 to 2 CO. Start with what's been given and set up the conversion factor to cancel "CO_2" out and give us what we're trying to find, CO:

$$30 \text{ molecules } CO_2 \text{ x } \frac{2 \text{ molecules } CO}{2 \text{ molecules } CO_2} = 30 \text{ molecules of } CO$$

b) We've been given moles of CO and are asked for moles of O_2. Set up a conversion factor, as we did in part "a":

$$6 \text{ moles } CO \text{ x } \frac{1 \text{ mole } O_2}{2 \text{ mole } CO} = 3 \text{ moles } O_2$$

Again, the chemical we started with will be canceled out just like a conversion factor, so we're left with units of "moles of O_2."

SOLUTION: a) 30 molecules of CO; b) 3 moles of O_2

Take-home messages from Exercise 3:
- When doing stoichiometry problems, always include chemical formulas as units to help track what chemical we have. Cancel out the chemicals, just like canceling out units.
- Always report the unit *and* the chemical in an answer to a stoichiometry problem.
- A stoichiometric ratio can be written with or without the word "moles" shown in the ratio. In the exercise, we showed the ratio with the word "moles," here it is shown without:

$$6 \text{ moles } CO \text{ x } \frac{1 O_2}{2 CO} = 3 \text{ moles } O_2$$

Notice that the words "mole" will cancel out in the ratio. In this guide, we will include the words "moles" (or "molecules) in stoichiometric ratios. In your own work, you may choose to omit them.

EXERCISE 4: Given the reaction: $2\,CO\,(g)\ +\ O_2\,(g)\ \rightarrow\ 2\,CO_2\,(g)$,
a) how many molecules of CO are needed to make 0.25 mole of CO_2?
b) how many grams of O_2 would react with 1.22 moles of CO?
c) how many kilograms of CO_2 can be made from 5.88 kg of CO?

STRATEGY: In each case, start with the information that has been given and use conversion factors to work towards the goal. To convert from one chemical to another, use the stoichiometric ratio as a conversion factor.

a) We're given 0.25 mol CO_2 and want molecules of CO:

$$0.25\,\text{mol}\,CO_2 \; x \; \frac{2\,\text{mol}\,CO}{2\,\text{mol}\,CO_2} \; x \; \frac{6.022 x 10^{23}\,\text{molecules}}{1\,\text{mol}} = 1.5 x 10^{23}\,\text{molecules of}\,CO$$

Notice that as you read the above calculation from left to right, the moles and the CO_2s cancel in the first conversion, then moles cancel in the next conversion. The units that remain are "molecules" and "CO." The answer seems reasonable because there are a lot of molecules in a portion of a mole. (2 sig figs – based on the data provided – 0.25.)

b) We're given 1.22 moles of CO and want grams of O_2:

$$1.22\,\text{mol}\,CO \; x \; \frac{1\,\text{mol}\,O_2}{2\,\text{mol}\,CO} \; x \; \frac{31.998\,\text{g}}{1\,\text{mol}} = 19.5\,\text{g}\,O_2$$

Notice that as you read the above calculation from left to right, the moles and the COs cancel after the first conversion to leave units of "moles" and "O_2." This indicates we need to use MM of O_2 to convert from moles to grams. After the MM conversion factor, the "mol"s cancel out, and we're left with units of "g" and "O_2." (3 sig figs in answer, based on data provided – 1.22)

c) We're given 5.88 kg of CO and want kg of CO_2:

$$5.88\,\text{kg}\,CO \; x \; \frac{1000\,\text{g}}{1\,\text{kg}} \; x \; \frac{1\,\text{mol}}{28.01\,\text{g}} \; x \; \frac{2\,\text{mol}\,CO_2}{2\,\text{mol}\,CO} \; x \; \frac{44.01\,\text{g}}{1\,\text{mol}} \; x \; \frac{1\,\text{kg}}{1000\,\text{g}} = 9.24\,\text{kg}\,CO_2$$

Notice that in reading the conversions from left to right, the first conversion converts the units from "kg" to "g," the second from "g" to "mol," the third from "CO" to "CO_2," the fourth from "mol" to "g," and the last from "g" to "kg." The final units are "kg" and "CO_2."

SOLUTION: a) $1.5 x 10^{23}$ molecules of CO; b) 19.5 g of O_2; c) 9.24 kg of CO_2

Take-home messages from Exercise 4, part c:
- Notice that we couldn't convert directly from 5.88 kg of CO to 5.88 kg of CO_2. The stoichiometric ratio is based on *mole* amounts, not masses - different chemicals have different molar masses. The ratio *must* be based on "counting" numbers, like moles or molecules.
- It is possible to sometimes take shortcuts. For example:

$$5.88\,\text{kg}\,CO \; x \; \frac{1\,\text{mol}}{28.01\,\text{g}} \; x \; \frac{2\,\text{mol}\,CO_2}{1\,\text{mol}\,CO} \; x \; \frac{44.01\,\text{g}}{1\,\text{mol}} = 9.24\,\text{kg}\,CO_2$$

Practice lots of stoichiometry problems to find shortcuts of your own.

Try It #3: Bituminous coal contains approximately 5% sulfur.
a) What mass of SO_2 (g) is produced from burning 2.24 tons of coal to provide electricity to a town for a day? The reaction for the combustion of sulfur is: S (s) $+ O_2$ (g) $\rightarrow$ SO_2 (g).
b) Given that $SO_2(g) + CaO(s) \rightarrow CaSO_3$ (s), what mass of CaO (s) would be needed inside an industrial scrubber to remove all of the SO_2 (g) generated?

4.3   YIELDS OF CHEMICAL REACTIONS

Key Terms:
*   **Theoretical yield** – maximum amount of product which can theoretically be obtained in a chemical reaction. This number can be calculated using stoichiometry.
*   **Actual yield** - amount of product actually formed in the reaction. A chemist would typically weigh the product after an experiment to determine the actual yield.

Key Concept:
*   Actual yield is usually less than 100%. Why?
    1) products may get lost during separation,
    2) other reactions may be taking place (side reactions), and
    3) as concentration of each reactant decreases, there is a lower probability of molecular collisions.

Useful Relationship:
* $\% \text{ Yield} = \dfrac{\text{Actual Yield}}{\text{Theoretical Yield}} \times 100$

Chemists want to know the percent yield for a reaction because:
*   it can act as a measure of what to expect the next time the reaction is performed.
*   it can be used to determine if a reaction is economically feasible. This is especially important to assess yield in multi-step syntheses because if the yield is low for each step then there won't be much product at the very end of the synthesis.

EXERCISE 5: If 42.77 kg of chlorine gas was formed during the decomposition of 98.0 kg of NaCl, what is the percent yield of this reaction?

STRATEGY: To find % yield, we need the "actual yield" (given: 42.77 kg) and the "theoretical yield." The theoretical yield is the maximum amount that could theoretically be made from the reactants provided. Write a balanced chemical equation first:

$$2 \text{ NaCl (s)} \rightarrow 2 \text{ Na (s)} + Cl_2 \text{ (g)}$$

$$\text{Theoretical Yield} = 98.0 \times 10^3 \text{ g NaCl} \times \frac{1 \text{ mol}}{58.443 \text{ g}} \times \frac{1 \text{ mol } Cl_2}{2 \text{ mol NaCl}} \times \frac{70.91 \text{ g}}{1 \text{ mol}} \times \frac{1 \text{ kg}}{1000 \text{ g}} = 59.4455 \text{ kg } Cl_2$$

$$\% \text{ Yield} = \frac{\text{Actual Yield}}{\text{Theoretical Yield}} \times 100 = \frac{42.77 \text{ kg } Cl_2}{59.4455 \text{ kg } Cl_2} \times 100 = 71.9482 \%$$

SOLUTION: The % yield for the reaction is 71.9% (3 sig figs). A yield of less than 100% is reasonable.

4.4   THE LIMITING REACTANT

In reality, all reactants are not consumed equally. One reactant will run out first. If one of the reactants is quite expensive or toxic or is difficult to separate out from the product mixture, a chemist will purposefully put in less than the stoichometric ratio of that reactant, just to ensure that it gets completely consumed in the reaction. This chemical would be called the "limiting reactant" because it limits the amount of product that can be made. The best way to understand how to track a limiting reactant in a chemistry problem is to work through an everyday example first. Exercise 6 will do this.

EXERCISE 6: Let's say that you have 36 quarters, 22 nickels and 30 dimes. You need to buy some sodas at the vending machine. One soda costs one dollar; the machine only takes combinations of three quarters, two dimes and one nickel, (3Q, 2D, 1N). a) Which coin runs out first? b) How many sodas can you buy? c) How many of each excess coins are left over?
   Take a few minutes to determine the answers to these questions before proceeding.

STRATEGY: For some students, the three questions are so easy to answer that it is almost difficult to pinpoint the strategy used. We're going to answer these questions from a chemistry perspective here, so that you can see how the stoichiometric thought processes are broken down.

Write a balanced equation to show the recipe or relationship between the coins and the soda. Then use that equation to determine how many sodas could be bought with each coin, assuming the other coins are in excess.

$$1Q + 2D + 3N \rightarrow 1\,\text{Soda}$$

a and b)

If the number of sodas depended only upon *quarters*: $36\,Q \times \dfrac{1\,\text{Soda}}{3\,Q} = 12\,\text{Sodas}$

If the number of sodas depended only upon *dimes*: $30\,D \times \dfrac{1\,\text{Soda}}{2\,D} = 15\,\text{Sodas}$

If the number of sodas depended only upon *nickels*: $22\,N \times \dfrac{1\,\text{Soda}}{1\,N} = 22\,\text{Sodas}$

The quarters will run out first, so we can buy a maximum of 12 sodas.

c) 12 sodas can be purchased. The quarters are completely used up, so no quarters are left. To determine how many dimes and nickels are left over, first determine how many were used to buy the 12 sodas, then subtract that from the number of each coin you had to start:

$12\,\text{Sodas purchased} \times \dfrac{2\,D}{1\,\text{Soda}} = 24\,D\text{ used.}$ 30 D to start – 24 D used = 6 D left over.

$12\,\text{Sodas purchased} \times \dfrac{1\,N}{1\,\text{Soda}} = 12\,N\text{ used.}$ 22 N to start – 12 N used = 10 N left over.

SOLUTION: a) Quarters run out first. b) 12 Sodas can be purchased. c) 6 dimes and 10 nickels will be left over.

Key Terms:
- **Limiting reactant (L.R.)** – reactant that gets used up first in a chemical reaction, and therefore limits the amount of product that can be produced.
- **Excess reactant** – reactant that is present in excess. Some of this chemical will remain in the reaction container after the reaction is complete.

Key Concepts:
- The balanced equation acts as the recipe, but the amounts of each reactant actually present will influence how much product is formed.
- The moles of product produced are determined by the limiting reactant.
- To identify the limiting reactant:
  1) Start with one reactant and determine the number of moles of product it can make.
  2) Repeat the same calculation for the other reactants.
  3) Compare the moles of products. The reactant that formed the least amount of product is the limiting reactant.

Helpful Hints
- To recognize a problem as a limiting reactant problem, look to see if starting amounts of more than one reactant have been given.
- The strategy we used in Exercise 6 is the strategy described in solving a limiting reactant problem. Refer back to the thought processes in Exercise 6 as an example of how to approach a limiting reactant problem.

EXERCISE 7: If you mix 1.00 mol of $K(s)$ with 1.00 mol of $I_2(s)$, how many moles of $KI(s)$ will form?

$$2 \, K(s) \; + \; I_2(s) \; \rightarrow \; 2 \, KI(s)$$

STRATEGY:

$$1.00 \, \text{mol K} \times \frac{2 \, \text{mol KI}}{2 \, \text{mol K}} = 1.00 \, \text{mol KI} \qquad 1.00 \, \text{mol I}_2 \times \frac{2 \, \text{mol KI}}{1 \, \text{mol I}_2} = 2.00 \, \text{mol KI}$$

SOLUTION: The K can make less KI so K is the limiting reactant; 1.00 mol of KI can be made.

EXERCISE 8: 1.5 g of carbon disulfide and 3.5 g of oxygen are placed in a flask. What mass of each product and reactant is in the container after the reaction between these chemicals is complete?

$$CS_2 \, (s) \; + \; 3 \, O_2 \, (g) \; \rightarrow CO_2 \, (g) \; + \; 2 \, SO_2 \, (g)$$

STRATEGY:
If the amount of SO_2 made depended only upon CS_2:

$$1.5 \, \text{g CS}_2 \times \frac{1 \, \text{mol}}{76.14 \, \text{g}} \times \frac{2 \, \text{mol SO}_2}{1 \, \text{mol CS}_2} = 0.039401 \, \text{mol SO}_2$$

If the amount of SO_2 made depended only upon O_2:

$$3.5 \, \text{g O}_2 \times \frac{1 \, \text{mol}}{31.998 \, \text{g}} \times \frac{2 \, \text{mol SO}_2}{3 \, \text{mol O}_2} = 0.072921 \, \text{mol SO}_2$$

CS_2 is the limiting reactant because less SO_2 would be produced from it than from O_2. CS_2 is completely used up, so none of it will remain after the reaction is complete.

$$\text{Mass of } SO_2 \text{ produced: } 0.039401 \text{ mol } SO_2 \times \frac{64.064 \text{ g}}{1 \text{ mol}} = 2.5242 \text{ g } SO_2$$

The amount of CO_2 formed also depends upon the CS_2:

$$1.5 \text{ g } CS_2 \times \frac{1 \text{ mol}}{76.14 \text{ g}} \times \frac{1 \text{ mol } CO_2}{1 \text{ mol } CS_2} \times \frac{44.01 \text{ g}}{1 \text{ mol}} = 0.86702 \text{ g } CO_2$$

The amount of O_2 that reacted can be calculated from the limiting reactant (CS_2) *OR* from the amount of SO_2 produced:

$$\text{Amount of } O_2 \text{ that } reacted \text{ with } CS_2\text{: } 1.5 \text{ g } CS_2 \times \frac{1 \text{ mol}}{76.14 \text{ g}} \times \frac{3 \text{ mol } O_2}{1 \text{ mol } CS_2} \times \frac{31.998 \text{ g}}{1 \text{ mol}} = 1.8911 \text{ g } O_2 \text{ used}$$

OR

$$\text{Amount of } O_2 \text{ } used \text{ } to \text{ } make \text{ } SO_2\text{: } 0.039401 \text{ mol } SO_2 \times \frac{3 \text{ mol } O_2}{2 \text{ mol } SO_2} \times \frac{31.998 \text{ g}}{1 \text{ mol}} = 1.8911 \text{ g } O_2 \text{ used}$$

$$\text{Mass of } O_2 \text{ left over} = \text{Mass of } O_2 \text{ to start} - \text{mass of } O_2 \text{ that reacted with } CS_2$$
$$= 3.5 \text{ g} - 1.8911 \text{ g} = 1.6 \text{ g}$$

SOLUTION: 0 g CS_2, 1.6 g O_2, 2.5 g SO_2, 0.87 g CO_2 (2 sig figs based on data provided)

Tables of Amounts

The text describes another method for doing limiting reactant problems, which is termed "Table of Amounts." This table organizes mole information for a limiting reactant problem into "Starting Amount," "Change in Amount," and "Final Amount" categories.

- To determine the limiting reactant, consider each starting material. Divide the starting amount of each substance by its corresponding coefficient. The one with the smallest value is the limiting reactant.
- The number in the "Change in Amount" line always corresponds to the stoichiometric ratio in the balanced equation.
- The sign in the "Change in Amount" line always corresponds to whether a chemical is being consumed (-) or produced (+).
- The final amount is found by comparing the "starting amount" to the "change in amount."

Exercise 8 is reworked here using the Table of Amounts method.

STRATEGY: Recognize that you have a limiting reactant problem by noticing that amounts for both starting materials are given. Determine moles of each starting material: for CS_2, 1.5 g = 0.01970 mol and for O_2, 3.5 g = 0.1094 mol. Determine the limiting reactant:

$$\frac{\text{Starting mol amount}}{\text{stoichiometric coefficient}} \qquad \text{for } CS_2 : \frac{0.01970}{1} = 0.01970 \qquad \text{for } O_2 : \frac{0.1094}{3} = 0.03647$$

CS_2 is the limiting reactant.

Make a Table of Amounts (in units of moles, based on the limiting reactant):

	CS_2 (s)	+	3 O_2 (g)	→	CO_2 (g)	+	2 SO_2 (g)
Moles:							
Starting Amount	0.01970		0.1094		0		0
Change in Amount	-0.01970		-0.05910*		+0.01970		+0.03940
Final Amount	0		0.05030		0.01970		0.03940

*Sample "Change in Amounts" calculation: $0.01970 \, mol \, CS_2 \times \dfrac{3 \, mol \, O_2}{1 \, mol \, CS_2} = 0.059160 \, mol \, O_2$ used

Convert the amounts remaining to grams using molar mass:
O_2: 0.05030 mol x 31.998 g/mol = 1.6 g
CO_2: 0.01970 mol x 44.01 g/mol = 0.87 g
SO_2: 0.03940 mol x 64.064 g/mol = 2.5 g

SOLUTION: 0 g CS_2, 1.6 g O_2, 2.5 g SO_2, 0.87 g CO_2

EXERCISE 9: The box shown below contains reactant molecules for the chemical reaction:
$$2 \, NO \, (g) + O_2 (g) \rightarrow 2 \, NO_2 \, (g)$$

Draw a molecular picture to depict the contents of the box after the reaction is complete.

Before Reaction

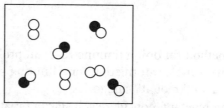

STRATEGY: The box contains 3 O_2 molecules and 4 NO molecules. This is a limiting reactant problem because amounts of both starting materials are provided. One way to analyze the problem is to just break apart an O_2 molecule and attach one O to a molecule of NO, repeating the process until one of the chemicals runs out. Another way to approach it is to set up calculations like in the Soda example. A third way to approach it is to use a Table of Amounts. Try the first two approaches on your own. We'll do a Table of Amounts here. (It will be in counting units of molecules rather than moles.)

Determine the limiting reactant: O_2: 3/1 = 3; NO: 4/2 = 2. NO is the limiting reactant.

	2 NO (g)	+	O_2 (g)	→	2 NO_2 (g)
Molecules:					
Starting Amount	4		3		0
Change in Amount*	-4		-2		+4
Final Amount	0		1		4

*Change in Amounts: $4 \, molec \, NO \times \dfrac{1 \, molec \, O_2}{2 \, molec \, NO} = 2$ molecules O_2 reacted

$$4 \text{ molec NO} \times \frac{2 \text{ molec NO}_2}{2 \text{ molec NO}} = 4 \text{ molecules NO}_2 \text{ formed}$$

SOLUTION:

After Reaction

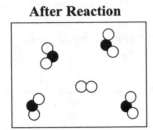

Try It #4: The thermite reaction shown below generates so much heat that the iron produced is molten (in the liquid phase). If 742 g of aluminum is allowed to react with 3315 g of Fe_2O_3, a) how much molten iron would be made? b) How many grams of unreacted starting material would remain?
$$Al (s) + Fe_2O_3 (s) \rightarrow Al_2O_3 (s) + Fe (l)$$

4.5 PRECIPITATION REACTIONS

Key Terms:
- **Soluble** – term used to describe a chemical that can dissolve in a given solvent. If the solvent is not specified, it will be water (the universal solvent).
- **Insoluble** – term used to describe a chemical that cannot dissolve in a given solvent.
- **Precipitation reaction** – reaction in which two soluble ions react to form an insoluble salt.
- **Net ionic equation** – balanced chemical equation which shows the *net* chemical change.
- **Spectator ion** – ion that does not participate in the net chemical reaction, but is simply floating around in solution (watching the reaction take place).

Key Concept:
- Most salts are water-insoluble.

Helpful Hints
- Most chemists know if a given salt is soluble or insoluble from experience in the lab.
- 4-1 in the text summarizes the soluble salts for your reference.
- Figure 4-6 in the text is a flow chart to help you determine whether or not a given salt is soluble.

EXERCISE 10: Determine whether or not a precipitate will form when aqueous calcium chloride is mixed with aqueous: a) sodium hydroxide, b) potassium bromide, or c) silver nitrate.

    STRATEGY: Use the flow chart in Figure 4-6 of the text to determine if any of the ions will combine to form a precipitate.
    a) Available ions: Ca^{2+} (aq), Cl^- (aq), Na^+ (aq), OH^- (aq)
    b) Available ions: Ca^{2+} (aq), Cl^- (aq), K^+ (aq), Br^- (aq)
    c) Available ions: Ca^{2+} (aq), Cl^- (aq), Ag^+ (aq), NO_3^- (aq)

    SOLUTION: a) $Ca(OH)_2$ (s); b) no precipitate; c) $AgCl$ (s).

EXERCISE 11: Write the net ionic equation and identify spectator ions for any reaction that took place in Exercise 10.

STRATEGY: The net ionic equation shows only the net reaction. Determine the net reaction from the precipitate that formed. The spectator ions are the ions that are not involved in the net reaction.

SOLUTION: a) $Ca(OH)_2$ formed, so the net ionic equation is: Ca^{2+} (aq) $+ 2$ OH^- (aq) $\rightarrow Ca(OH)_2$ (s). The spectator ions are: Cl^- (aq) and Na^+ (aq).
b) no reaction
c) AgCl formed, so the net ionic equation is: Ag^+ (aq) $+ Cl^-$ (aq) $\rightarrow$ AgCl (s). The spectator ions are NO_3^- (aq) and Ca^{2+} (aq).

Precipitation Stoichiometry

EXERCISE 12: 30.00 mL of 0.100 M $AgNO_3$ (aq) is mixed with 20.00 mL of 0.200 M Na_2CO_3 (aq). a) How much precipitate will form? b) What are the concentrations of silver ions and carbonate ions after the reaction?

STRATEGY: This is a limiting reactant problem because amounts of both starting materials were given. We first need a balanced chemical equation. Follow the flow chart in Figure 4-6 of the text to see that the net ionic equation for this process is between the silver ions and the carbonate ions. Nitrate and sodium ions are the spectator ions.

$$2\,Ag^+ (aq) + CO_3^{2-} (aq) \leftrightarrow Ag_2CO_3 (s)$$

If the amount of Ag_2CO_3 produced depended only upon Ag^+:

$$0.03000\,L \times \frac{0.100\,mol\,Ag^+}{L} \times \frac{1\,mol\,Ag_2CO_3}{2\,mol\,Ag^+} = 0.001500\,mol\,Ag_2CO_3$$

If the amount of Ag_2CO_3 produced depended only upon CO_3^{2-}:

$$0.02000\,L \times \frac{0.200\,mol\,CO_3^{2-}}{L} \times \frac{1\,mol\,Ag_2CO_3}{1\,mol\,CO_3^{2-}} = 0.00400\,mol\,Ag_2CO_3$$

Ag^+ is the limiting reactant.

The maximum amount of precipitate that will form is based on the limiting reactant:

$$0.00150\,mol\,Ag_2CO_3 \times \frac{275.748\,g}{1\,mol} = 0.4136\,g\,of\,Ag_2CO_3$$

b) There will be no Ag^+ left in solution at the end of the reaction. Moles of carbonate remaining in solution = moles to start – moles used.

Moles of carbonate to start: $$0.020\,L \times \frac{0.200\,mol\,Na_2CO_3}{L} \times \frac{1\,mol\,CO_3^{2-}}{1\,mol\,Na_2CO_3} = 0.00400\,mol\,CO_3^{2-}$$

Moles of carbonate used: $$0.00150\,mol\,Ag_2CO_3 \times \frac{1\,mol\,CO_3^{2-}}{1\,mol\,Ag_2CO_3} = 0.00150\,mol\,CO_3^{2-}$$

Moles of CO_3^{2-} remaining in solution = moles to start – moles used = $0.00400 - 0.00150 = 0.00250$

$$[CO_3{}^{2-}] = \frac{\text{moles of } CO_3{}^{2-}}{\text{total volume}} = \frac{0.00250 \text{ mol } CO_3{}^{2-}}{(20.00 \text{ mL} + 30.00 \text{ mL})} \times \frac{1000 \text{ mL}}{1 \text{ L}} = 0.0500 \text{ M } CO_3{}^{2-}$$

SOLUTION: a) 0.414 g of Ag_2CO_3 (3 sig figs); b) $[Ag^+] = 0$ M and $[CO_3{}^{2-}] = 0.0500$ M (3 sig figs). The answers seem reasonable because one reactant was consumed and the other is present in a lower concentration relative to its initial concentration.

4.6   ACID-BASE REACTIONS

Key Terms:
- **Acid** – substance that donates protons.
- **Base** – substance that accepts protons.
- **Neutralization reaction** – reaction of an acid with a base to produce water and a salt.
- **Strong acid** – substance that quantitatively donates protons to water (a strong acid ionizes 100% in water).
- **Weak acid** – substance that quantitatively donates protons to hydroxide ion, but not to water (a weak acid only ionizes a small amount in water).

Key Concepts:
- Most acids are weak.
- Acids and bases will be covered in greater detail in Chapter 17.
- When an acid donates a proton to water, hydronium ion, H_3O^+, is formed.

EXERCISE 13: Write a balanced equation for the neutralization of sulfuric acid and sodium hydroxide.

STRATEGY: In a neutralization reaction, an acid and a base will form a salt and water. The acid is H_2SO_4 (aq) and the base is NaOH (aq). The salt that will form is based on the counter-ions from the acid and the base: Na^+(aq) and $SO_4{}^{2-}$(aq). The ratio for these ions is 2:1, so the salt's formula is Na_2SO_4. This is a soluble salt, so the phase will be (aq). The unbalanced equation is:
$$H_2SO_4 \text{ (aq)} + NaOH \text{ (aq)} \rightarrow Na_2SO_4 \text{ (aq)} + H_2O \text{ (l)}$$

SOLUTION: H_2SO_4 (aq) + 2 NaOH (aq) $\rightarrow$ Na_2SO_4 (aq) + 2 H_2O (l)

Titration

Key Terms:
- **Titration** – systematic addition of a known concentration of acid to a base of unknown concentration until the exact stoichiometric ratio of acid to base has reacted. (This can also be reversed so that one knows the concentration of base, but not the acid.)
- **Titrant** – chemical of known concentration that is being added to the unknown.
- **Stoichiometric point** – point in titration at which the exact stoichiometric ratio of moles of acid and base have been mixed.
- **Standard solution** – a solution whose concentration has been carefully determined (usually to four sig figs). Standard solutions are usually used as titrants.

Key Concept:
- At the stoichiometric point, the moles of acidic hydrogen = moles of OH⁻.

Units and Conversions:
- Use Molarity as a conversion factor in aqueous calculations: M = moles of solute/L of solution.

EXERCISE 14: What volume of standardized 0.2004 M HCl is needed to neutralize 0.485 g of KOH that has been dissolved in 26 mL of water?

STRATEGY: This is a neutralization reaction, so we need a balanced equation.

$$HCl \, (aq) \, + \, KOH \, (aq) \, \rightarrow \quad H_2O \, (l) \, + \, KCl \, (aq)$$

$$0.485 \text{ g KOH} \times \frac{1 \text{ mol}}{56.11 \text{ g}} \times \frac{1 \text{ mol HCl}}{1 \text{ mol KOH}} \times \frac{1 \text{ L of solution}}{0.2004 \text{ mol HCl}} = 0.04313 \text{ L} = 43.13 \text{ mL of solution}$$

Reading the calculation from left to right, we convert mass of KOH to moles, then use the stoichiometric ratio to convert to moles of HCl. Notice that here we used the molar concentration of HCl as a conversion factor to cancel units of moles and be left with units of "volume of solution." Notice, too, that the volume measurement of 26 mL was not needed in the calculation.

SOLUTION: 43.1 mL of 0.2004 M HCl (aq) will be needed. (3 sig figs based on data provided)

4.7   OXIDATION-REDUCTION REACTIONS

Key Terms:
- **Oxidation** – loss of electrons.
- **Reduction** – gain of electrons.
- **"Redox"** – nickname for "oxidation/reduction."
- **Half-reaction** – balanced chemical equation that shows either an oxidation or a reduction process.
- **Activity series** – list of metals in order of increasing ease of oxidation.

Key Concept:
- Oxidation and reduction go together; one process cannot happen without the other. If one chemical loses electrons (oxidation), then another chemical must pick up those electrons (reduction) because all electrons must be accounted for in the overall process.

Common Types of Redox Reactions

Type	Definition	Sample Net Ionic Equation
Metal Displacement	A metal cation being displaced by another metal	Cu^{2+} (aq) + Mg (s) → Mg^{2+} (aq) + Cu (s)
Metal and Acid	An active metal will react with an acid to make $H_2(g)$ & $H_2O(l)$	$2H_3O^+(aq)$ + Zn(s) → $H_2(g)$ + $2H_2O(l)$ + $Zn^{2+}(aq)$
Metal and Water	An active metal will react with water to form $H_2(g)$ & $OH^-(aq)$	Na(s) + $2H_2O(l)$ → $H_2(g)$ + $2OH^-$ (aq) + Na^+(aq)
Oxidation by O_2 (g)	Reaction of a substance with $O_2(g)$	2Mg (s) + O_2 (g) → 2MgO (s)

EXERCISE 15: The reaction that occurs to make table salt is: Na (s) + Cl$_2$ (g) → NaCl (s).
a) Write the half reactions for this reaction. b) Which half reaction is the oxidation portion of the
reaction? c) Balance the overall equation.

    STRATEGY: a) Determine the charges on each substance to see what gained and lost electrons.

    Na (s) and Cl$_2$ (g) are both elements, so they each have a "0" charge.
    NaCl is an ionic compound made up of Na$^+$ ions and Cl$^-$ ions.

    SOLUTION: a) Na → Na$^+$ + e$^-$; Cl$_2$ + 2 e$^-$ → 2 Cl$^-$;
    b) Na is undergoing oxidation because it lost an electron; c) 2 Na (s) + Cl$_2$ (g) → 2 NaCl (s)

EXERCISE 16: A steel propeller on a boat is usually fitted with a zinc collar, which is a ring of zinc
attached to the neck of the propeller. Zinc oxidizes more easily than iron. Use this fact to write the net
ionic equation that occurs when the propeller reacts with ocean water.

    STRATEGY: Since Zn oxidizes more easily than Fe, the Zn will react with the water. A metal reacting
    with water will make hydrogen gas and hydroxide ion. Balance the atoms and the charges.

    SOLUTION: Zn (s) + 2 H$_2$O (l) → Zn^{2+} (aq) + H$_2$ (g) + 2 OH$^-$ (aq)

    Practical note: None of the Fe will react with the water until all of the Zn is gone. This protects the
    propeller; the zinc collar can be replaced before it completely dissolves.

EXERCISE 17: All fuels undergo a redox reaction when they combust. Combustion is classified as
"oxidation by O$_2$" because O$_2$ is needed as a reactant. When butane undergoes combustion, it produces
carbon dioxide and water. Write a balanced chemical equation to illustrate this reaction. Butane is C$_4$H$_{10}$
and it is a liquid at room temperature.

    STRATEGY: Butane and oxygen are reactants, and carbon dioxide and water are products. The phases
    for each substance must be included too.
$$C_4H_{10} \text{ (l)} + O_2 \text{ (g)} \rightarrow CO_2 \text{ (g)} + H_2O \text{ (l)}$$

    Balance the Cs and the Hs: C$_4$H$_{10}$ (l) + O$_2$ (g) → 4 CO$_2$ (g) + 5 H$_2$O (l)

    We need 6.5 O$_2$ molecules to balance the Os: C$_4$H$_{10}$ (l) + 6.5 O$_2$ (g) → 4 CO$_2$ (g) + 5 H$_2$O (l)

    Multiply through by 2 to get rid of the fraction: 2 C$_4$H$_{10}$ (l) + 13 O$_2$ (g) → 8 CO$_2$ (g) + 10 H$_2$O (l)

    SOLUTION: 2 C$_4$H$_{10}$ (l) + 13 O$_2$ (g) → 8 CO$_2$ (g) + 10 H$_2$O (l)

Try It #5: Silver tarnish is composed of Ag$_2$S, which is a black, insoluble solid. Tarnish can be removed
by setting the piece of silver on a piece of aluminum foil in warm salt water. The aluminum and the silver
will undergo a metal displacement redox reaction. (Given: Al cations have a +3 charge.) a) Write a redox
reaction to illustrate this process. b) Which metal is more easily oxidized, silver or aluminum?

Chapter 4 Self-Test

You may use a periodic table and a calculator for this test.

1. How many grams of H_2 are needed to produce 4.00 mol of NH_3?

$$N_2 \, (g) \; + \; H_2 \, (g) \; \rightarrow \; NH_3 \, (g)$$

2. Write the net ionic equation for the reaction which occurs when aqueous $Pb(NO_3)_2$ is mixed with aqueous $MgCl_2$. Show all phases.

3. Draw molecular pictures to illustrate all solute particles in Question #2. Draw one picture for reactants and one picture for products.

4. What volume of 0.500 M aqueous nitric acid is needed to react with 3.00 g of sodium metal?

5. One step in the production of titanium metal is shown by this reaction.

$$TiCl_4 \; + \; Mg \; \rightarrow \; Ti \; + \; MgCl_2$$

a) 45.00 kg of titanium was isolated when 189.0 kg of $TiCl_4$ was reacted with 52.5 kg of Mg. What is the percent yield of the reaction?

b) If the reaction had gone to completion (100% yield), what mass of the excess reactant would have been left over?

c) In this chemical reaction, which is true of Ti?
it's an acid   it's a base   it's being oxidized   it's being reduced   it's a precipitate

6. What is the concentration of Na^+ ions in solution after 125 mL of 0.100 M HCl (aq) react with 15.00 mL of 1.00 M sodium hydroxide (aq)?

7. Use this information to help answer the question.

Relative ease with which a metal undergoes oxidation: Mg>Fe>Cu

Consider two blocks of iron that a person is trying to preserve. One piece of iron is coated with copper and the other piece of iron is coated with magnesium. If both pieces get scratched so that the iron is exposed, in which block will the Fe stay preserved longer? Explain.

Answers to Try Its:

1. $2 \, CO \, (g) \; + \; O_2 \, (g) \rightarrow \; 2 \, CO_2 \, (g)$ (remember to show simplest ratio)
2. a) $2 \, NH_3 \, (g) \; + 3 \, N_2O \rightarrow \; 3 \, H_2O \, (g) \; + 4 \, N_2 \, (g)$,
   b) $CS_2 \, (s) \; + \; 3 \, O_2 \, (g) \rightarrow CO_2 \, (g) \; + \; 2 \, SO_2 \, (g)$,
   c) $H_3PO_4 \, (aq) \; + 3 \, KOH \, (aq) \; \rightarrow \; 3 \, H_2O \, (l) \; + \; K_3PO_4 \, (aq)$
3. a) 203 kg SO_2; b) 178 kg of CaO
4. a) 1.54 kg Fe could be made; b) 1.12×10^3 g Fe_2O_3 left over
5. a) $3 \, Ag_2S(s) \; + \; 2 \, Al \, (s) \rightarrow \; 6 \, Ag \, (s) \; + \; Al_2S_3 \, (aq)$
   b) Al is more easily oxidized than Ag. If Ag was more easily oxidized, then no reaction would occur because the Ag in Ag_2S is already in its oxidized form.

Answers to Self-Test:

1. 12.1 g H_2
2. $Pb^{2+} \, (aq) \; + \; 2 \, Cl^- \, (aq) \rightarrow \; PbCl_2 \, (s)$

3.

Before Reaction   **After Reaction**

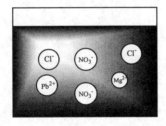 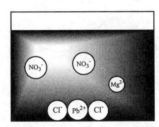

4. 261 mL
5. a) 94.32% (the limiting reactant data contains four sig figs, so the answer should too);
   b) 4.1 kg Mg left over, no $TiCl_4$ left over;
   c) it's being reduced
6. 0.107 M Na^+ (the sodium ion is a spectator ion!)
7. The Mg-coated piece would stay preserved longer, because none of the Fe would react until all of the Mg has undergone oxidation. (Mg oxidizes more easily than Fe). In the copper coated piece, the Fe would react before the copper would because Fe oxidizes more easily than Cu.

Chapter 5: The Behavior of Gases

Learning Objectives

In this chapter you will learn how to:

- Relate speed, kinetic energy, and temperature of gaseous substances.
- Qualitatively and quantitatively determine pressure effects on a gas.
- Calculate and conceptualize what happens to a gas when its environment (pressure, volume, and/or temperature) changes.
- Quantitatively assess gas mixtures using Dalton's Law of Partial Pressures.
- Use the Ideal Gas Law in solving stoichiometry problems.
- Calculate weather information (relative humidity and dew point).

Practical Aspects

This chapter lays the foundation for molecular behavior in the gaseous phase. The concepts in this chapter will be applied in numerous situations in later chapters.

5.1 PRESSURE

Key Terms:
- **Pressure** – force per unit area. Pressure depends upon the force and quantity of molecules colliding with the walls of the container.
- **Barometer** – device used to measure atmospheric pressure.
- **Manometer** – device used to measure the difference in pressure between two gases.
- **Atmosphere** – pressure that will support a column of mercury 760 mm high.

Units and Conversions
- 1 atm = 760 mmHg = 760 torr = 101.325 kPa

EXERCISE 1: Atmospheric pressure decreases as elevation increases. The typical atmospheric pressure at sea level is 1.00 atm. A typical barometer reading in Denver, Colorado is 0.83 atm. Calculate the atmospheric pressure in Denver in units of mmHg, torr, kPascals and inches of Hg.

STRATEGY: Use conversion factors. Start with what's given and set up conversion factors to work towards desired units.

$$0.83 \text{ atm} \times \frac{760 \text{ torr}}{1 \text{ atm}} = 630.8 \text{ torr} = 630 \text{ torr} = 630 \text{ mmHg}$$

$$0.83 \text{ atm} \times \frac{101.325 \text{ kPa}}{1 \text{ atm}} = 84 \text{ kPa}$$

$$630 \text{ mm Hg} \times \frac{1 \text{ cm}}{10 \text{ mm}} \times \frac{1 \text{ in}}{2.54 \text{ cm}} = 25 \text{ inches of Hg}$$

SOLUTION: 0.83 atm = 630 torr = 630 mmHg = 84 kPa = 25 inches of Hg (all to 2 sig figs)

5.2   DESCRIBING GASES

Key Terms:
- **Ideal gas** – a gas that meets these requirements: 1) The size of the molecule is negligible compared to the size of the container, and 2) The kinetic energy of the gas is enormous compared to the attractive forces between molecules, making all molecular collisions completely elastic.
- **Ideal gas equation** – mathematical equation that relates all of the factors that influence an ideal gas: PV = nRT

Key Concepts:
- The Kelvin scale *must* be used in gas law calculations.
- The pressure each molecule exerts depends upon temperature (temperature affects both the frequency and the forcefulness of collisions).

Useful Relationships:
- PV = nRT. (The derivation of the ideal gas law is provided in the text.)

Units and Conversions:
- R = 8.314 J/mol K = 0.08206 L atm/mol K. R is called the universal gas constant.

EXERCISE 2: Standard temperature and pressure conditions ("STP") for gases are 1.00 atm and 273 K. What volume does exactly one mole of gas occupy under these conditions?

STRATEGY: This is a direct application of the ideal gas law: PV = nRT. We have all variables except for V, so we can solve for V. (This problem is simple enough that we'll short-cut the 7-step method.)

$$V = \frac{nRT}{P} = \frac{1.000 \text{ mol}}{1.00 \text{ atm}} \times \frac{0.08206 \text{ L atm}}{\text{mol K}} \times 273 \text{ K} = 22.4 \text{ L}$$

SOLUTION: One mole of gas occupies 22.4 L at STP.

Take-home message from Exercise 2:
- Multiplication and division has no "order of operations" requirements so the "1.00 atm" value for P could be written under n (like it is), under T, or on its own as "x 1/P." The answer will be the same.

Key Concept:
- One mole of an ideal gas occupies 22.4 L at STP. Use this as a basis for comparison when determining if a quantitative gas law problem seems reasonable.

EXERCISE 3: What effect will each change have on the pressure inside the container? What effect will it have on the container's volume? Assume ideal gas behavior.

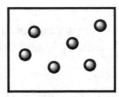

a) Removing three molecules from the container. (Condition: the container walls can move.)
b) Removing three molecules from the container. (Condition: the container walls can NOT move.)
c) Adding three molecules to the container. (Condition: the container walls can move.)

STRATEGY AND SOLUTION: $PV = nRT$, so if n is cut in half, then one of four things must occur: T must double OR P must be cut in half OR V must be cut in half OR a combination of changes must occur.

a) At the instant the three molecules are removed, the pressure will decrease by half. The walls of the container are movable, so as soon as pressure decreases, volume will decrease. The volume will continue to decrease until the pressure is reestablished. The net result is that the volume will decrease by ½ and the pressure will stay the same. If this is difficult to picture, visualize it as an air-filled balloon: if you were to let out half the air, the balloon's volume would decrease by half.

Key Concept:
- For a container with movable walls, volume will change as needed to make $P_{inside} = P_{outside}$.

b) If the walls can't move, then V is constant. P will decrease by ½. (There are ½ as many collisions on the container's walls.)
c) This is similar to part "a." This time V will increase to re-establish a constant P. The net result is no change in pressure and the new volume is 1.5 times the original volume.

Try It #1: How could the conditions in Exercise 3 be altered so that the force with which each molecule hits the walls of the container is increased?

Working with One Gas Under Varying Conditions

Consider one gas under two different conditions. $PV = nRT$; solve for R (a constant): $R = \dfrac{PV}{nT}$; the equations for the two different conditions can be set equal to each other because both equal R:

$$\frac{P_1 V_1}{n_1 T_1} = \frac{P_2 V_2}{n_2 T_2}$$

Notice that if any of the components (P, V, n, or T) stays constant, then that term will drop out of the equation. For example, if V stays at 5 L, then the "5"s would cancel out.

The laws developed from early experiments with gases were given names to honor their discoverers:
- Boyle's Law – (at constant n and T): as P increases, V decreases.
- Charles' Law – (at constant P and n): as T increases, V increases.

Helpful Hints

- Notice that the *only* relationship you need to memorize here is PV=nRT. If you have two sets of conditions, simply solve PV=nRT for R and establish the relationship shown above. Then cross out any term (P, V, n, T) that does not change.
- The mathematical relationship for a gas under varying conditions is sometimes subscripted "i" for "initial" and "f" for "final" in place of Condition "1" and Condition "2."

EXERCISE 4: 0.50 mole of nitrogen gas is heated from 290. K to 500. K inside a rigid container. If the original pressure in the container was 700. torr, what was the final pressure in units of atm?

STRATEGY: This question contains a lot of data, so the 7-step method will be used to evaluate it.
1. **Determine what is asked for.** Final pressure.
2. **Visualize the problem.** "Rigid" indicates the size of the container cannot change. A specific quantity of gas is inside the container and the container is heated. The molecules will speed up when heated, so pressure will increase.
3. **Organize the data.** We've been given the amount of gas, the initial pressure and temperature, and the final temperature.
4. **Identify the process to solve the problem.** This is an application of the ideal gas law: one gas under varying conditions. V and n are constant.
5. **Manipulate the equations.**

$$\frac{P_1 V_1}{n_1 T_1} = \frac{P_2 V_2}{n_2 T_2} \text{ simplifies to } \frac{P_1}{T_1} = \frac{P_2}{T_2} \text{ because } V \text{ and } n \text{ are constant.}$$

$$\textit{Rearrange the equation to solve for the unknown : } P_2 = \frac{P_1 T_2}{T_1}$$

Notice: We don't even need the "0.50 mol" of gas information to solve the problem!

6. **Substitute and calculate.**

$$P_2 = \frac{P_1 T_2}{T_1} = 700. \text{ torr } x \frac{500. \text{ K}}{290. \text{ K}} x \frac{1 \text{ atm}}{760 \text{ torr}} = 1.59 \text{ atm}$$

The answer is reported to 3 sig figs because the data used in the calculation contained 3 sig figs.

7. **Does the Result Make Sense?** The initial pressure was under 1 atm. The final pressure is greater than the initial pressure, which is reasonable.

SOLUTION: The final pressure is 1.59 atm.

EXERCISE 5: 0.50 mole of nitrogen gas is heated from 290. K to 500. K inside a rigid container. If the original pressure in the container was 700. torr, what is the volume of the container?

STRATEGY: At first glance, this looks a lot like the previous exercise. When it is broken down into pieces, you'll see it's treated differently.
1. **Determine what is asked for.** Volume of the container.
2. **Visualize the problem.** "Rigid" indicates the size of the container cannot change. A specific quantity of gas is inside the container.
3. **Organize the data.** In order to find the container's volume, we need P, n and T for one set of conditions (P = 700. torr, n = 0.50 mol, T = 290. K). "500 K" is not needed to solve the problem.

4. **Identify the process to solve the problem.** This is an application of the ideal gas law: one gas under one set of conditions: $PV = nRT$.

5. **Manipulate the equations.** $PV = nRT$, so $V = nRT/P$

6. **Substitute and calculate.**

$$V = \frac{nRT}{P} = \frac{0.50\,\text{mol}}{700.\,\text{torr}} \times \frac{0.08206\,\text{L atm}}{\text{mol K}} \times 290.\,\text{K} \times \frac{760\,\text{torr}}{1\,\text{atm}} = 12.92\,\text{L} = 13\,\text{L}$$

The value "0.50 mol" limits the sig figs in the answer to 2.

7. **Does the Result Make Sense?** 1 mol of gas occupies 22.4 L at STP, so the answer seems reasonable.

SOLUTION: The volume of the container is 13 L.

Try It #2: A balloon with a volume of 0.25 L is inside a freezer at –12°F. If the balloon is removed from the freezer and placed in a room at 21°C, what will be the new volume of the balloon?

5.3 MOLECULAR VIEW OF GASES

Key Concepts:
- At a given temperature, all gases have the same kinetic energy distribution.
- When referring to a single molecule, one must discuss *average* kinetic energy.
- Temperature is a measure of kinetic energy. At 0 Kelvin, there is no kinetic energy. As temperature increases above 0 Kelvin, so does kinetic energy. (KE is proportional to T.)
- The Kelvin scale is *always* used in calculations involving temperature.

Useful Relationships:

Key Concept At a given temperature...	Mathematical Relationship to Support Concept	Notice this in the Mathematical Relationship:
All gaseous objects have the same average kinetic energy	$KE_{molar} = \dfrac{3RT}{2}$ (units = J/mol)	KE depends *only* upon temperature
"	$KE = \frac{1}{2}\,mu^2$	At a given temperature, KE is constant: as an object's mass increases, u decreases (and vice versa)

EXERCISE 6: a) What is the kinetic energy of 0.500 mole of nitrogen molecules on a hot day in Phoenix (112°F)? b) What is the kinetic energy of one nitrogen molecule at this temperature?

STRATEGY: Convert to Kelvin first: °F = (9/5)°C + 32; $T_K = T_C + 273$; so 112°F = 317 K

a) Use the kinetic energy equation for gases. We've been given a specific number of moles, so we'll need to multiply by that quantity at the end to determine KE for the set of molecules.

$$KE = \frac{3RT}{2} = \frac{3}{2} \times \frac{8.314\,J}{mol\,K} \times 317\,K \times 0.500\,mol = 1.98 \times 10^3\,J$$

b) We can't calculate the kinetic energy of one nitrogen molecule because molecules exhibit a range of kinetic energies at a given temperature. The best we can do is calculate the *average* kinetic energy.

$$KE_{ave} = \frac{3RT}{2} = \frac{3}{2} \times \frac{8.314\,J}{mol\,K} \times 317\,K \times \frac{1\,mol}{6.022 \times 10^{23}\,molecules} = 6.56 \times 10^{-21}\,J/molecule$$

SOLUTION: a) 1.98×10^3 J; b) 6.56×10^{-21} J/molecule (on average). Both answers seem reasonable: an individual molecule will have a much, much smaller KE than the collective KE of 0.500 mole.

Take-home messages from Exercise 6:
- Always use units of Kelvin for gas calculations involving temperature.
- Watch units! Remember to convert between kg and grams when converting Joules.
- Notice that each problem was set up so that unwanted units would cancel and we'd be left with the desired units.
- *There's often more than one correct way to attack a chemistry problem.* As long as you apply the concepts correctly and watch your units, you should obtain the same answer via different methods.

Ideal Gases

Key Concepts:
- The volume occupied by molecules of an ideal gas is negligible compared to the volume of the container.
- All molecular collisions between molecules of an ideal gas are elastic, meaning that the forces between molecules are negligible compared to the molecules' kinetic energies.
- When working with ideal gases in PV=nRT calculations, the identity of the gas does not matter; only the number of gas molecules, n, matters.

EXERCISE 7: Consider the two boxes shown below. Box 1 contains argon atoms and Box 2 contains helium atoms. If both gases behave as ideal gases and if both boxes have the same pressure, what is the temperature in Box 2 in units of °C?

Box 1 (T = 21°C)   **Box 2**

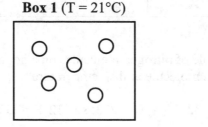
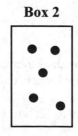

STRATEGY: The gases are ideal gases, so identity of each gas does not matter – only the quantity matters. This problem can be treated as if it were one gas under two sets of conditions.

$$\frac{P_1V_1}{n_1T_1} = \frac{P_2V_2}{n_2T_2} \text{ simplifies to } \frac{V_1}{T_1} = \frac{V_2}{T_2}; \; T_2 = \frac{V_2T_1}{V_1} \text{ because P and n are constants}$$

Notice that V_2 is ½ of V_1. Notice, too, that if we forgot to convert to Kelvin (21°C = 294 K), we'd think the answer is 10.5°C. This is not the case:

$$T_2 = \frac{V_2T_1}{V_1} = \frac{0.5V_1}{V_1} \text{ x } 294 \text{ K} = 147 \text{ K}, \text{ and } 147 - 273 = -126°C$$

SOLUTION: The temperature in Box 2 must be -126°C.

Take-home messages from Exercise 7:
- Ideal gas calculations are greatly simplified (relative to non-ideal gases, later in the text) because we don't need to factor in the identity of the gas molecules.
- Watch for relationships between two conditions – we used *relative* volumes in this calculation. The relationships between the two containers in this exercise allowed us to make mathematical comparisons as if we were comparing conditions for one gas. This does not work for every case.

5.4   ADDITIONAL GAS PROPERTIES

Variations of Working with One Gas Under One Set of Conditions

Two common calculations performed that utilize the ideal gas law are molar mass and gas density calculations.

- **Molar mass**: MM = mass/moles, so MM can be related to the ideal gas law through n
- **Gas density**: Density = mass/Volume, so density can be related to the ideal gas law through n/V (because n=m/MM)

EXERCISE 8: 5.52 g of an unknown gas is placed in a 10.00 L vessel. It is found that at 32°C the gas exerts a pressure of 655 mmHg. Which gas is it: N_2, C_2H_2, or CH_4?

STRATEGY: We'll do a shortcut version of the 7-step method. We need to find the identity of the gas and can use molar mass to achieve this. The respective molar masses of the gases are 28.0, 26.0, and 16.0 g/mol. The real question is, "what is the molar mass of the gas?"

$$PV = nRT; \;\; MM = m/n, \text{ so } n = m/MM$$

$$PV = \frac{mRT}{MM} \text{ so } MM = \frac{mRT}{PV}$$

Remember to convert temperature to Kelvin: 32°C = 305 K

$$MM = \frac{mRT}{PV} = \frac{5.52 \text{ g}}{655 \text{ torr}} \text{ x } \frac{0.08206 \text{ L atm}}{\text{mol K}} \text{ x } \frac{305 \text{ K}}{10.00 \text{ L}} \text{ x } \frac{760 \text{ torr}}{1 \text{atm}} = 16.0 \text{ g/mol}$$

SOLUTION: The unknown gas is CH_4. The MM matched one of the choices, so the answer seems reasonable.

Take-home message from Exercise 8:
- Sometimes it helps to work backwards on the problem and come up with a simpler question or "mini-goal" for the problem.

Rates of Gas Movement

Key Terms:
- **Diffusion** – the movement of a collection of molecules of one type through another (without stirring).
- **Effusion** – movement of molecules as they escape from a container and into a vacuum.

Key Concept:
- When referring to a single molecule, one must discuss *rms* speed or *average* kinetic energy.

Key Concept *At a given temperature...*	Mathematical Relationship to Support Concept	Notice this in the Mathematical Relationship:
Smaller objects move faster than larger objects	$u_{ave} = u_{rms} = \sqrt{\dfrac{3RT}{MM}}$	Molecular speed depends upon 2 things: 1. temperature (directly) 2. MM (inversely)

Alternate approach to Exercise 6:

In exercise 6, we were asked to find the kinetic energy associated with 0.500 mole of nitrogen molecules at 317 K. This problem could also be approached using the new equation in this section. First, find "u", and then plug it into KE = ½ mu²:

$$u_{ave} = \sqrt{\frac{3RT}{MM}} = \sqrt{3 \text{ x} \frac{8.314\text{x}10^3 \text{ g m}^2}{\text{sec}^2 \text{ mol K}} \text{ x } 317\,\text{K x} \frac{\text{mol}}{28.01\,\text{g}}} = 5.314\text{x}10^2\,\text{m/sec}$$

$$KE = \frac{1}{2}mu^2 = \frac{1}{2} \text{ x } 0.500\,\text{mol x} \frac{28.01\,\text{g}}{\text{mol}} \text{ x} \left(\frac{5.314\text{x}10^2 \text{ m}}{\text{sec}}\right)^2 \text{ x} \frac{1\,\text{kg}}{1000\,\text{g}} = 1.98\text{x}10^3\,\text{J}$$

*Notice that we had to convert grams to kg to obtain units for Joules: "kg m²/sec²."

Take-home messages from this approach to Exercise 6:
- Watch units! Remember to convert between kg and grams when converting Joules.
- Notice that each problem was set up so that unwanted units would cancel and we'd be left with the desired units.
- *There's often more than one correct way to attack a chemistry problem.* As long as you apply the concepts correctly and watch your units, you should obtain the same answer via different methods.

EXERCISE 9: Here are two containers of neon. Both are at the same temperature. The rms speed of neon at this temperature is 1000 m/sec.

Box 1   Box 2

a) Which container's molecules are moving faster?
b) True or false: the speed of each Ne atom is 1000 m/sec.
c) Neon weighs five times as much as helium. Draw a molecular picture of helium atoms that will have the same average kinetic energy as the neon atoms shown in Box 2. Specify the conditions.

    SOLUTION: a) neither – both containers are at the same temperature, so their rms speeds are the same. Both containers DO have a distribution of speeds, but collectively their speeds are said to be equivalent.
    b) false – the *average* speed is 1000 m/sec, but an individual molecule's speed may vary from this.
    c) KE depends only upon temperature. As long as T is the same, we could draw any size box with any number of atoms in it.

Try It #3: Calculate the average speed of these gas molecules at 0°C: a) carbon dioxide; b) carbon monoxide.

EXERCISE 10: How could the concepts in this section be used to separate a mixture of hydrogen gas and carbon dioxide gas?

    STRATEGY AND SOLUTION: At any given temperature, lighter gases move faster than heavier gases. Put the gas mixture into Container A. Set up a series of empty containers next to Container A as shown here:

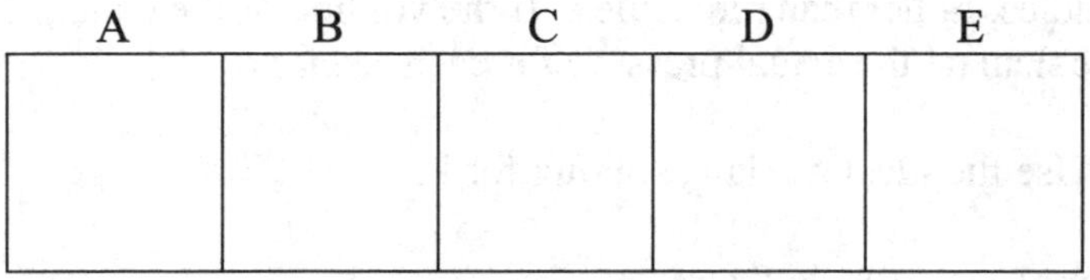

    Prick a small hole in each connecting wall. The lighter gas (hydrogen) will have a greater probability of going through the hole in the wall between A and B. After a short amount of time, Box B will contain a higher proportion of hydrogen than Box A. This process will continue; each box will become more enriched in hydrogen than the previous box. It may take more than 5 boxes to completely separate the two chemicals, but this scheme illustrates the concept.

    This method is frequently used to separate gas mixtures.

5.5   GAS MIXTURES

Key Concept:
- Gas behavior at a given temperature depends only upon the *number* of gaseous particles.

Dalton's Law of Partial Pressures: The total pressure of a gas mixture will equal the sum of the partial pressures of each gas within the mixture. (Each gaseous substance will exert its own pressure.)

$$P_T = p_1 + p_2 + p_3 \ldots$$

Notice partial pressure is designated with lower case italics to distinguish it from P_{total}.

EXERCISE 11: Consider a completely empty container with fixed walls. Draw three molecular pictures to illustrate the concept of Dalton's Law. The picture should show this: 1) 4 molecules of Gas #1 inside the container; 2) 8 molecules of Gas #2 (and nothing else) inside the container; 3) 4 molecules of Gas #1 AND 8 molecules of Gas #2 inside the container. The temperature is the same inside each container.

SOLUTION:

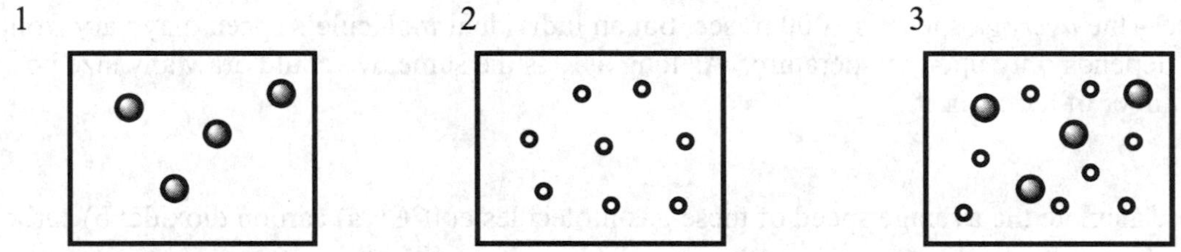

At the same V and T, pressure changes depend only upon n. The pressure inside the second box is twice the pressure inside the first box because there are twice as many molecules in it. The pressure inside Box 3 is triple that of Box 1 because it contains 3 times the number of molecules as Box 1. Quantitatively, if the pressure in Box 1 is x, then the pressure in Box 2 is 2x. The pressure in Box 3 will be "x + 2x" or 3x – this is Dalton's Law.

EXERCISE 12: Consider the container from Exercise 11. Assume that each scenario presented in Exercise 11 is referring to *moles* of particles, rather than molecules. If the volume of the container is 3.00×10^2 L and the temperature is 325 K, calculate the actual pressure for each scenario.

STRATEGY AND SOLUTION: Use the ideal gas law, solving for P:

$$P = \frac{nRT}{V} = \frac{\text{"n" mol}}{3.00 \times 10^2 \text{ L}} \times \frac{0.08206 \text{ L atm}}{\text{mol K}} \times 325 \text{ K} = \text{pressure inside the container (atm)}$$

Plug in the correct number of moles for each scenario and solve for P.

Scenario	Moles of Gas	Pressure
1	4	0.356 atm
2	8	0.711 atm
3	12	1.07 atm

Notice that in Scenario 3, $p1 + p_2 = P_T$, or 0.356 atm + 0.711 atm = 1.067 atm = 1.07 atm.

Describing Gas Mixtures

The concentration of a gas within a mixture of gases is typically described in one of these ways:

Way to measure a gas mixture:	Symbol	Mathematical Relationship	Units
Mole Fraction	X	$X_A = \dfrac{\text{moles of A}}{\text{total moles in sample}}$	None
Partial Pressure	p	$p_A = X_A\, P_{total}$	Pressure units
Parts per Million	-	$[A] \text{ in ppm} = \dfrac{\#\text{ of molecules of A}}{\#\text{ of molecules in total sample}} \times 10^6$	ppm
Parts per Billion	-	$[A] \text{ in ppb} = \dfrac{\#\text{ of molecules of A}}{\#\text{ of molecules in total sample}} \times 10^9$	ppb

EXERCISE 13: A system at 912 torr contains a gas with a partial pressure of 501 torr. What is the mole fraction of this gas?

STRATEGY: Use the relationship for partial pressure and mol fraction.

$$p_A = X_A P_T, \text{ so } X_A = \frac{p_A}{P_T} = \frac{501 \text{ torr}}{912 \text{ torr}} = 0.549$$

SOLUTION: The mol fraction of the gas is 0.549. The answer is between 0.5 and 1, which seems reasonable, because this gas is accounting for a little over ½ of the total pressure.

EXERCISE 14: Here are the symptoms associated with a one hour exposure to carbon monoxide: 1000 ppm = mild headache; 1300 ppm = throbbing headache, cherry-red skin; 2000 ppm = death. What symptoms would the average person experience after one hour in a room that contained these levels of carbon monoxide: a) 200,000 ppb, and b) $X_{CO} = 0.0040$?

STRATEGY: Convert to units of ppm and compare to CO data.

a) The "ppb" scale is # of CO molecules in 10^9 total molecules and "ppm" scale is # of CO molecules in 10^6 total molecules, so 1000 ppb = 1 ppm. 200,000 ppb will be 200 ppm.

b) "mol fraction" is moles of CO in 1 mol total. We can multiply this by 10^6 to get units of ppm: just like scaling up to moles of CO in 1 million moles total. $(0.0040)(10^6) = 4000$ ppm CO. Another way to show this is:

$$x_{CO} = \frac{n_{CO}}{n_{total}} = 0.0040; \quad n_{CO} = 0.0040 \text{ x } n_{total} = 0.0040 \text{ x } 1{,}000{,}000 = 4000 \text{ ppm}$$

SOLUTION: a) no apparent symptoms; b) death. This exercise illustrates the importance of scale.

Try It #4: Answer Exercise 14, given that the partial pressure of CO in the room is 0.912 mmHg (at sea level).

5.6   GAS STOICHIOMETRY

Key Concepts:
- The mole is the currency of chemistry - all calculations involving chemical changes go through moles.
- To convert from one substance to another, use the stoichiometric ratio from the balanced equation.

EXERCISE 15: The chemical reaction that makes an airbag inflate is: $2 \text{ NaN}_3 \text{ (s)} \rightarrow 2 \text{ Na (s)} + 3 \text{ N}_2 \text{ (g)}$. How many grams of NaN_3 are needed to make an air bag for a car that will inflate to 4.75 gallons at 72.0°F with a pressure of 1.10 atm?

STRATEGY: This is a stoichiometry problem that contains gas law information. The mini-goal of this question is to find moles of NaN_3. The key to this mini-goal is the stoichiometric ratio. Remember, the mole is the currency of chemistry, so when converting from one chemical to another, always use the stoichiometric ratio. We are given P, V, and T for nitrogen gas, so n for nitrogen gas can be found. Be sure to convert temperature to Kelvin first: 295K.

$$\text{For the nitrogen gas}: PV = nRT, \text{ so } n = \frac{PV}{RT} = \frac{1.10 \text{ atm}}{295 \text{ K}} \text{ x } \frac{\text{mol K}}{0.08206 \text{ L atm}} \text{ x } 4.75 \text{ gal x } \frac{3.785 \text{ L}}{1 \text{ gal}} = 0.81696 \text{ mol}$$

Now use stoichiometry to convert to NaN_3: $0.81696 \text{ mol N}_2 \text{ x } \frac{2 \text{ mol NaN}_3}{3 \text{ mol N}_2} \text{ x } \frac{65.02 \text{ g}}{\text{mol}} = 35.4 \text{ g NaN}_3$

SOLUTION: 35.4 g of NaN_3 are needed for the airbag. The final answer was rounded to 3 sig figs to correlate with the data's precision.

Take-home message from Exercise 15:
- If a stoichiometry problem provides or asks for units of "grams," simplify the problem by thinking "moles."

5.7   CHEMISTRY OF THE ATMOSPHERE

Key Terms:
- **Troposphere** – region of the atmosphere that is closest to earth. N_2 and O_2 make up over 99% of the gases in the troposphere.
- **Vapor** – gaseous substance that arises from evaporation.
- **Dynamic equilibrium** – condition in which the rate of the forward process equals the rate of the reverse process, so that no net observable change is observed.
- **Vapor pressure (*vp*)** – pressure that vapor exerts above a solid or liquid inside a closed system at equilibrium. (Rate of evaporation equals rate of condensation.) Vapor pressure depends on temperature and the identity of the substance.
- **Dew point** – temperature at which the partial pressure of water equals the vapor pressure of water, resulting in condensation (fog or dew).

Key Concept:
- Vapor pressure depends upon the identity of the substance and temperature.

Helpful Hint
- Table 5-4 in the text contains vapor pressures of water at various temperatures.

Useful Relationship:
- $\text{Relative Humidity} = \dfrac{p_{H_2O}}{vp_{H_2O}} \times 100$

EXERCISE 16: The temperature outside is 72°F. The relative humidity is 82%. a) What is the partial pressure of water vapor in the air? b) What is the dew point?

STRATEGY: a) Use the relationship for relative humidity. We will need to look up vapor pressure of water at this temperature. Table 5-4 in the text lists the vapor pressure of water at various temperatures in units of °C.

Convert °F to °C using °F = (9/5)°C + 32. 72°F = 22°C. Here's the information available from Table 5-4 in the text: H_2O's *vp* is 17.535 torr at 20°C and 23.756 torr at 25°C.

Interpolate to find *vp* at 22°C: The temperature range between the two numbers is 5°C and the *vp* range is 6.221 torr, which translates to 1.244 torr/°C for this range. The vp at 22°C will be:
$$17.535 + 2(1.244) = 20.023 \text{ torr}$$

$\text{R.H.} = \dfrac{p_{H_2O}}{vp_{H_2O}} \times 100, \text{ so } p_{H_2O} = \dfrac{\text{R.H.} \times vp_{H_2O}}{100} = \dfrac{82 \times 20.023 \text{ torr}}{100} \times \dfrac{1 \text{ atm}}{760 \text{ torr}} = 0.0216 \text{ atm} = 0.022 \text{ atm}$

b) The dew point depends upon this partial pressure of water in the air. We need to find the temperature at which the vapor pressure of water equals 0.0216 atm or 16.42 torr (100% R.H).

Look at Table 5-4 in the text. 16.42 torr is between 15°C and 20°C. Interpolate to determine the exact temperature of the dew point: 15°C to 20°C is a 5°C range and the range for the two vapor pressures is (17.535 – 12.788) = 4.747 torr, making the vapor pressure change for this range is 0.9494 torr/°C.

The temperature that corresponds to 16.42 torr is:

$$17.535 - 16.42 = 1.115 \text{ torr lower than at } 20°C$$

$$1.115 \text{ torr}(1°C/0.9494 \text{ torr}) = 1.0586°C \text{ lower than } 20°C = 18.9°C \text{ or } 66°F.$$

SOLUTION: a) 0.022 atm; b) 19°C or 66°F (2 sig figs). The answers seem reasonable. Even at a high relative humidity, the partial pressure of water vapor in the atmosphere should be low (0.022 atm compared to 1.00 atm total). The dew point should be below the original temperature of 72°F and it is.

Chemistry in the Troposphere

Key Terms:
- **Photochemical smog** – smog that originates from 1) the combustion of nitrogen and 2) the chemicals produced from UV light reacting with these combustion products. Major components in photochemical smog include various nitrogen oxides and ozone.
- **"London" smog** – smog that originates from the combustion of coal. (Coal contains varying amounts of sulfur.) The sulfur is converted to SO_2, which can be further oxidized to SO_3. SO_2 and SO_3 can react with water to form H_2SO_3 and H_2SO_4, respectively, both components of acid rain.

Environmental effects of acid rain:
- Dead lakes – water becomes too acidic for plants and animals to survive.
- Eggshells, marble, and limestone dissolve – acids react with carbonate compounds.
- Kills plants – many toxic elements in soil don't dissolve in water, but do dissolve in acidic solutions.

Chapter 5 Self-Test

You may use a periodic table and a calculator for this test. This water vapor pressure table may be helpful.

Temperature (°C)	10	15	20	25
Vapor pressure of H$_2$O (torr)	9.209	12.788	17.535	23.756

1. A completely empty 5.0 gallon BBQ propane tank with a mass of 4.25 kg was taken to a gas station to be filled. The mass of the tank after filling was 8.86 kg. What is the pressure inside the tank after filling if the temperature that day was 83°F? (Recall that propane's molecular formula is C$_3$H$_8$.)

2. Mexico City is 2240 m above sea level. Its average atmospheric pressure is 25% lower than at sea level. On a given day in Mexico City, the ground-level concentration of ozone was 0.19 ppm. Calculate this concentration in terms of partial pressure in units of Pascals.

3. Here's part of the weather report for Mexico City: The temperature is 20.0°C and the dew point is 10.0°C. What is the relative humidity?

4. 0.55 mole of a gas is placed in a 2.00 L container at 24.3°C. The container is heated to 400 K. The walls of the container cannot move. a) What is the pressure inside the container at this elevated temperature? b) What is the volume of the container at this elevated temperature?

5. Consider the three boxes shown. Each box is the same size and each gas is behaving as an ideal gas. The temperature inside each container is the same.

1 mole of He	1 mole of Ar	1 mole of Xe
Box 1	Box 2	Box 3

a) What can you say about the relative kinetic energies of the three gases?

b) What can you say about the pressures that each of the three gases exert on the walls of each box?

The wall between Box 1 and Box 2 is now removed. (We'll call this the "new" box.)

c) What will the He and the Ar do when the wall is removed?

d) What will happen to the pressure inside the new box (relative to the isolated boxes)?

e) What will happen to the pressure of the He in the new box (relative to when it was in Box 1)?

6. What volume of air at 757 mmHg and 21°C must a person inhale for 525 g of glucose to be metabolized? Air is composed of about 21% oxygen.

$$C_6H_{12}O_6 \text{ (s)} + O_2 \text{ (g)} \rightarrow CO_2 \text{ (g)} + H_2O \text{ (l)}$$

Answers to Try Its:

1. The only way to increase force per molecule is to increase the molecule's speed by increasing temperature.
2. 0.30 L (the balloon expanded, as would be predicted)
3. a) carbon dioxide: 393 m/sec; b) carbon monoxide: 493 m/sec
4. 1200 ppm = headache, somewhere between mild and throbbing

Answers to Self-Tests

1. 140 atm (2 sig figs)
2. 0.014 Pa (notice question asked for Pa, not kPa)
3. 52.5%
4. a) 9.0 atm; b) 2.00 L (no change in volume)
5. a) They're the same.
    b) They're the same.
    c) They'll diffuse until they're uniformly distributed in the new volume.
    d) The same (no change)
    e) The pressure of He in new box is ½ what it was in Box 1 before the wall was removed. (The new volume is twice the original volume.)
6. 2.01×10^3 L

Chapter 6: Energy and Its Conservation

Learning Objectives

In this chapter, you will learn how to:

- Identify a system, its surroundings, its state, and changes that it may undergo.
- Calculate the energy flow within a given system by assessing heat and work.
- Use calorimetry to evaluate heat flow for a given process.
- Calculate energy changes for reactions from bond energies.
- Compare enthalpy and energy changes for a given process.
- Use standard heats of formation to calculate enthalpy changes for reactions.

Practical Aspects

The study of chemical energetics is one of the most practical everyday chemistry topics to study. A basic understanding of how energy flows is crucial to understanding most scientific phenomena. As you will learn, energy is composed of heat and work.

At its most basic level, chemists study chemical energetics to understand the relative strengths of chemical bonds. From a practical standpoint, chemists want to know whether or not a given chemical reaction will produce/release heat energy. An applications chemist would use the knowledge from this chapter to assess a given physical or chemical process and attempt to determine conditions which would yield the greatest amount of usable energy for that particular application. An engineer would be interested in constructing systems that would maximize either heat or work for a given process. Engineers would also want a conceptual understanding of why different materials absorb heat differently, so that they could choose the correct material for a given job. A biologist might be interested in studying the various caloric values of foods that provide our bodies with energy.

The concepts covered in this chapter will be used again in Chapter 14.

6.1   TYPES OF ENERGY

Key Terms, Key Concepts, and Useful Relationships
- These are summarized in a table on the next page.

Units and Conversions:
- Energy is measured in Joules (J).
- $1\ J = 1\ kg\ m^2/sec^2$

Summary Table of Various Forms of Energy

Form	Definition	Useful Relationships/Information
Kinetic Energy, KE	Energy of Motion	$KE = \frac{1}{2} mu^2$, where m = mass and u = speed
Potential Energy, PE	Energy of Position, Stored Energy	Both definitions of PE are used frequently in chemistry.
Chemical Energy	Energy stored inside a molecule resulting from attractive forces between the atoms within the molecule (a type of potential energy)	When a substance undergoes a chemical transformation, chemical energy is either released or absorbed by the substance.
Thermal Energy	Constitutes a variety of energy types, depending on the substance analyzed: kinetic, rotational, vibrational, translational. Heating an object increases these energies.	Example: Kinetic energy of a monoatomic gas, as seen in Chapter 5: $$KE_{molar} = \frac{3RT}{2}$$
Electrical Energy	Energy arising from forces between two charged particles (a type of potential energy)	$$E_{electrical} = k\frac{q_1 q_2}{r}$$ where $k = 2.31 \times 10^{-16}$ J pm, q_1 and q_2 = charges on the respective objects, r = distance between the objects.
Radiant Energy	Light Energy	UV light, visible light, Infrared light are some examples. (covered in detail in 7)

EXERCISE 1: List the energy transformations that occur during each process: a) moving a notebook from the floor to a high shelf; b) providing energy to heat your home with an electrical heater.

SOLUTION:

a) Kinetic energy (moving the notebook) is converted into potential energy (the notebook now has potential to fall to the floor). Technically, chemical energy was transformed into the kinetic energy because chemical reactions had to take place inside your body to provide you with energy to move.

b) Electrical energy from a power plant (or several power plants) is converted into thermal energy in your home. The electrical energy could have come from a variety of sources: 1) a waterfall or a windmill – kinetic energy from the moving water or wind gets converted to electrical energy; 2) a nuclear power plant would convert nuclear chemical energy into thermal energy by heating up water to make steam, which would be converted to kinetic energy by making the steam move a turbine, which would be converted to electrical energy; 3) a coal plant or other fuel plant would convert chemical energy into thermal energy then kinetic energy, then electrical energy.

Try It #1: List the energy transformations that occur when you burn a log in a campfire.

EXERCISE 2: Determine the kinetic energy of an electron traveling at 65 miles per hour.

STRATEGY: Use the equation: "KE = ½ mu²" to determine the kinetic energy. The value for speed will have to be in units of "m/sec;" we can do that first or incorporate it into the problem. Both methods will be shown to illustrate that the same result is obtained.

Method 1: Convert rate units to m/sec first, then apply that to the KE equation in a second step.

$$\frac{65 \text{ miles}}{\text{hr}} \times \frac{1 \text{ hr}}{3600 \text{ sec}} \times \frac{1609 \text{ m}}{\text{mile}} = 29.05 \text{ m/sec}$$

$$\text{KE} = \text{½ mu}^2 = \frac{1}{2} \times 9.1 \times 10^{-31} \text{ kg} \times \left(\frac{29.05 \text{ m}}{\text{sec}}\right)^2 \times \frac{1 \text{ J}}{1 \text{ kg m}^2 / \text{sec}^2} = 3.8 \times 10^{-28} \text{ J}$$

Method 2: Do calculation set-up and conversion in one step.

$$\text{KE} = \text{½ mu}^2 = \frac{1}{2} \times 9.1 \times 10^{-31} \text{ kg} \times \left(\frac{65 \text{ miles}}{\text{hr}}\right)^2 \times \left(\frac{1 \text{ hr}}{3600 \text{ sec}}\right)^2 \times \left(\frac{1609 \text{ m}}{1 \text{ mile}}\right)^2 \times \frac{1 \text{ J}}{1 \text{ kg m}^2 / \text{sec}^2} = 3.8 \times 10^{-28} \text{ J}$$

Notice in the second method that the conversion factors had to be squared in order for the units to cancel out. When punching numbers into a calculator, be sure to account for the squared terms.

SOLUTION: The kinetic energy of one electron traveling at 65 miles/hour is 3.8×10^{-28} J. (2 sig figs from data)

Take-home messages from Exercise 2:
- There's often more than one correct way to approach a chemistry problem.
- When doing a multi-step mathematical process – as in Method 1 – wait until *the end of the last step* to round off to sig figs.

6.2   THERMODYNAMICS

Thermodynamics is the study of energy flow. Energy flows from one object to another, but is never created or destroyed. In order to discuss and quantitate energy flow, several terms must be defined:

Key Terms:
- **Law of Conservation of Energy** – Energy cannot be created nor destroyed. It can only be transferred from one body to another or converted from one form to another.

- **System** – the material(s) being studied.
- **Surroundings** – everything else in the universe except for the system.
- **Boundary** – that which separates the system from the surroundings.

Key Concept:
- Depending upon what you want to analyze, you might define a particular system differently than another person. The "system" is simply "what is being studied."

EXERCISE 3: Let's say we'd like to study the thermodynamics of a cup of coffee. Define the system, surroundings, and boundary in this situation.

SOLUTION: We would define the system as the coffee itself (i.e., the contents of the cup). The boundary would be the cup, and the surroundings would be every other thing in the entire universe.

Heat and Energy

Key Terms:
- **Temperature (T)** - measure of kinetic energy of molecular motion.
- **Heat (q)** – transfer of thermal energy between a system and its surroundings.
- **Molar heat capacity (C)** – amount of heat energy required to raise one mole of substance by one degree Celsius.

Key Concepts:
- Heat always flows *from* hot *to* cold.
- ΔT depends upon four factors:
  1) amount of heat transferred – the more heat transferred, the greater the ΔT;
  2) direction of heat transferred – $\Delta T = T_f - T_i$, so initial vs. final T will predict sign of q;
  3) amount of material – the more substance present, the more heat it can absorb or release;
  4) identity of material – some materials absorb heat better than others.
- ΔT in units of Kelvin is the same as ΔT in units of degrees Celsius.
    For example: raising the temperature from 10°C to 25°C gives a ΔT of +15°C.
        In Kelvin it would be: $(25 + 273) - (10 + 273) = +15$ K.
- When energy flows *into* a system *from* the surroundings, the system gains energy.
- When energy flows *from* a system *into* the surroundings, the system loses energy.
- *Energy changes are measured from the point of view of the system.* If no subscript is noted then the value is referring to the system.

Helpful Hints
- Metals have low heat capacities; they are good conductors of heat.
- Water has an unusually high heat capacity.

Useful Relationships:
- $q = nC\Delta T$
- $q_{surroundings} = -q_{system}$; heat flows between the system and the surroundings, so their values will be equal in magnitude, but opposite in sign.

EXERCISE 4: If you have 10 moles each of copper and water – one is very hot and the other is very cold – and you simultaneously toss them into the same insulated container,
   a) which will gain or lose more heat?
   b) what will be true of the magnitudes of the ΔTs?
   c) If the copper was originally the hotter object, what will be the sign of ΔT for each substance?
   d) If the copper is considered the system and the water is considered the surrounding, what is the relationship between q for each of these?

STRATEGY: Assess what is occurring. The two substances are initially at different temperatures, so when they come in contact, heat will flow from one to the other until their temperatures are equal.

SOLUTION:
a) The heat lost by one will equal the heat gained by the other. Neither will gain nor lose more than the other.
b) ΔT_{Cu} will not equal ΔT_{H_2O} because the substances have different molar heat capacities. Copper, with a lower heat capacity, will undergo a greater temperature change than water.
c) Cu: temperature is decreasing, so T_f is smaller than T_i. The sign for ΔT_{Cu} will be negative. The sign for ΔT_{water} will be positive.
d) q for copper will be equal in magnitude but opposite in sign to q for water. We could write this as: $q_{copper} = - q_{water}$ or $-q_{copper} = q_{water}$. Either equation is valid; notice that both equations simply illustrate these values are opposite in sign from each other.

EXERCISE 5: A 2.0 kg cast iron skillet at room temperature (19°C) is placed in an oven at 175°C. How much heat energy does the skillet absorb when it heats up to the oven's temperature? Assume that the skillet is pure iron for the calculation. C_{iron} is 25.10 J/mol K.

STRATEGY: We'll use a shortcut of the 7-step method to analyze this problem. We're asked to find q_{iron}. We've been given the mass of iron and its temperature change. We can use "$q=nC\Delta T$." Here are some considerations:
   • ΔT is always final minus initial conditions: $175° - 19° = 156°C$; $\Delta T = 156$ K.
   • Mass can be converted to moles using molar mass.

$$q_{iron} = n_{iron}C_{iron}\Delta T_{iron} = 2.0\,kg \times \frac{1000\,g}{1\,kg} \times \frac{1\,mol}{55.845\,g} \times \frac{25.10\,J}{mol\,K} \times 156\,K = 1.4 \times 10^5\,J$$

SOLUTION: The heat energy required is 1.4×10^5 J. This answer seems reasonable because it is a positive number, which indicates heat was absorbed by the iron.

Try It #2: If a 0.46 kg hot copper pan lost 9.8×10^3 J of heat as it cooled down to 72°F, what was its original temperature?

Work

Key Term:
• **Work (w)** – energy used to move an object against an opposing force.

Key Concepts:
- When a system does work on the surroundings, the system loses energy, so the sign of w is negative.
- When the surroundings do work on a system, the system gains energy, so the sign of w is positive.
- Like heat, work is *transferred* between the system and the surroundings, so work of the system will be equal in magnitude but opposite in sign from work of the surroundings.

Useful Relationships:
- $w_{surroundings} = -w_{system}$

First Law of Thermodynamics: Law of Conservation of Energy

Key Term:
- **Law of Conservation of Energy** (also known as the First Law of Thermodynamics) – Energy cannot be created nor destroyed, just transferred. (In other words, the energy of the universe is constant.)

Key Concepts:
- The only two types of energy that can be exchanged between a system and its surroundings are heat (q) and work (w): $\Delta E = q + w$.
- The energy gained by the system is lost by the surroundings and vice versa.

Useful Relationships:
- The relationship: $\Delta E = q + w$ can be incorporated with what we already know about heat, work and energy to give us many useful relationships. We use the subscripts "sys" and "surr" for "system" and "surroundings", respectively, to keep track. Convince yourself that these relationships are true:
  - $-\Delta E_{surr} = q_{sys} + w_{sys}$
  - $\Delta E_{sys} + \Delta E_{surr} = \Delta E_{total} = 0$
  - $q_{sys} = \Delta E_{sys} - w_{sys}$

EXERCISE 6: If a system does 15 J of work and receives 18 J of heat, what is the value of ΔE for the system?

STRATEGY: The key to answering this question is to remember what the signs for q and w indicate: if the system *receives* heat, the value for q is "+," and if it *does* work the value for w is "-."

SOLUTION: $\Delta E_{sys} = q_{sys} + w_{sys} = (18 \text{ J}) + (-15 \text{ J}) = 3 \text{ J}$ (1 sig fig from subtraction rules)

State Functions and Path Functions

Key Terms:
- **State** – the conditions that describe the system (For example: pressure, volume, temperature, amount of substance).
- **Change of state** – a change in conditions of the system.
- **State function** – a function or property whose value depends only on the present state (condition) of the system, not on the path used to arrive at that condition.
- **Path function** – a function that depends upon *how* the change occurs.

Important features of state functions:
- A state function can be analyzed from its initial and final conditions. The *process* in going from the initial to final conditions does not need to be known.
- For any state function, the overall change is zero if the system returns to its original condition.
- Reversibility: if a state function has a numerical value of "+5" in one direction, it will have a value of "-5" in the reverse direction.
- Values for many state functions are tabulated in reference tables and books.
- State changes can be calculated by the most convenient means available.
- Since state functions can be analyzed from initial and final conditions only, they often contain a "Δ" in calculations.

Key Concepts:
- Change in energy (ΔE) of a system is a state function, so it only depends upon initial and final conditions, and not the path taken.
- Work and heat are not state functions; they are path-dependent.

EXERCISE 7: Which is a state function? a) The change in volume of a balloon that was popped. b) The amount of calories you burn when you run two miles.

STRATEGY: A state function is independent of the path taken.

SOLUTION: a) This is a state function because a balloon at a given starting volume will have a final, deflated volume of zero regardless of how the air was let out of it. b) This is not a state function because the amount of calories burned depends upon more than the starting and ending point of the run. For example, how fast did you run? Were there any hills? Was the ground hard or sandy? This is a path function, not a state function.

6.3   ENERGY CHANGES IN CHEMICAL REACTIONS

Features of Reaction Energies

Key Term:
- **Molar energy change** (ΔE_{molar}) – change in energy for a process that is based on one mole of a specific reactant.

Key Concepts:
- Bond breaking requires energy; bond formation releases energy.
- A net negative reaction energy indicates that:
    - the products are more stable than the reactants, and
    - the reaction releases energy.
- A net positive reaction energy indicates that:
    - the reactants are more stable than the products, and
    - the reaction absorbs energy
- If a reaction releases energy when it proceeds in the forward direction, it will gain that same amount of energy if it proceeds in the reverse direction, and vice versa.
- The amount of energy released or absorbed in a chemical reaction depends upon the quantity of materials undergoing reaction; the more chemicals reacting, the more energy released or absorbed.

- The energy change associated with a chemical reaction can be calculated by using the balanced chemical equation for that reaction as a stoichiometric conversion factor. For example, in the reaction:

$$Fe_2O_3 \text{ (s)} + 3 \text{ CO (g)} \rightarrow 2 \text{ Fe (s)} + 3 \text{ CO}_2 \text{ (g)} \qquad \Delta E = -24.8 \text{ kJ}$$

we see that three moles of CO (g) reacting with plenty of Fe_2O_3 (s) will release 24.8 kJ of energy (remember a negative sign means energy released). This means that:

- six moles of CO (g) reacting with plenty of Fe_2O_3 (s) will release twice that amount of energy:

$$6.00 \text{ mol CO } \times \frac{-24.8 \text{ kJ}}{3 \text{ mol CO}} = -49.6 \text{ kJ}$$

- one mole of CO (g) reacting with plenty of Fe_2O_3 (s) will release one third that amount of energy:

$$1.00 \text{ mol CO } \times \frac{-24.8 \text{ kJ}}{3 \text{ mol CO}} = -8.30 \text{ kJ}$$

- 14.0 g of CO reacting with plenty of Fe_2O_3 (s) will release one sixth that amount of energy:

$$14.0 \text{ g CO } \times \frac{1 \text{ mol CO}}{28.0 \text{ g CO}} \times \frac{-24.8 \text{ kJ}}{3 \text{ mol CO}} = -4.13 \text{ kJ}$$

Helpful Hint:
- Figure 6-12 in the text nicely diagrams energy flows in reactions.

EXERCISE 8: Given that: $8 \text{ Mg(s)} + \text{Mg(NO}_3)_2\text{(s)} \rightarrow \text{Mg}_3\text{N}_2\text{(s)} + 6 \text{ MgO(s)}$; $\Delta E = -3884 \text{ kJ}$
Calculate the amount of energy released or absorbed for the following processes:

a) 4.000 moles of $Mg(NO_3)_2$ (s) and plenty of Mg (s) react
b) the molar energy of reaction with respect to Mg(s)
c) 88.5 g of Mg(s) and plenty of $Mg(NO_3)_2$ (s)
d) 2.00 moles of Mg_3N_2(s) react with excess MgO(s).

STRATEGY: Use appropriate stoichiometric relationships to perform conversions.
   a) Start with moles of what you're given, $Mg(NO_3)_2$ (s), and create a conversion factor between that and energy of reaction.

$$4.000 \text{ mol Mg(NO}_3)_2 \times \frac{-3884 \text{ kJ}}{1 \text{ mol Mg(NO}_3)_2} = -15,540 \text{ kJ}$$

   b) Start one mole of Mg(s), and create a conversion factor between Mg(s) and energy of reaction.

$$1.000 \text{ mol Mg } \times \frac{-3884 \text{ kJ}}{8 \text{ mol Mg}} = -485.5 \text{ kJ}$$

c) Start with 88.5 g of Mg (given), convert grams to moles, then create a conversion factor between moles of Mg and energy of reaction.

$$88.5 \, g \, Mg \quad x \quad \frac{1 \, mole \, Mg}{24.305 \, g \, Mg} \quad x \quad \frac{-3884 \, kJ}{8 \, mol \, Mg} = -1768 \, kJ$$

d) Notice that the materials given are product in our reference reaction, so this question is asking for what happens in the reverse reaction. Therefore, our stoichiometric ratio will involve the reverse sign for energy: +3884 kJ instead of −3884 kJ. Start with moles of what you're given, Mg_3N_2 (s), and create a conversion factor between that and energy of the reverse reaction.

$$2.00 \, mol \, Mg_3N_2 \quad x \quad \frac{3884 \, kJ}{1 \, mol \, Mg_3N_2} = 7768 \, kJ$$

SOLUTION:
a) −15,540 kJ. We're starting with four times as much starting material as the balanced equation specifies, so we'll release four times as much energy in the process.
b) −485.5 kJ. We have much less than 8 moles of Mg, so the process will release much less than 3884 kJ of energy.
c) −1770 kJ (round to 3 sig figs). 88.5 g of Mg is less than 8 moles of Mg, so it is reasonable that this will release less than 3884 kJ of energy.
d) 7770 kJ (round to 3 sig figs). This answer seems reasonable because energy is *absorbed* in the reverse reaction, and the answer has a positive sign. Furthermore, we are reacting twice as many moles of Mg_3N_2 as in the balanced equation, so the magnitude of the energy change will be double that of the reference reaction.

Try It #3: Given the reaction: Ni (s) + 4 CO (g) → Ni(CO)₄ (g)   ΔE = −154 kJ, how many kilograms of CO are required for this reaction to release 756 kJ of energy?

Bond Energy and Reaction Energy

Key Terms:
- **Bond energy (BE)** – amount of energy required to break a covalent bond.
- **Average bond energy** – amount of energy required to symmetrically break a given bond within one mole of gaseous molecules. When a bond breaks symmetrically, one electron goes to each atom, resulting in two neutral fragments.

Key Concepts:
- Bond energies are assessed in many different ways. The type of bond energy we will use is called "average bond energy."
- Average bond energy applies to a generic bond between two atoms, so actual bond energies may deviate slightly from the tabulated values.
- The overall energy for a reaction can be approximated by calculating the difference between the energy required to break bonds in reactant molecules and the energy released by bonds formed in product molecules.

Helpful Hint
- Table 6-2 in the text lists average bond energies for several bonds.

Useful Relationship:
- $\Delta E_{reaction} = \Delta E_{\text{bond breaking}} + \Delta E_{\text{bond formation}} = \Sigma BE_{reactants} - \Sigma BE_{products}$

EXERCISE 9: Ethanol can be made by fermenting sugar in corn. Corn ethanol can be used as an alternate fuel to gasoline. Here is the combustion reaction for ethanol:

$$C_2H_6O \ (l) + \ 3 \ O_2 \ (g) \rightarrow 2 \ CO_2 \ (g) + 3 \ H_2O \ (g)$$

Use Average Bond Energies from Table 6-2 in the text to calculate the energy of the reaction. Note: assume that you've been given the structure of each molecule.

STRATEGY: Use the relationship: $\Delta E_{reaction} = \Sigma BE_{reactants} - \Sigma BE_{products}$ to calculate the reaction energy.

We must take a tally of all the bonds broken and formed. Structures of all molecules in the reaction have been provided.

Structures:	1 Ethanol 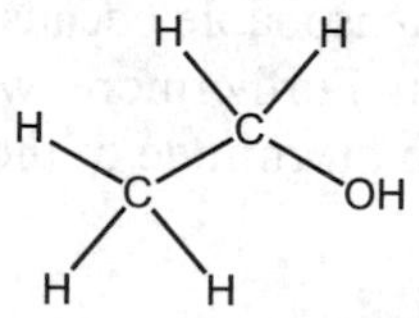	3 Oxygens 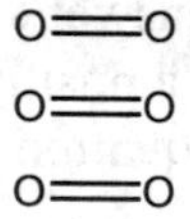	2 Carbon dioxides 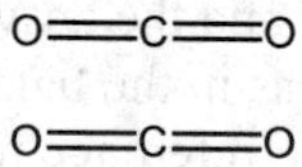	3 Waters 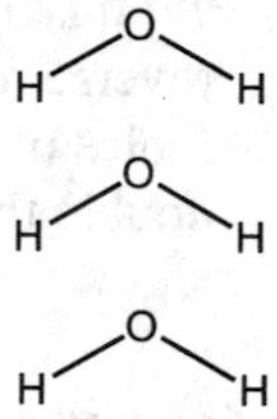
Tally: Bond Energies in parentheses (kJ/mol)	5 C-H = 5(415) 1 C-O = 1(360) 1 O-H = 1(460) 1 C-C = 1(345)	3 O=O = 3(495)	4 C=O = 4(800) *the BE of C=O in CO_2 is 800 kJ/mol	6 O-H = 6(460)
Tally Subtotal: (kJ/mol)	3240	1485	3200	2760

$$\Delta E_{reaction} = \Sigma BE_{reactants} - \Sigma BE_{products} = (3240+1485) - (3200+2760)$$
$$= -1235 \text{ kJ/mol}$$

Try It #4: Use Average Bond Energies from Table 6-2 in the text to calculate the energy of the reaction:
$$H_2(g) + \ Cl_2(g) \leftrightarrow \ 2 \ HCl \ (g)$$

6.4   MEASURING ENERGY CHANGES: CALORIMETRY

Calorimeters

Key Terms:
- **Calorimeter** – container designed to create an insulated system, so that heat measurements can be carefully made without any loss of heat to the surrounding environment. Two main types of calorimeters exist:
    - **Constant-pressure calorimeter** – typically used to assess reactions that involve aqueous solutions. One can be made from a Styrofoam cup and a thermometer.
    - **Constant-volume calorimeter** – typically used to assess reactions that involve gases.
- **Endothermic** – gain in energy. (ΔE is "+.")
- **Exothermic** – loss of energy. (ΔE is "-.")
- **Heat capacity of a calorimeter (C_{cal})** – amount of heat required to raise the temperature of the entire calorimeter by one degree Celsius.

Key Concepts:
- A calorimeter is a highly insulated system. Energy cannot flow in or out of the calorimeter; energy is only transferred between the chemicals in the calorimeter and the calorimeter walls.
- A calorimeter is typically calibrated prior to use to determine its heat capacity.
- In constant-pressure calorimetry, it is typical to assume that the heat capacity of the calorimeter equals the heat capacity of the water inside the calorimeter.

Useful Relationships:
- $q_{chemicals} = -q_{calorimeter}$. Heat is only transferred between the chemicals and the calorimeter.
- $q_{calorimeter} = C_{cal}\Delta T$

Units and Conversions:
- SI unit for energy is the Joule (J), $1\ J = 1\ N\ m\ = 1\ kg\ m^2/sec^2$
- 1 calorie = 1 cal = 4.184 J
- 1 Cal = 1 kcal = expression used for food (banana = 70 Cal that your body can use as fuel)

EXERCISE 10: An unknown metal was believed to be chromium. The heat capacity of chromium is 23.35 J/molK. The following laboratory experiment was performed in hopes of verifying the metal's identity. A 11.56 g piece of the metal was cooled to 5.68°C. It was then transferred to a Styrofoam cup calorimeter containing 30.00 g of water at 25.000°C. The temperature of the water dropped to 24.240°C. Is the unknown metal chromium?

> *STRATEGY:* Use the 7-step method.
> 1. **Determine what is asked for.** The identity of the metal.
> 2. **Visualize the problem.** Consider the concepts involved in this particular calorimetry problem. The Styrofoam cup acts as a constant-pressure calorimeter, meaning that all of the energy gained by the metal (system) must be released by the calorimeter (surroundings). In constant-pressure calorimetry, one can assume that the heat capacity of the calorimeter equals the heat capacity of the water inside the calorimeter. Furthermore, the heat will flow between the two objects until their temperature is the same (24.240°C).
> 3. **Organize the data.** The data that has been provided includes:
>   Water: mass$_{water}$ = 30.00 g, and ΔT_{water} = 24.240°C – 25.000°C = -0.760°C = -0.760 K.

Metal: $mass_{metal} = 11.56$ g, $\Delta T_{metal} = 24.240°C - 5.68°C = 18.56°C = 18.56$ K.

4. **Identify the process to solve the problem.** The appropriate mathematical reasoning is therefore:

$$q_{metal} = -q_{calorimeter} = -C_{cal}\Delta T_{cal} = -n_{water}C_{water}\Delta T_{water}$$

which simplifies to:

$$q_{metal} = -n_{water}C_{water}\Delta T_{water}$$
$$\text{and we know that } q_{metal} = n_{metal} C_{metal} \Delta T_{metal}$$
$$\text{so, } n_{metal} C_{metal} \Delta T_{metal} = -n_{water}C_{water}\Delta T_{water}$$

Every substance has its own heat capacity. We can determine the heat capacity of the metal from the experiment, and see if it matches the heat capacity for chromium.

5. **Manipulate the equations.** Solve for C_{metal}: $n_{metal} C_{metal} \Delta T_{metal} = -[n_{water} C_{water} \Delta T_{water}]$

$$C_{metal} = \frac{-n_{water} C_{water} \Delta T_{water}}{n_{metal}\Delta T_{metal}}$$

6. **Substitute and calculate.** For this calculation to work, we'll need to put in the molar mass of Cr to convert grams of metal to moles of metal. If the metal is indeed Cr, then the heat capacity will match with Cr's.

$$C_{metal} = -30.00 \text{ g} \times \frac{1\,\text{mole}}{18.015\,\text{g}} \times \frac{75.291\,\text{J}}{\text{mol K}} \times (-0.760 \text{ K}) \times \frac{1}{11.56\,\text{g}} \times \frac{51.996\,\text{g}}{1\,\text{mol}} \times \frac{1}{18.56\,\text{K}} = 23.09 \text{ J}/\text{molK}$$

The heat capacity of the metal did not match the heat capacity of Cr, so the metal is not Cr.

7. **Do the results make sense?** It is difficult to assess reasonableness in this question, other than to say the metal is not Cr.

Try It #5: Is nickel the unknown metal in Exercise 10?

Calculating Energy Changes: $\Delta E = q + w$

Key Concepts:
- Work is force applied across a distance, so if no volume change occurs, then no work is done.
- In a constant-volume calorimeter, $\Delta V = 0$ so no work is done; $\Delta E_{(constant\ volume)} = q_v$, where the subscript "v" denotes constant-volume conditions. (For example: a closed container with rigid walls is a constant-volume situation.)
- In a constant-pressure calorimeter, volume can change, so work may be done. For constant-pressure conditions (denoted by a subscript "p"), $\Delta E_{(constant\ pressure)} = q_p + w_p$.

EXERCISE 11: When 3.66 g of methane is combusted inside a constant-volume calorimeter, the temperature of the calorimeter rises 19.19°C. The heat capacity of the calorimeter is 9.56 kJ/°C. Calculate the molar energy of combustion for methane.

STRATEGY: We're asked to find the energy released by one mole of methane when it undergoes combustion. $\Delta E = q + w$, but this is a constant-volume situation, so no work is done. The energy change can therefore be calculated directly from q.

$$\Delta E_{combustion} = q_{combustion} = -q_{calorimeter} = -C_{calorimeter}\Delta T_{calorimeter}$$

$$\Delta E_{combustion} = -\frac{9.56\,kJ}{°C} \times 19.19°C = -183.46\,kJ$$

This is the ΔE observed for the 3.66 g of methane used. Convert to units of kJ/mol:

$$\Delta E_{combustion,\,molar} = \frac{-183.46\,kJ}{3.66\,g} \times \frac{16.0426\,g}{1\,mol} = -802.2\,kJ/mol$$

SOLUTION: The molar energy of combustion for methane is –802 kJ/mol. (3 sig figs) Combustion is an exothermic process, so this value should be negative, and it is.

6.5   ENTHALPY

Expansion Work

Key Term:
- **Pressure** – force per unit area.

Key Concepts:
- If there is no volume change then no work is done.
- If work is being done *by* the system *on* the surroundings then E is leaving the system, so the sign of w must be negative. This corresponds to an expansion of a gas because $V_f > V_i$.
- At constant pressure $\Delta E = q_p - P\Delta V$. (Two examples of constant-pressure situations: 1) an open container or 2) a closed container that can expand or contract.)

Useful Relationship:
- $w = -P_{ext}\Delta V$ This equation is derived from the relationship between work, pressure and volume.

Units and Conversions:
- 1 L atm = 101.325 J

EXERCISE 12: Calculate the work done by the gas molecules in a container as the gas expands from 1.0 L to 2.5 L against a constant external pressure of 3.00 atm. Report your answer in units of Joules.

STRATEGY: Work can be calculated from the relationship "$w = -P_{ext}\Delta V$," where $\Delta V = V_f - V_i$. All variables have been provided.

$$w = -P_{ext}\Delta V = -[3.00\,atm\,(2.5\,L - 1.0\,L)] = -4.500\,L\,atm$$

$$-4.500\,L\,atm \times 101.325\,J\,/1\,L\,atm = -456.0\,J$$

SOLUTION: The work done by the gas molecules is –460 J. (2 sig figs from volume data) The number seems reasonable because a system loses energy during an expansion, so the value should be negative.

EXERCISE 13: Consider the process: Z_2 (g) → 2 Z (g), which takes place inside a rigid, sealed container at constant temperature. Fill in the blank with "+," "-," or "0":

a) q is: ____ b) w is _____ ; c) ΔE is ____; d) ΔE_{total} is ___.

STRATEGY: Visualize the process. A diatomic molecule fragments into two atoms. The walls of the container cannot move (rigid), so no change in volume occurs.

SOLUTION:

a) q in this process must be "+" because energy is required to break bonds.

b) w is "0," because no volume change occurred.

c) ΔE must be "+" because $\Delta E = q+w$. q was "+" and w was 0.

d) ΔE_{total} is zero because energy cannot be created nor destroyed. The energy absorbed by the system was lost by the surroundings.

Summary of Signs

The positive and negative signs can get overwhelming in this chapter. Here is a summary of what the sign on each thermodynamic function indicates. In each case, we're talking about the system. Remember that the sign for the surroundings would be opposite and have the opposite meaning. If you had difficulty with Exercise 13, this table should be especially helpful.

Function	If the sign is	It means:
ΔE	"+"	The system is gaining energy. (The system is taking energy from the surroundings.)
"	"-"	The system is losing energy. (The system is transferring energy to the surroundings.)
q	"+"	The system is gaining heat energy. (This is an endothermic process.)
"	"-"	The system is losing heat energy. (This is an exothermic process.)
w	"+"	The system is receiving work energy from the surroundings. (A contraction is occurring.)
"	"-"	The system is doing work on the surroundings. (An expansion is occurring.)

Try It #6: Consider this process occurring inside a closed, but flexible, system at constant pressure:
$$CO_2 (s) \rightarrow CO_2 (g)$$
What is the appropriate sign for each (+ , - , or 0)? a) w; b) ΔE_{total}; c) q_{surr}

Enthalpy

Key Term:
- **Enthalpy (ΔH)** – heat for a process under constant pressure conditions.

Key Concepts:
- Enthalpy is a *defined* thermodynamic property that simplifies heat calculations.
- Chemists usually measure ΔH because most reactions are done at constant P.
- Standard thermodynamic conditions are: temperature = 298 K, gas pressure = 1 atm, solution concentration = 1 M for aqueous solutes. Standard conditions are denoted with a "°."
- It is necessary to specify pressure and temperature conditions for enthalpy, so standard conditions are typically used. $\Delta H°$ = enthalpy change under standard temperature and pressure conditions.
- ΔH and ΔE will only differ significantly if there is a large change in the number of gas molecules present (which would result in PV work). For solids and liquids, $P\Delta V \sim 0$ so $\Delta H = \Delta E$.

Useful Relationship:
- $\Delta H = q_p = \Delta E + P\Delta V$

EXERCISE 14: It requires 64.1 kJ of heat energy to evaporate 218 g of mercury at 630. K (its boiling point) inside a container whose pressure is a constant 1.00 atm. Determine ΔH and ΔE for this process.

> *STRATEGY:* In a constant pressure environment, $\Delta H = q_p$ and ΔE is related to ΔH in the equation:
> $$\Delta H = q_p = \Delta E + P\Delta V$$
>
> ΔH is simply 64.1 kJ. In order to find ΔE, we must determine $P\Delta V$. In this process, the mercury is going from a liquid to a gas. The initial volume of the liquid will be negligible compared to the final volume in the gaseous state. Use the ideal gas law to determine the final volume of the container. We'll convert grams to moles in one step in the ideal gas law:
>
> $$PV = nRT, \text{ so } V = \frac{nRT}{P} = \frac{218\,g}{1.00\,atm} \times \frac{mol}{200.59\,g} \times \frac{0.08206\,L\,atm}{mol\,K} \times 630.\,K = 56.185\,L$$
>
> $$P\Delta V = 1.00\,atm\,(56.185\,L - 0\,L) \times \frac{101.325\,J}{1\,L\,atm} = 5.6929 \times 10^3\,J$$
>
> $$\Delta E = \Delta H - P\Delta V = 64.1\,kJ - 5.6929\,kJ = 58.4\,kJ$$
> (be sure to convert to similar units prior to subtracting!)

SOLUTION: $\Delta H = 64.1$ kJ and $\Delta E = 58.4$ kJ. The answers seem reasonable. $P\Delta V$ is typically quite small compared to ΔH.

Try It #7: In Exercise 5, we saw that it required 1.4×10^5 J of heat energy to raise the temperature of a 2.0 kg iron skillet from 19°C to 175°C in the oven. What are ΔH, ΔE, and $P\Delta V$ for this process?

Ways to Analyze Enthalpy

Key Terms:
- **Standard enthalpy of formation** ($\Delta H°_f$) – $\Delta H°_{rxn}$ to create 1 mole of product from its elements in their standard states. Note: fractional coefficients are allowed.
- **Hess' Law** – the overall heat gain/loss in a chemical process equals the sum of the enthalpy changes in each step of the process: $\Delta H_T = \Delta H_1 + \Delta H_2 + \Delta H_3...$

Useful Relationships:
- The text shows how Hess' Law can be manipulated to yield this very useful relationship,
$$\Delta H°_{rxn} = \Sigma \text{ (coeff) } \Delta H°_f(\text{products}) - \Sigma \text{ (coeff) } \Delta H°_f(\text{reactants})$$

Helpful Hint
- Appendix D in the text lists standard heats of formation ($\Delta H°_f$) for many substances.

EXERCISE 15: Which have a $\Delta H°_f$ of zero? a) Ag (s); b) NaCl (s); c) Xe (g); d) CO_2 (g); e) Al (l)

STRATEGY: $\Delta H°_f$ will be zero if the initial and final conditions are the same. A formation reaction shows the formation of one mole of a substance from its elements in their standard states. If the chemical in question *is* an element in its standard state, then no change is required to form it.

SOLUTION: a) Ag(s) and c) Xe(g) are the only substances shown that are elements in their standard states, so these are the only choices that will have a $\Delta H°_f$ of zero.

Try It #8: Which have a $\Delta H°_f$ of zero? a) N_2 (g); b) O_2 (l); c) Na (s)

EXERCISE 16: Calculate $\Delta H°_{rxn}$ for each reaction, given that:
$$3 H_2 (g) + N_2 (g) \rightarrow 2 NH_3 (g); \Delta H_1° = -91.8 \text{ kJ}$$

a) $2 NH_3 (g) \rightarrow 3 H_2 (g) + N_2 (g)$ $\Delta H_a° = ?$
b) $3/2 H_2 (g) + 1/2 N_2 (g) \rightarrow NH_3 (g)$ $\Delta H_b° = ?$
c) $NH_3 (g) \rightarrow 3/2 H_2 (g) + 1/2 N_2 (g)$ $\Delta H_c° = ?$

STRATEGY: Compare the original reaction to each reaction shown. Use the fact that $\Delta H°$ is a state function to solve each problem.
a) This reaction is the reverse of the original reaction. State functions depend upon final minus initial conditions, so $\Delta H_a°$ will be the reverse of $\Delta H_1°$: 91.8 kJ.
b) The coefficients in this reaction have been halved relative to the original reaction. The amount of heat energy depends upon the amount of material present, so $\Delta H_b°$ will be half of $\Delta H_1° = \frac{1}{2}(-91.8$ kJ) = -45.9 kJ.
c) This reaction has been reversed and the coefficients have been halved relative to the original reaction. $\Delta H_c° = -\frac{1}{2} \Delta H_1° = -\frac{1}{2}(-91.8$ kJ) = 41.9 kJ.

SOLUTION: $\Delta H_a° = 91.8$ kJ; $\Delta H_b° = -45.9$ kJ; $\Delta H_c° = 45.9$ kJ.

Key Concepts:
- The concepts applied in Exercise 16 apply to all state functions. Using ΔH as an example:
  - If you add two reactions, then add their ΔHs.
  - If you reverse a reaction, then reverse the sign of its ΔH.
  - If you multiply/divide the coefficients of the chemical equation, then multiply/divide the ΔH.

Summary of Ways to Find ΔH for a Reaction:
(ΔH is a state function so it can be calculated by the most convenient means available)

1. Standard heats of formation.
2. $\Delta H^\circ_{rxn} = \Sigma$ (coeff) ΔH°_f(products) - Σ (coeff) ΔH°_f(reactants)
3. Hess' Law
4. Calorimetry experiment at constant pressure.

EXERCISE 17: Given the following information, calculate ΔH°_3.

$$Fe_2O_3 (s) + 3 CO (g) \rightarrow 2 Fe (s) + 3 CO_2 (g) \qquad \Delta H^\circ_1 = -24.8 \text{ kJ}$$

$$CO (g) + 1/2 \ O_2 (g) \rightarrow CO_2 (g) \qquad \Delta H^\circ_2 = -283.0 \text{ kJ}$$

$$2 Fe(s) + 3/2 \ O_2 (g) \rightarrow Fe_2O_3 (s) \qquad \Delta H^\circ_3 = ?$$

STRATEGY: Use Hess' law to add reactions 1 and 2 together in such a way that they result in reaction 3. We want 1 Fe_2O_3 (s) as a product, so reversing reaction 1 will provide us with that:

$$2 Fe (s) + 3 CO_2 (g) \rightarrow Fe_2O_3 (s) + 3 CO (g) \qquad \Delta H^\circ_{1 \text{ reversed}} = +24.8 \text{ kJ}$$

We need to get rid of the 3 CO *and* the 3 CO_2 in our reversed reaction 1. If we multiply reaction 2 by three, and add it to reversed reaction 1, then all of the chemicals we want will remain and all that we don't want will cancel out:

$$3 CO (g) + 3/2 \ O_2 (g) \rightarrow 3 CO_2 (g) \qquad 3\Delta H^\circ_2 = 3(-283.0 \text{ kJ})$$
$$2 Fe (s) + 3 CO_2 (g) \rightarrow Fe_2O_3 (s) + 3 CO (g) \qquad \Delta H^\circ_{1 \text{ reversed}} = +24.8 \text{ kJ}$$

Adding the above two reactions together results in ΔH°_3.
~~3 CO (g)~~ + 3/2 O_2 (g) +2 Fe (s) + ~~3 CO₂ (g)~~ → ~~3 CO₂ (g)~~ + Fe_2O_3 (s) + ~~3 CO (g)~~

SOLUTION: $\Delta H^\circ_3 = \Delta H^\circ_{1 \text{ reversed}} + 3\Delta H^\circ_2 = (+24.8 \text{ kJ}) + 3(-283.0 \text{ kJ}) = -824.2 \text{ kJ}$

EXERCISE 18: Calculate ΔH°_{rxn} for the reactions shown below from thermodynamic data in Appendix D in the text.
a) $CaCl_2 (s) \rightarrow Ca^{2+} (aq) + 2 Cl^- (aq)$; (this reaction can be used to make hot packs)
b) $2 Fe (s) + 3 CO_2 (g) \rightarrow Fe_2O_3 (s) + 3 CO (g)$

STRATEGY: Look up values of ΔH°_f for each substance and apply it to the equation:

$$\Delta H°_{rxn} = \Sigma \text{ (coeff) } \Delta H°_f \text{(products)} - \Sigma \text{ (coeff) } \Delta H°_f \text{(reactants)}$$

a) $CaCl_2(s) \rightarrow Ca^{2+} (aq) + 2 Cl^- (aq)$:
$$\Delta H°_{rxn} = [(1 \text{ mol } Ca^{2+})(-543.0 \text{ kJ/mol}) + (2 \text{ mol } Cl^-)(-167.1 \text{ kJ/mol})]$$
$$-[(1 \text{ mol } CaCl_2)(-795.4 \text{ kJ/mol})]$$
$$= -81.8 \text{ kJ}$$

b) $2 Fe (s) + 3 CO_2 (g) \rightarrow Fe_2O_3 (s) + 3 CO (g)$:
$$\Delta H°_{rxn} = [(3 \text{ mol } CO)(-110.525 \text{ kJ/mol}) + (1 \text{ mol } Fe_2O_3)(-824.2 \text{ kJ/mol})]$$
$$- [(2 \text{ mol } Fe)(0 \text{ kJ/mol}) + (3 \text{ mol } CO_2)(-393.509 \text{ kJ/mol})]$$
$$= 24.8 \text{ kJ}$$

SOLUTION: a) –81.8 kJ; b) 24.8 kJ. The answers seem reasonable – reaction "a" has a negative value for $\Delta H°_{rxn}$, indicating the reaction releases heat energy, which is what a hot pack does. Reaction "b" was seen in the previous Exercise, and we obtained the same value for $\Delta H°_{rxn}$ as shown there.

Try It #9: Use thermodynamic data from Appendix D in the text to calculate $\Delta H°_{rxn}$ for:
$$N_2H_4 (l) + H_2 (g) \rightarrow 2 NH_3 (g)$$

EXERCISE 19: How much heat is needed to react 550 kg of solid iron with an excess of carbon dioxide?
Given: $2 Fe (s) + 3 CO_2 (g) \rightarrow Fe_2O_3 (s) + 3 CO (g)$; $\Delta H°_{rxn} = 24.8 \text{ kJ}$

STRATEGY: The amount of heat released or required depends upon the amount of material reacted. The above equation shows that 24.8 kJ of energy are required for every 2 moles of Fe (or every 3 moles of CO_2, every 1 mole of Fe_2O_3, or every 3 moles of CO). Use the $\Delta H°_{rxn}$ in a stoichiometric conversion factor with Fe.

$$550 \times 10^3 \text{ g Fe} \times \frac{1 \text{ mol}}{55.845 \text{ g}} \times \frac{24.8 \text{ kJ}}{2 \text{ mol Fe}} = 1.2 \times 10^5 \text{ kJ}$$

SOLUTION: 1.2×10^5 kJ of heat energy is needed. The number is reasonable, because this is a large quantity of Fe, and the value is positive, which indicates energy is required.

6.6 ENERGY SOURCES

Fast Facts:
- Figure 6-21 in the text shows that energy needs per capita have increased by more than 100-fold as civilization has progressed.
- By the mid nineteenth century, work done by machines exceeded work done by animals.
- In 2001, 85% of the total world energy production came from petroleum, coal, and natural gas, all of which are non-renewable energy sources.
- Photosynthesis is the process by which plants react carbon dioxide and water to form glucose and oxygen. This reaction requires energy from the sun, and the products of the reaction store chemical energy to be used at a later time.
- The combustion process releases large amounts of energy.
- Nuclear energy is another important energy source, accounting for approximately 6.5% of human energy production.

Chapter 6 Self-Test

You may use a periodic table, a calculator, Table 6-2 in the text, and Appendix D in the text for this test. The heat capacity for water is 75.291 J/mol K.

1. Consider the process occurring inside a container whose sides cannot move:

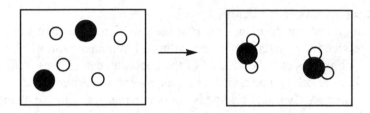

   Determine the sign of ΔE for the system (+, -, or 0). Briefly explain your choice.

2. Given this information: NH_4^+ (aq) + NO_3^- (aq) → NH_4NO_3 (s) ΔH = -21.1 kJ/mol
   Calculate q_{water} when 35.0 g of solid ammonium nitrate dissolves in 300.0 mL of water.

3. Given: $2 (NH_2)_2CO$ (s) → $2 C$ (s) + O_2 (g) + $2 N_2$ (g) + $4 H_2$ (g) ; $\Delta H°$ = 666.2 kJ

   a) Calculate $\Delta H_f°$ of $(NH_2)_2CO$ (s)

   b) Calculate the amount of heat energy absorbed or released when 15.0 g of $(NH_2)_2CO$ (s) decomposes to its elements in their standard states.

4. A 5.263 g strip of magnesium was combusted inside a constant-pressure calorimeter. The temperature of the calorimeter rose from 24.2°C to 61.7°C. Given that the heat capacity of the calorimeter was 3.475 kJ/°C, calculate:

   a) the amount of heat released during the combustion.
   b) the molar heat of combustion for magnesium.

5. Use Table 6-2 in the text to determine ΔE for the reaction: N_2H_4 (l) + H_2 (g) → $2 NH_3$ (g). (Hint: each structure contains only single bonds. Nitrogen atoms are the central atoms.)

6. Consider the reaction: $Ni(CO)_4$ (g) → Ni (s) + $4 CO$ (g) $\Delta H°$ = 161 kJ. During this process, 1.07 moles of $Ni(CO)_4$ (g) cause the reaction container to expand from 7.0 L to 28.0 L against a constant external pressure of 5.000 atm.

   a) Calculate work for this process.
   b) Calculate ΔE for the process.
   c) Did the system or the surroundings do the work?
   d) Is the system gaining energy or losing energy?

Answers to Try Its:

1. A chemical transformation occurs when a log burns. Chemical energy (potential energy stored in oxygen molecules and in the chemicals that make up the log) is converted into radiant energy (we observe light) and thermal energy (we feel heat).
2. 77.4°C (The °F is not linear with Kelvin scale, so convert °F to °C first.)
3. 0.550 kg CO
4. −185 kJ/mol
5. Yes, Nickel's heat capacity is 26.07 J/molK.
6. Before answering any questions here, let's analyze the situation. We have a closed system, so energy can be exchanged between the system and surroundings, but matter cannot. We have a constant-pressure system, which means that the volume of the container can change in order to maintain a constant pressure. Thus, work can be done. The process shows carbon dioxide undergoing a phase change from solid to gas. A chemical change is not occurring, just a phase change. Energy will have to be added to get the molecules to move enough to go into the gas state. To answer the questions:
   a) w_{sys} is "-" because when the gas forms, the container will expand. The sign for "$-P_{ext}\Delta V$" is "-."
   b) ΔE_{total} is "0" because of the first law of thermodynamics. It is ALWAYS "0."
   c) q_{surr} must be "-." One must first assess q_{sys}, which is "+" because energy must be added to the system to convert the solid to a gas (increased molecular movement/overcome intermolecular forces.) Since q_{sys} is "+," q_{surr} must be "-." The increased q of the system was taken from the surroundings.
7. $\Delta H = q_p = 1.4 \times 10^5$ J, $\Delta E = 1.4 \times 10^5$ J. (A kitchen oven is a constant pressure environment, so $\Delta H = q_p$.) The skillet is a solid, so $P\Delta V = 0$.
8. N_2 (g) and Na (s). O_2 is a gas under standard conditions, so O_2 (l) won't have a $\Delta H°_f$ of zero.
9. $\Delta H° = -142.4$ kJ

Answers to Self-Test:

1. $\Delta E = q + w$. "w" is zero because the sides of the container cannot move. (ΔV is zero.) "q" is negative because bonds are being formed, which releases energy. The sign for ΔE is therefore negative.
2. -9.23 kJ (Notice that you didn't even need the volume of water! Notice, too, that the process in question is the reverse of the reaction provided.)
3. a) −333.1 kJ; b) 83.2 kJ (heat absorbed)
4. a) −130. kJ; b) −602 kJ/mol
5. −185 kJ
6. a) w = −105 L atm or −10.6 kJ; b) $\Delta E = 161.7$ kJ; c) the system did the work, as indicated by work's negative sign. (The system lost work energy.); d) the system gained energy, as indicated by ΔE's positive sign.

Chapter 7: Atoms and Light

Learning Objectives

In this chapter, you will learn how to:

- Explain the behavior of light energy using the wave-particle duality of light.
- Relate light's energy to its wavelength and frequency.
- Use Einstein's explanation for the photoelectric effect to calculate the binding energy of an electron in a metal.
- Use Bohr's theory of quantized energy levels in atoms to calculate atomic emission lines in hydrogen.
- Interpret an energy level diagram for an element.
- Generate a set of quantum numbers to describe a given atomic orbital.
- Draw pictures of atomic orbitals.

Practical Aspects

In Chapter 2, you learned about several experiments that probed at the structure of the atom, as well as several theories that were developed from these experiments. This chapter picks up where those experiments left off, and introduces experiments that further probe the structure of the atom. This body of work in understanding atomic structure ranks as one of the great achievements in science, as is evidenced by the fact that a high percentage of these scientists were Nobel Prize winners.

Sections 7.5 and 7.6 introduce quantum mechanical theory – a theory which contributes to the understanding of electron arrangement in atoms by defining orbitals, or regions where a given electron is likely to be. You will learn that different orbitals have different shapes, sizes, and orientations. As you will see in Chapters 8, 9, and 10, these shapes, sizes and orientations of an atom's orbitals will define how that atom connects to other atoms to make molecules. Develop a clear visual picture of these orbitals now, and it will help you with the upcoming chapters, as well as any future chemistry courses you take.

7.1 CHARACTERISTICS OF ATOMS

Here's a quick review of what you should know about atoms up to this point:

- Atoms contain a positively-charged nucleus (made up of protons and neutrons) surrounded by negatively-charged electrons.
- In a neutral atom, the number of protons and electrons is equal.
- Most of the atom is empty space; the volume of an atom is significantly larger than the volume of its nucleus.
- Atoms of different elements have different physical and chemical properties.
- Atoms attract one another.
- Atoms can combine with one another to make molecules.

7.2   CHARACTERISTICS OF LIGHT

Light Has Wave Aspects

Light acting as a wave is characterized by its:

Characteristic	Symbol	Unit	Definition
Frequency	ν (pronounced "nu")	$\sec^{-1}$ or Hz	the number of wave crests that pass a reference point in one second
Wavelength	λ (pronounced "lambda")	nm	the distance between any two similar points on a wave
Amplitude	A	-	height of the wave (This is a measurement of intensity.)

Key Terms:
- **Electromagnetic radiation** – light energy.
- **Spectroscopy** – the study of the interaction between light and matter.
- **White light** – a mixture of all the wavelengths of visible light.
- **Intensity** – the *number* of waves of a given wavelength.

Key Concepts:
- The electromagnetic spectrum is a range of all the types of light energy that exist, arranged by frequency or wavelength.
- The types or regions of light energy that exist are: Gamma, X-rays, UV, Visible, Infrared, Microwave, and Radio.
- Our eyes can only detect visible light, which contains wavelengths of light between 400 and 700 nm.
- White light can be separated into its individual wavelengths by shining it through a prism.
- Each wavelength of visible light is associated with a color of the rainbow, with red showing up at the 700 nm end and violet at the 400 nm end. (See Figure 7-4 in the text for a detailed illustration.)

Units and Conversions:
- c = speed of light = 2.997925×10^8 m/sec
- 1 nm = 10^{-9} m; (sample shortcut unit conversion: 600 nm = 600×10^{-9} m)

Useful Relationship:
- $\lambda\nu = c$

Helpful Hint
- The colors of the rainbow are: "Roy G. Biv" or "Red, Orange, Yellow, Green, Blue, Indigo, Violet."

EXERCISE 1: Light of 432 nm has what frequency in units of Hertz? What color of light is this?

STRATEGY: Use the relationship "$\lambda\nu = c$" and solve for frequency. Use "Roy G. Biv" (or Figure 7-4 in the text) to approximate the color of the light.

$$\nu = \frac{c}{\lambda} = \frac{2.9979 \times 10^8 \text{ m}}{\text{sec}} \times \frac{1}{432 \times 10^{-9} \text{ m}} = 6.94 \times 10^{14} \text{ sec}^{-1} = 6.94 \times 10^{14} \text{ Hz}$$

SOLUTION: The light's frequency is 6.94×10^{14} Hz and its color will be purple.

Try It #1: Estimate the wavelength and corresponding frequency of orangish-red light.

The Photoelectric Effect

Key Terms:
- **The Photoelectric Effect** – light shining on a metal surface can remove an electron from the metal if the light is of sufficient energy. Note that no matter how bright (i.e. intense) the light is, the electron will not be removed unless the light energy exceeds a minimum (threshold) frequency.
- **Photon** – a "packet" or "bundle" of light energy.
- **Binding energy** – measurement of how tightly an electron is attached to a given atom. Different atoms have different binding energies.
- **Threshold frequency (ν_o)** – minimum frequency that a photon of light needs to overcome the atom's binding energy.

Einstein's explanation of the photoelectric effect:
Light can exist as "photons" (bundles of energy). If a single photon's energy is strong enough to overcome the binding energy of the electron, then the photoelectric effect will be observed. Any photon energy that is in excess of the binding energy will be converted into kinetic energy of the electron. This can be shown mathematically as:

$$KE_{electron} = E_{photon} - E_{binding}$$

Key Concepts:
- In order for an electron to be removed from an atom, the atom must absorb enough energy from a photon to overcome the binding energy of the electron. This minimum amount of energy is called the "threshold frequency."
- No matter how intense the light is, the photoelectric effect will not happen unless each photon can overcome the binding energy of the electron.
- Einstein's explanation of the photoelectric effect indicated that light also can act as particles.
- The wave-particle duality of light: light can behave like waves and/or particles.

Units and Conversions:
- h = Planck's constant = 6.626×10^{-34} J sec

Useful Relationship:
- $E = h\nu$ This equation is used to relate light energy and frequency.

EXERCISE 2: Consider a pool table with three billiard balls on it (Balls A, B, and C). You are supposed to shoot the three balls very lightly, using the same amount of energy each time. What you don't know is that Ball B is stuck to the pool table with a piece of gum and Ball C has been glued to the table with super-bonding glue. a) Which ball will have the greatest kinetic energy when you shoot it? b) Which ball will have zero kinetic energy? c) Use this situation as an analogy of the photoelectric effect. Use terms from this section to describe what happens to each ball. d) For which scenario(s) is the photoelectric effect observed?

SOLUTION:
a) Ball A will have the greatest kinetic energy.
b) Ball C will have zero kinetic energy.
c) If this is an analogy of the photoelectric effect, then the balls must be electrons on different metals, and the energy with which you hit the balls must be the energy of the photon.
Ball A: has no binding energy, so all of the energy of the photon is converted into kinetic energy.
Ball B: The ball is able to move, but its kinetic energy is less than that of Ball A. The difference in kinetic energies between Balls A and B is the energy that went into overcoming the binding energy of Ball B. ($KE = E_{photon} - E_{binding}$). The fact that the billiard ball moved is evidence that the frequency of the photon is greater than the threshold frequency, v_o, for this billiard ball.
Ball C: The ball did not move, so the photon's energy was not strong enough to overcome the high binding energy of the ball. In other words, the photon's energy was less than hv_o for Ball C.
d) The photoelectric effect was observed for balls A and B only.

EXERCISE 3: A photon of light of 545 nm strikes a metal surface. Given that the binding energy for the metal is 3.27×10^{-19} J, will an electron be ejected from the metal, and if so, what will its kinetic energy be?

STRATEGY: In order for the photoelectric effect to occur, the energy of the photon must exceed the binding energy of the electron to the metal. Calculate the energy of the photon and compare the two numbers.

$$E_{photon} = \frac{hc}{\lambda_{photon}} = \frac{6.626 \times 10^{-34} \text{ J sec}}{545 \times 10^{-9} \text{ m}} \times \frac{2.9979 \times 10^8 \text{ m}}{\text{sec}} = 3.64 \times 10^{-19} \text{ J}$$

The photon's energy is greater than the binding energy, so an electron can be removed from an atom of the metal. The electron's kinetic energy will be:

$$KE_{electron} = E_{photon} - E_{binding} = 3.64 \times 10^{-19} \text{ J} - 3.27 \times 10^{-19} \text{ J} = 0.37 \times 10^{-19} \text{ J} = 3.7 \times 10^{-20} \text{ J}$$

SOLUTION: An electron can be removed from the metal if a photon of 545 nm light hits it. The kinetic energy of the electron will be 3.7×10^{-20} J.

EXERCISE 4: Let's look at the metal in Exercise 3 again. a) Would a photon of 615 nm be able to remove an electron from this metal? b) Would two photons of 615 nm be able to remove an electron from the metal?

STRATEGY: Use the same procedure as in Exercise 3, substituting in the new wavelength.
a) $E_{photon} = 3.23 \times 10^{-19}$ J, which is less than the binding energy, so 615 nm light could not remove an electron from the metal.
b) Two photons would have $2(3.23 \times 10^{-19}$ J) of energy or 6.46×10^{-19} J, which is greater than the binding energy. BUT! The photoelectric effect depends upon the energy *per photon*, which in this case is not enough to overcome the electron's binding energy. The number of photons will make the *intensity* greater, but has no effect on the energy per photon, and thus no effect on the photon's ability to remove the electron from the metal.

SOLUTION: a) One 615 nm photon cannot remove this electron. b) Two photons can't remove it either. (A thousand photons of this wavelength couldn't either.)

Try It #2: What is the longest wavelength of light energy that could produce the photoelectric effect in the metal described in Exercise 3?

7.3   ABSORPTION AND EMISSION SPECTRA

Light and Atoms

Key Terms:
- **Ground state** – the lowest energy state, or most stable state, for an atom.
- **Excited state** – a higher energy state than the atom's ground state.

Key Concepts:
- Energy must be absorbed by an atom for it to move to an excited state.
- When an atom returns from an excited state to its ground state, it must lose the exact amount of energy it originally gained. Most of this energy is emitted as electromagnetic radiation (light energy). Some of this energy could be emitted as heat.

Helpful Hints
- Energy-level diagrams are used to illustrate electronic transitions in atoms. See Figure 7-9 in the text for an example of one.
- This topic will be covered in great detail in the next section.

Atomic Spectra

Key Terms:
- **Spectrum** – plot of intensity as a function of energy, wavelength, or frequency (typically focuses on a specific region of the electromagnetic spectrum, for example, a "visible spectrum" or an "infrared spectrum").
- **Absorption spectrum** – plot of intensity of the absorption of light energy over a range of wavelengths.
- **Emission spectrum** – plot of intensity of the emission of light energy over a range of wavelengths.
- **Intensity** – measure of *how many* photons of a given wavelength are being absorbed or emitted.
- **Quantized** – restricted to specific values.
- **Energy level diagram** – plot of absorption and emission lines for an element, with Energy on the vertical axis. Figures 7-13 and 7-14 in the text show sample energy level diagrams.

Key Concepts:
- Atomic energy levels are quantized, or restricted to specific values (first proposed by Niels Bohr).
- When an atom emits energy in the visible region, it will show up as a specific color.
- Each element has its own unique emission spectrum that can act as a fingerprint for that element.
- A visible atomic absorption spectrum looks like a rainbow with some colors blacked out; the blacked out colors are the wavelengths of light absorbed by the element.
- A visible atomic emission spectrum looks like a black background with colored bars of light added to it. The bars of light are the photons of light emitted by the element at those wavelengths.

Useful Relationship:

- $E_n = -\dfrac{2.18 \times 10^{-18}\ \text{J}}{n^2}$ Use this relationship to calculate the energy level occupied by an electron in a hydrogen atom.

EXERCISE 5: Which are quantized? a) The rungs on a ladder; b) a slope on a hill; c) the gears in a car.

STRATEGY: The term "quantized" means "restricted to certain values."

SOLUTION: Rungs on a ladder are located only in certain places, so they can be said to be quantized. The gears in a car are also quantized: first gear, second gear, etc. You need to shift into a specific (quantized) gear for the car to run. A slope on a hill is *not* quantized because it has a continuous range of elevations; for example, a tree is not limited to quantized locations where it can grow on the slope.

EXERCISE 6: Given that the wavelength of the red line in hydrogen's atomic emission spectrum is 656 nm, what is the energy associated with one mole of photons that cause the red line?

STRATEGY AND SOLUTION:

$$E = h\nu = \frac{hc}{\lambda} = \frac{6.626 \times 10^{-34}\ \text{J sec}}{656 \times 10^{-9}\ \text{m}} \times \frac{2.9979 \times 10^8\ \text{m}}{\text{sec}} \times \frac{6.022 \times 10^{23}\ \text{photons}}{1\ \text{mol}} = 1.82 \times 10^5\ \text{J/mol}$$

EXERCISE 7: Calculate the wavelength of light that is emitted when electrons move from the *n*=6 state to *n*=2 state within hydrogen.

STRATEGY: Use the hydrogen atom energy-level relationship: $E_n = -\dfrac{2.18 \times 10^{-18}\ \text{J}}{n^2}$ to find the energy associated with the transition, then calculate wavelength from energy.

$$E_6 = -\frac{2.18 \times 10^{-18}\ \text{J}}{6^2} = -6.056 \times 10^{-20}\ \text{J} \quad \text{and} \quad E_2 = -\frac{2.18 \times 10^{-18}\ \text{J}}{2^2} = -5.45 \times 10^{-19}\ \text{J}$$

The energy emitted when the electron goes from *n*=6 to *n*=2:

$$E_{emitted} = E_f - E_i = E_2 - E_6 = (-5.45 \times 10^{-19}\ \text{J}) - (-6.056 \times 10^{-20}\ \text{J}) = -4.84 \times 10^{-19}\ \text{J}$$

$$E = h\nu = \frac{hc}{\lambda}; \text{so}\ \lambda = \frac{hc}{E} = \frac{6.626 \times 10^{-34}\ \text{J sec}}{4.84 \times 10^{-19}\ \text{J}} \times \frac{2.9979 \times 10^8\ \text{m}}{\text{sec}} \times \frac{10^9\ \text{m}}{1\ \text{m}} = 410.\ \text{nm}$$

SOLUTION: The light emitted when an electron falls from energy level 6 to 2 within a hydrogen atom is purple light at 410. nm. This value coincides with the energy level diagram shown in Figure 7-13 of the text, so the answer seems reasonable.

EXERCISE 8: The atomic emission spectrum for a given element shows a very bright, intense bar of green light and a faint bar of purple light. a) Which bar of light corresponds to a greater number of electronic transitions occurring? b) Which bar of light corresponds to a greater amount of energy being released per transition?

STRATEGY AND SOLUTION:
a) Wavelength depends on the particular transition that is occurring. Intensity of light at that wavelength depends upon *how many* photons are making that transition. The bright, intense green bar of light is therefore related to the electronic transition that is occurring the greatest number of times.

b) When one electron falls from an excited state to its ground state within an atom, the energy is emitted as light energy. Wavelength is related to energy by $E = hc/\lambda$. The shorter the wavelength, the greater the energy emitted. Purple light is of shorter wavelength, and therefore higher energy, than green light. The purple line corresponds to greater energy emitted per atom.

7.4 PROPERTIES OF ELECTRONS

Key Terms:
- **Heisenberg Uncertainty Principle** – it is impossible to simultaneously know the position and the motion of an electron; the more that is known about its position, the less is known about its motion, and vice versa.
- **Delocalized** – not localized; not in one location.

Key Concepts:
- Each electron has a mass of 9.109×10^{-31} kg and a charge of -1.602×10^{-19} C.
- Electrons behave like magnets – they may be aligned with a magnetic field or opposite a magnetic field (this will be covered in detail in Chapter 8).
- Electrons have wave properties (this was first proposed by Louis de Broglie).
- Electrons are delocalized; they are constantly moving around inside the atom.
- Photons and electrons both have wave-particle duality, but their properties are treated differently mathematically. See Table 7-1 in the text for a summary of equations to use for photons vs. electrons.

Useful Relationships:
- $\lambda_{particle} = \dfrac{h}{mu}$ This is called the "de Broglie" equation. It relates λ (a wave property) to m (a particle property). This equation is only used for very small particles, like electrons.

EXERCISE 9: Compare the speed of an electron that has a wavelength of 340 nm to the speed of a photon whose wavelength is 340 nm.

STRATEGY: Use the de Broglie equation to find the speed of the electron. A photon is light energy, so the speed of the photon is the speed of light: 2.9979×10^8 m/sec.

$$\lambda_{particle} = \frac{h}{mu}, \text{ so } u = \frac{h}{m\lambda} = \frac{6.626 \times 10^{-34} \text{ J sec}}{340 \times 10^{-9} \text{ m}} \times \frac{1 \text{ kg m}^2 / \text{sec}^2}{1 \text{ J}} \times \frac{1}{9.109 \times 10^{-31} \text{ kg}} = 2.1 \times 10^3 \text{ m/sec}$$

SOLUTION: At 340 nm, a photon has a speed of 2.9979×10^8 m/sec and an electron has a speed of 2.1×10^3 m/sec. These numbers seem reasonable because light has no mass so it should be much faster than an electron (which has mass).

7.5   QUANTIZATION AND QUANTUM NUMBERS

Key Terms:

- **Quantum mechanical theory** – theory that provides a mathematical explanation which links quantized energies to the wave characteristics of atoms.
- **Schrödinger equation** – mathematical equation developed by Erwin Schrödinger that describes the kinetic and potential energies associated with an atom.
- **Orbital** – a three-dimensional shape that maps where a given electron has a relatively high probability of being found.
- **Node** – a region in space where there is zero probability of finding an electron. Most orbitals contain nodes within them.
- **Quantum number** – a number that describes a quantized property of an electron. There are four different quantum numbers, and each one describes a different feature of the electron.

Key Concepts:

- Only energies of electrons that are bound to atoms are quantized.
- For each quantized energy, the Schrödinger equation will generate a 3-dimensional wave function to describe where the electron can be located. (This is an orbital.)
- Different electrons within a given atom have different quantized energies, and therefore will be located in different orbitals.
- An orbital is described by its size, shape and 3-D orientation in space.

The Four Quantum Numbers

These numbers are derived from the Schrödinger equation and are so named because they are each "quantized" - only certain values can exist for each. A complete description of an electron within an atom must be described using all four quantum numbers.

Quantum Number	Name	Information It Provides	Possible Numerical Values
n	Principal Quantum Number	- describes energy level or shell that the electron is occupying - describes size of the orbital: the larger the number, the larger the orbital	1, 2, 3, 4, …
l	Azimuthal Quantum Number Or Angular Momentum Quantum Number	- describes the orbital's shape - defines subshells (see below)	Integers from zero to $(n-1)$
m_l	Magnetic Quantum Number	- describes the orbital's orientation	Integers from $-l$ thru $+l$
m_s	Spin Quantum Number	- describes the electron's spin	$+\frac{1}{2}$ or $-\frac{1}{2}$ (sometimes called "spin up" or "spin down")

Possible Subshells (Orbitals):

If l =	0	1	2	3
then you have this subshell:	s	p	d	f

Helpful Hint

- Quantum mechanical theory is quite abstract, and at this point in the chapter, a complete picture has not yet been presented. For now, just think of the quantum numbers as a game or puzzle with a very specific set of rules that must be followed. Learn the rules now. By the end of the chapter you'll be given a visual perspective on what these quantum numbers mean to a chemist.

EXERCISE 10: If the principal quantum number for an electron is $n=2$, a) what possible values can l be? b) what possible values can m_l be?

 STRATEGY: Use the rules in the above quantum numbers tables to answer these questions.
 SOLUTION:
 a) If $n=2$, then l can be 0 or 1.
 b) If $l=0$ then m_l can be 0. If $l=1$ then m_l can be -1 or 0 or 1.

EXERCISE 11: List one valid set of quantum numbers for an electron that is in a "d" subshell.

 STRATEGY: Use the rules in the above tables as guidelines.
 • A "d" subshell corresponds to $l=2$.
 • n must be at least 3 because if n is 3, then $n-1=2$. This is from the possible values for $l=0$ to $n-1$.
 • m_l can be -2, -1, 0, 1, or 2.
 • m_s can be $+\frac{1}{2}$ or $-\frac{1}{2}$.

 SOLUTION: One set of quantum numbers for an electron in a "d" subshell is: $n=3$, $l=2$, $m_l=1$, $m_s=-\frac{1}{2}$. Many other possible sets exist.

Try It #3: List another set of quantum numbers for an electron in a d subshell.

Try It #4: a) What are the possible values for l if $n=1$? b) What are the possible values for m_l if $l=4$? c) What are the possible values for m_s if $l=2$?

EXERCISE 12: Which set(s) of quantum numbers is/are not permissible? For those sets that are not permissible, indicate the quantum number that makes it invalid.

	n	l	m_l	m_s
a)	1	2	0	$+\frac{1}{2}$
b)	4	3	0	$+\frac{1}{2}$
c)	4	3	4	$-\frac{1}{2}$
d)	3	1	-2	$-\frac{1}{2}$

STRATEGY: Use the rules in the quantum numbers tables as guidelines.

SOLUTION:
- Set "a" is not valid because if $n=1$, then l can only be 0. l cannot be 2 unless n is at least 3.
- Set "b" is valid.
- Set "c" is not valid because if $l=3$, then m_l can only be –3, -2, -1, 0, 1, 2, or 3. m_l cannot be 4 unless l is at least 4 (which would mean n would have to be at least 5).
- Set "d" is not valid because m_l can only be –2 if l is at least 2. In this case with $l=1$, m_l must be –1, 0, or 1.

Try It #5: Is it possible to have an $n=3$ electron in an f orbital? Explain.

7.6   SHAPES OF ATOMIC ORBITALS

Orbital Depictions

Figure 7-19 in the text shows the three basic ways to depict orbitals. Here is a summary:

Type	Benefits	Drawbacks
Electron density plot - graph of electron density as a function of distance from the nucleus	• Simple to draw • Several plots can be superimposed to give one overall picture for the atom • Shows nodes	• Shows only 2 dimensions
Orbital density picture	• 3-Dimensional • shows nodes	• Time consuming
Electron contour drawing (this is most frequently used)	• Simplified orbital density picture to show overall contour	• Can only see the surface of the orbital, so nodes are not visible

Details of Orbital Shapes

Key Concepts:
- An orbital can be described using a specific shorthand method. For example, an electron in an orbital that has $n=2$ and $l=1$ is a "2p" orbital.
- As n increases, orbital size increases.
- A p orbital has a nodal plane running through the center of the orbital. A d orbital has two nodal planes. (See the drawings in the chart on the next page.)
- All orbitals with the same principal quantum number have roughly the same size (in a given atom).
- For orbitals with the same principal quantum numbers in *different* atoms, orbitals become smaller as nuclear charge increases because a large positive charge pulls on the electron more strongly than a small positive charge.

Helpful Hint
- Figures 7-20, 7-22, and 7-23 in the text illustrate the respective contour plots of all s, p, and d orbitals. See the next page for a summary of these plots.

Summary Chart of Orbital Shapes

Orbital (as indicated by l)	Shape (nodal planes are shown with arrows)	Orientations (as indicated by m_l)
s ($l=0$)	Spherical	1 orientation, based on 1 value for m_l (for $l=0$, $m_l = 0$)
p ($l=1$)	Dumbbell	• 3 orientations, based on 3 values for m_l (for $l=1$, $m_l = -1$, 0, or 1) • The orientations are: p_x, p_y, or p_z • Subscripts indicate the dumbbell lies along the x, y, or z-axis.
d ($l=2$)	Cloverleaf or Dumbbell w/ donut	• 5 orientations, based on 5 values for m_l (for $l=2$, $m_l = -2$, -1, 0, 1, or 2) • There are 4 cloverleaves in various orientations and 1 dumbbell with donut. • The text describes the subscripts for these orbitals.

EXERCISE 13: Draw contour plots to show the comparison between each pair of orbitals: a) 3py vs. 4p$_y$; b) 1s vs. 3s; c) 4s vs. 4p$_x$ vs. 4d$_{xy}$

a) 3p$_y$ 4p$_y$ **b)** 1s   3s   **c)** 4s   4p$_x$ 4d$_{xy}$

Notice that for a given value of n, the orbital size stays about the same. Here are the answers to Exercise 13c superimposed to illustrate this.

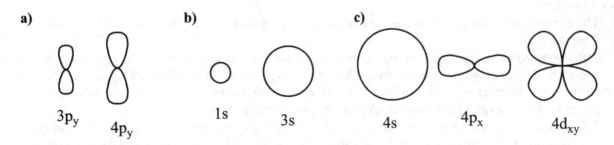

c) 4s   4p$_x$ 4d$_{xy}$ =

Try It #6: How many 4d orbitals exist for a given atom?

EXERCISE 14: Here are drawings (contour plots) of several orbitals shown below. The possible designations for these orbitals (in no particular order) are: a 3s orbital in Si, a 2s orbital in Si, a 2s orbital in C, a 2s orbital in O. Match the picture to the orbital.

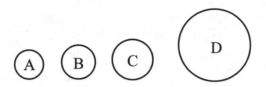

STRATEGY: The orbital size is determined primarily by the principal quantum number, but also by the nuclear charge.

SOLUTION: The largest orbital, D, must correspond to the Si 3s, which has the largest principal quantum number. All of the other orbitals are 2s orbitals for atoms with different nuclear charges: C = +6, O = +8, and Si = +14. The element with the greatest nuclear charge will have the strongest attractions to the 2s electrons. This element will pull the electrons towards itself the most intensely, and therefore have the smallest orbital. A must be the silicon 2s orbital, B must be the oxygen 2s orbital and C must be the carbon 2s orbital.

7.7   SUNLIGHT AND THE EARTH

Key Concepts:
- The atmosphere is separated into several regions: troposphere, stratosphere, mesosphere, and thermosphere.
- The **thermosphere** (the region of our atmosphere closest to the sun) shields us from most of the high-energy ionizing radiation from the sun. This high-energy radiation breaks apart O_2 molecules into atoms and it fragments N_2 molecules. These atoms and fragments later re-form into molecules, releasing heat energy in the process. As a result, the thermosphere is quite hot.
- The **stratosphere** contains the ozone layer, which shields us from most of the UV light from the sun. UV light breaks the ozone molecules into atomic and molecular oxygen and breaks molecular oxygen into atoms of oxygen.
- Depletion of the ozone layer has caused increased amounts of UV light to reach the earth.
- The **troposphere** is the region closest to the earth's surface. The types of light that reach the earth's surface are visible light, infrared light, and some UV light. The earth absorbs this light and releases it in the form of infrared light, which makes its way back through our atmosphere and out to space.
- The troposphere contains a small amount of CO_2, water vapor, and CH_4 gas, (and minor amounts of some other gases), each of which is quite good at absorbing infrared light. These gases are termed "Greenhouse Gases" because they can absorb the infrared light energy given off by the earth, thus trapping the energy close to earth.

Chapter 7 Self-Test

You may use a periodic table and a calculator for this test.
$c = 2.9979 \times 10^8$ m/sec; $h = 6.626 \times 10^{-34}$ J sec; mass of an electron $= 9.109 \times 10^{-31}$ kg

1. A radio station in San Diego, CA, broadcasts at a frequency of 100.7 MHz. a) What is the wavelength of this electromagnetic radiation? b) Is this electromagnetic radiation more or less energetic than visible light?

2. A photon of UV light has a frequency of 1.000×10^{15} Hz.
   a) Calculate the energy of one mole of these photons.
   b) This light generated a photoelectric effect on a metal. The speed of the ejected electron was 1.200×10^6 m/sec. What is the metal's binding energy?

3. Use this emission spectra data to answer the questions below.

Wavelength (nm)	Relative Intensity
655	Bright
582	Medium-bright
501	Weak

   a) Each spectral line indicated in the table corresponds to one of these electronic transitions:

   $n=7$ to $n=2$, $n=2$ to $n=7$, $n=5$ to $n=2$, $n=2$ to $n=5$, $n=3$ to $n=2$, $n=2$ to $n=3$, $n=2$ to $n=2$

   Match the lines to the appropriate transitions and defend your choice with a brief explanation.

   b) Which spectral line is emitting the most photons? Explain briefly.

4. a) What is the symbol for the azimuthal quantum number?
   b) What information does this quantum number provide?

5. Which set(s) of quantum numbers is(are) not allowed? Which quantum number(s) makes the set(s) unallowable?

	n	l	m_l	m_s
a)	3	2	0	$+\frac{1}{2}$
b)	4	4	0	$+1$
c)	2	1	2	$-\frac{1}{2}$

6. Draw pictures of all of the orbitals that have a principal quantum number of 2.

7. Draw a contour picture of each orbital. Be sure to illustrate relative sizes of the orbitals.

   a) a $3p_x$ orbital for Se   b) a $3p_y$ orbital for S   c) a 3d orbital for S (you choose the orientation)

   d) a 2s orbital

Answers to Try Its:

1. An estimate of this light's wavelength is 650 nm, and its corresponding frequency is 4.6×10^{14} Hz.
2. The binding energy corresponds to a 607 nm wavelength, so just less than 607 nm could generate the photoelectric effect.
3. $n=3$, $l=2$, $m_l = 2$, $m_s = -\frac{1}{2}$ (lots of other possibilities exist too)
4.

   a) If $n=1$, then $l=0$.
   b) If $l=4$, m_l can be $-4, -3, -2, -1, 0, 1, 2, 3,$ or 4.
   c) $m_s = -\frac{1}{2}$ or $+\frac{1}{2}$ regardless of the other quantum numbers.

5. If $n=3$, then the maximum value for l is 2, so it is not possible to have an f orbital ($l=3$ for an f orbital) if $n=3$.
6. For a given n value, there will be five d orbitals, each with a different orientation.

Answers to Self-Test:

1. a) 2.977 m; b) less energetic (visible light is 400-700 nm, which is much shorter wavelength, higher energy than this radio wave).
2. a) 3.990×10^5 J; b) 6.8×10^{-21} J
3. a) 501 nm is the $n=7$ to $n=2$ transition, 582 nm is the $n=5$ to $n=2$ transition, and 655 nm is the $n=3$ to $n=2$ transition. These transitions were chosen because in an emission spectrum, each line is generated when electrons fall from an excited state to a ground state. Furthermore, the shortest wavelength corresponds to the greatest energy released, which is $n=7$ to $n=2$, while the longest wavelength corresponds to the least energy released, which is $n=3$ to $n=2$.
   b) The 655 nm line is the one with the most photons because it is the most intense line.
4. a) l; b) orbital shape
5. a) is OK; b) is not allowed. If $n=4$, then l can be at most 3. Also, m_s cannot be 1 because its only possible values are $+\frac{1}{2}$ or $-\frac{1}{2}$. c) is not allowed. If $l=1$, then m_l cannot be 2.
6. If $n=2$, then l can be 0 (an s orbital) or 1 (a p_x, p_y, or p_z orbital). The p_z orbital is perpendicular to the plane of the paper (shown at an angle to illustrate this).

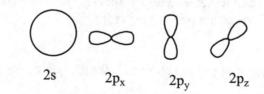

2s   2p$_x$ 2p$_y$ 2p$_z$

7.

   a)   b)   c)   d)

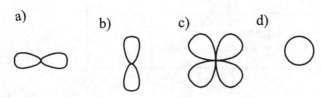

Notice that "a" is slightly smaller than "b". Also b & c are the same size. d is significantly smaller than the others.

Chapter 8: Atomic Energies and Periodicity

Learning Objectives

In this chapter, you will learn how to:

- Use the Aufbau Principle, Pauli Exclusion Principle, and Hund's Rule to determine the manner in which electrons fill into atomic orbitals.
- Draw electron configurations and energy-level diagrams to provide a detailed illustration of how electrons are arranged in atoms.
- Determine if a given element has magnetic properties (paramagnetism).
- Determine the number of valence electrons within an atom.
- Predict periodic trends based on atomic structure.

Practical Aspects

This chapter builds on Chapter 7 concepts by completing the picture of how electrons are arranged within the atom. Electron arrangement has a huge impact on an atom's properties, so it is essential to understand the concepts in this chapter in order to understand chemical bonding and reactivity. The next two chapters in the text (on chemical bonding) will build on the concepts covered here.

8.1   ORBITAL ENERGIES

A given orbital's energy (and thus size) is determined by a combination of two opposing forces:
1. The attraction of an orbital electron to the nucleus.
2. Repulsive forces of other electrons.

Key Concepts:
- Every different element or ion has a unique combination of protons and electrons, and thus will have a unique set of orbital energies.
- Attractive forces lower the energy of the orbital (making the orbital smaller) and repulsive forces raise the energy (making it larger).

Screening

Key Terms:
- **Screening** (also called shielding) – partial canceling of the full attractive forces between a nucleus and an electron by any electron that is between the nucleus and that electron.
- **Effective nuclear charge (Z_{eff})** – the net charge attraction to the nucleus that an electron feels.

EXERCISE 1: In the picture shown below, a) which electron is the most screened from the nucleus? b) Which electron is not screened at all? c) Which electron has the lowest effective nuclear charge?

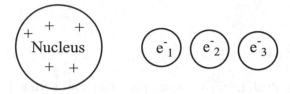

STRATEGY AND SOLUTION:
a) Electron #3 is screened the most from the nucleus – there are two electrons between it and the nucleus.
b) Electron #1 is not screened at all from the nucleus. There are no electrons between it and the nucleus.
c) The two electrons sitting between the nucleus and Electron #3 cancel out 2 of the "+" charges of the nucleus, so that Electron #3 can only feel a net attraction to a "3+" charge. Electron #3, therefore, has the lowest Z_{eff}.

Try It #1: In the above picture, what is the effective nuclear charge of electron #2?

EXERCISE 2: Now, let's link the concept of screening to what we know about orbitals. Each set of orbitals is inside one atom. One orbital in each set feels more attraction to the nucleus and therefore screens the other(s) from the nucleus. In each set of orbitals, identify which orbital acts as a screen for the other(s). a) 1s vs. 3s   b) 4s vs. 4p vs. 4d

STRATEGY: Pictures may help to visualize this. Draw pictures to illustrate each orbital set. Compare the pictures to determine which orbital has a higher probability of electron density near the nucleus – that orbital will be the one that screens the other orbital(s) from the nucleus.

a)   **b)**

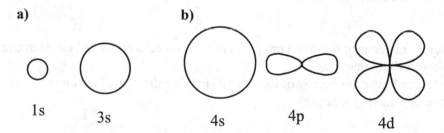

1s   3s   4s   4p   4d

SOLUTION:
a) We can see from the pictures that the 1s orbital is significantly smaller than the 3s orbital. If both of these orbitals are in one atom, then the electrons in the 1s orbital will have a higher probability of being closer to the nucleus than the 3s orbitals. The 1s orbital electrons will screen the 3s electrons from the nucleus.
b) All orbitals are the same size, so we must compare orbital shapes to see if one orbital will screen the others. Comparing the three orbitals shown, we can see that the 4d orbital (with its two nodal planes) has the lowest electron density near the nucleus and the 4s orbital has the highest electron density near the nucleus. The 4s orbital will therefore screen both the 4p and 4d orbitals. Furthermore, the 4p orbital will screen the 4d orbital. The screening in these orbitals is much less significant than the screening in part a) where the orbitals had different *n* values.

Several generalizations can be made from the results of Exercise 2, as shown in the Key Concepts here.

Key Concepts:
- Complete screening – Orbitals with lower n values (i.e. lower principal quantum numbers) completely screen orbitals with higher n values. For example, $n=1$ orbitals completely screen $n=2$ orbitals and $n=3$ orbitals from the nucleus.
- Partial screening – Within a given principal quantum number, the greater the probability of electron density near the nucleus, the more it will shield the other orbitals. For example, within a given n, an s orbital partially screens a p orbital.

Try It #2: The two orbitals shown are in one atom. Which one screens the other from the nucleus?

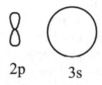

2p   3s

8.2 STRUCTURE OF THE PERIODIC TABLE

Key Terms:
- **Pauli Exclusion Principle** – Each electron within an atom will have a unique set of quantum numbers. (In other words, no two electrons within an atom can have the same four quantum numbers.)

EXERCISE 3: Create sets of quantum numbers to determine how many electrons can fit into: a) a 2s orbital, b) a $4p_x$ orbital, c) a $3d_{xy}$ orbital.

    STRATEGY: Use the Pauli Exclusion Principle to make the sets.
    a) 2s: If $n=2$ and $l=0$, then m_l can only be zero and m_s can be +/- ½.
    b) $4p_x$: $n=4$ and $l=1$, one orientation has been specified, so m_l can only be one value, m_s can be +/- ½. (We'll specify the one value for m_l as –1, but it could just as well be 0 or +1.)
    c) $3d_{xy}$: $n=3$ and $l=2$, one orientation has been specified, so m_l can only be one value, m_s can be +/- ½. (We'll specify the one value for m_l as –2, but it could just as well be –1, 0, +1 or +2.)

    SOLUTION:
    a) There are 2 sets of quantum numbers: $n=2$, $l=0$, $m_l=0$, $m_s=+$½ and $n=2$, $l=0$, $m_l=0$, $m_s=-$½.
    b) There are 2 sets of quantum numbers: $n=4$, $l=1$, $m_l=-1$, $m_s=+$½ and $n=4$, $l=1$, $m_l=-1$, $m_s=-$½.
    c) There are 2 sets of quantum numbers: $n=3$, $l=2$, $m_l=-2$, $m_s=+$½ and $n=3$, $l=2$, $m_l=-2$, $m_s=-$½.

Exercise 3 establishes more generalizations, as outlined below in the Key Concepts.

Key Concepts:
- Every orbital can hold a maximum of two electrons.
- Two electrons within one orbital must have opposite spins.

The Aufbau Principle

Key Terms:
- **The Aufbau Principle** – principle that describes how electrons fill within an atom: electrons will fill the lowest energy orbitals first, so that the overall atom can be in the lowest possible energy state.
- **Ground-state configuration** – the lowest energy combination of electrons within orbitals.

Key Concepts:
- The periodic table is arranged by shells and subshells. It can be used as a tool to predict how the electrons fill within the atom.
- Columns on the periodic table indicate orbital type: "s-block" and "p-block"(representative elements), "d-block" (transition metals), and "f-block" (inner transition metals).
- Rows on the periodic table indicate n value:
  - "s-block" electrons: n = row number
  - "p-block" electrons: n = row number
  - "d-block" electrons: n = row number – 1
  - "f-block" electrons: n = row number – 2
- When arranging electrons within an atom, you must consider: 1) the energy of the shell (n value), and 2) the energy of the subshell (s<p<d...).

Helpful Hints
- Figures 8-8 and 8-10 in the text show how the periodic table is sectioned into "s-block," "p-block," "d-block," and "f-block" elements.
- To predict the order in which electrons fill into a given atom: 1) find the element on the periodic table; 2) starting with atomic number 1, read from left to right along the periodic table until you reach the desired element.

EXERCISE 4: Use the periodic table to determine which orbitals are filled and partially filled for these elements: a) S; b) Na; c) Ni

    STRATEGY: Locate the element on the periodic table. Read along the periodic table, until you reach the element. Using S as an example: S is in the p-block of row 3. If we read along the periodic table, we see that row 1 is full of electrons (i.e., 1s orbital is full), row 2 is full (i.e., 2s and 2p orbitals are full), and row 3 is partially full (i.e., 3s is full, and the 3p orbitals are partially full). Notice that the 3p orbitals are of the same energy, so they are mentioned as a set.

    SOLUTION:
a) S: filled orbitals –1s, 2s, 2p, 3s; partially-filled orbitals: 3p.
b) Na: filled orbitals –1s, 2s, 2p; partially-filled orbital: 3s.
c) Ni: filled orbitals –1s, 2s, 2p, 3s, 3p, 4s; partially-filled orbitals: 4d.

Try It #3: List the filled and partially filled orbitals for an atom of Zr (atomic number 40).

EXERCISE 5: How many 6d orbitals of the same energy are in a given atom? What is the maximum number of electrons that they can accommodate?

STRATEGY: The number of possible orientations for a given subshell will determine the number of orbitals that have the same energy in that subshell. For any d subshell, there are five orbitals with different orientations:

$$6d_{xy}, 6d_{xz}, 6d_{yz}, 6d_{z^2}, \text{ and } 6d_{x^2-y^2}$$

SOLUTION: There are five 6d orbitals of equal energy, as shown above. Each orbital can hold a maximum of two electrons, so collectively these orbitals can accommodate a maximum of 10 electrons.

Summary of Orbitals That Have the Same Energy:

Orbital Type	# of Orbitals That Have the Same Energy (as predicted by number of orbital orientations)	Maximum number of electrons this set of orbitals can accommodate
s	1	2
p	3	6
d	5	10
f	7	14

EXERCISE 6: Consider all of the possible existing orbitals that are less energetic than a 3p orbital. a) Arrange those orbitals from least to most energetic. b) Of the orbitals listed, which one will fill with electrons first? Which one will fill last?

STRATEGY: All orbitals with $n=1$ or $n=2$ are less energetic than an $n=3$ orbital. Also, a 3s orbital is lower in energy than a 3p orbital. $n=1$ contains only an s orbital: 1s; $n=2$ contains these orbitals: 2s, $2p_x$, $2p_y$, and $2p_z$

SOLUTION: a) The orbitals, arranged from lowest to highest energy are: 1s, then 2s, then $2p_x$, $2p_y$ and $2p_z$ (the 2 p orbitals as a set have equal energy), then 3s. b) Electrons will be placed in the lowest energy orbital first, which is the 1s orbital. The last orbital to fill will be the 3s orbital, which is the highest energy orbital of the set.

Try It #4: a) What is the maximum number of electrons the 4p subshell can hold? b) Which of these orbitals – 3s, 4s, $3d_{xy}$, $4d_{xy}$, $3d_{xz}$ – have the same energy?

Valence Electrons

Key Terms:
- **Valence electrons** - electrons in the outermost shell and any partially-filled "d" and "f" subshells.
- **Core electrons** - electrons in inner shells and any completely filled "d" and "f" subshells.

Key Concepts:
- *Valence electrons are the electrons that are involved in chemical reactions.* If two atoms come near each other, the first things to come in contact are the valence electrons of each atom.
- Core electrons are not involved in chemical reactions.

EXERCISE 7: Determine the number of valence electrons for each atom as well as the orbitals in which the valence electrons are located. a) P; b) K; c) Fe; d) Al; e) Kr; f) Hg

STRATEGY: Valence electrons are the electrons in the outermost shell (*n*) plus any electrons in unfilled "d" and "f" orbitals.

SOLUTION:

Atom:	P	K	Fe	Al	Kr	Hg
# of valence electrons:	5	1	8	3	8	2
The valence electrons are in this/these orbital(s):	3s, 3p	4s	4s, 3d	3s, 3p	4s, 4p	6s

Try It #5: Determine the number and location of the valence electrons in: a) Ni; and b) Sr.

8.3   ELECTRON CONFIGURATIONS

Key Terms:

- **Electron configuration** – description of how the electrons within an atom are arranged.
- **Hund's Rule** - within a set of orbitals that have the same energy, each orbital must receive one electron before any orbital can have two. (This spreads the electrons out as far as possible to minimize electron-electron repulsions.) All unpaired electrons will have the same spin.
- **Diamagnetic** – term used to describe an atom in which all electron spins are paired.
- **Paramagnetic** – term used to describe an atom that contains unpaired spins. Paramagnetic substances act as magnets.
- **Net spin** – the sum of all unpaired electron spins within an atom.

Key Concepts:

- The more unpaired spins present in an atom, the more intensely it will respond to a magnetic field.
- Paramagnetism or diamagnetism can be predicted from an atom's energy level diagram (see below).

Summary of ground state electron configurations:
1. Each electron occupies the most stable available orbital.
2. No two electrons can have the same four quantum numbers.
3. Each orbital can have a maximum of 2 electrons.
4. The higher the *n*, the less stable the orbital.
5. The higher the *l*, the less stable the orbital (within one *n*).

Three ways to illustrate electron configurations:

1. **List the quantum number sets for every electron in the atom.**
   (We saw this in Chapter 7.) This is time consuming to do. The two methods shown below are quicker and provide more useful information.

2. **Write an electron configuration**.
   This method shows how many electrons are in orbitals that don't have the same energy. In Exercise 4, we determined where the electrons in an atom of sulfur were located by reading the information from the periodic table: a filled 1s orbital, a filled 2s orbital, a filled 2p set of orbitals, a filled 3s orbital, and 4 electrons in the 3p orbitals. The electron configuration for this can be shown in two ways:
   - **a full electron configuration**: $1s^2\, 2s^2\, 2p^6\, 3s^2\, 3p^4$.

   - **a shorthand electron configuration**: $[Ne]\, 3s^2\, 3p^4$. This abbreviates a full electron configuration by substituting a bracketed noble gas symbol to represent as much of the configuration as possible. $[Ne] = 1s^2\, 2s^2\, 2p^6$.

3. **Draw an energy-level diagram**.
   This shows even more detail than the electron configuration by illustrating *how* the electrons are distributed among non-filled orbitals that are of the same energy. This is also sometimes called an **orbital diagram**. Here is the energy-level diagram, or orbital diagram, for sulfur:

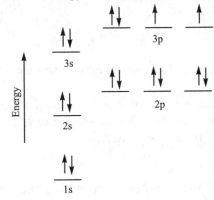

Notice these features in the diagram: a) The Aufbau Principle is obeyed as illustrated by arranging energy vertically (1s then 2s, then the set of 2p orbitals, etc.); b) the Pauli Exclusion Principle is obeyed; and c) Hund's Rule is obeyed. Notice too, that any of these designations for the 3p orbitals would be equally as valid as the one above:

$$\underline{\uparrow}\ \underline{\uparrow\downarrow}\ \underline{\uparrow}\qquad \underline{\uparrow}\ \underline{\uparrow}\ \underline{\uparrow\downarrow}\qquad \underline{\downarrow}\ \underline{\downarrow}\ \underline{\uparrow\downarrow}\qquad \underline{\downarrow}\ \underline{\uparrow\downarrow}\ \underline{\downarrow}\qquad \underline{\uparrow\downarrow}\ \underline{\downarrow}\ \underline{\downarrow}$$
$$3p3p3p3p3p$$

Each obeys the rules for drawing orbital diagrams. It is customary, but not essential, to record unpaired spins as "spin up." The net spin in each diagram would therefore be 2 electrons$(+\,½\,) = +1$.

EXERCISE 8: Draw the full and shorthand electron configuration for: a) Na; b) Ni; c) As; d) Li

   SOLUTION: a) Na: $1s^2\, 2s^2\, 2p^6\, 3s^1$ and $[Ne]\, 3s^1$
           b) Ni: $1s^2\, 2s^2\, 2p^6\, 3s^2\, 3p^6\, 4s^2\, 3d^8$ and $[Ar]\, 4s^2\, 3d^8$
           c) As: $1s^2\, 2s^2\, 2p^6\, 3s^2\, 3p^6\, 4s^2\, 3d^{10}\, 4p^3$ and $[Ar]\, 4s^2\, 3d^{10}\, 4p^3$
           d) Li: $1s^2\, 2s^2$ and $[He]\, 2s^1$

Notice that lithium and sodium are both alkali metals and that they both contain 1 electron in an s orbital. All group I metals have 1 valence electron in an s orbital.

EXERCISE 9: Draw the energy-level diagram (orbital diagram) for: a) oxygen, b) lithium, and c) sodium.

STRATEGY AND SOLUTION: Follow the Aufbau Principle, Pauli Exclusion Principle, and Hund's Rule to construct the diagrams.

a) Oxygen (8 electrons)   b) Lithium (3 electrons)   c) Sodium – 11 electrons

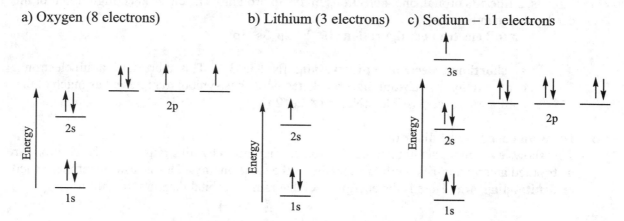

EXERCISE 10: Classify oxygen, lithium, and sodium as either paramagnetic or diamagnetic.

STRATEGY AND SOLUTION: Diamagnetic substances have all spins paired, while paramagnetic substances have at least one unpaired spin. The net spin on oxygen is +1, lithium is + ½, and sodium is + ½. Therefore all of these atoms are paramagnetic.

Try It #6: For Germanium, Ge, draw the a) full electron configuration, b) the shorthand electron configuration, and c) the energy-level diagram. d) Determine the net spin on the atom. e) Is it diamagnetic or paramagnetic?

Special Considerations

Key Terms:
- **Isoelectronic** – term used to describe two different substances that have the same number of electrons (for example: Na^+ and Ne each contain 10 electrons, so they are isoelectronic).
- **Excited state** – electron that is not in its ground state.

Key Concepts:
- For transition metal *cations*, the $(n-1)d$ orbitals are more stable than the ns orbital. Always place electrons into the $(n-1)d$ orbitals before the ns orbital for these cations. (For example, fill 3d orbitals before 4s orbitals when writing configurations for transition metal cations.)
- It becomes difficult to predict the filling order of electrons in transition metals, due to subtle energy differences in their orbitals. These predictions are beyond the scope of the course; so in general, just fill electrons according to the periodic table. However, there are some elements that are used frequently, so their somewhat odd configurations are worth remembering: Cr and Mo: s^1d^5, Cu, Ag, and Au: s^1d^{10}.

EXERCISE 11: a) What 2+ cation is isoelectronic with Ti? b) Write the electron configuration for each.

STRATEGY: a) When it is a neutral atom, this element in question will have 2 more total electrons than Ti. The element is therefore Cr. b) For Ti, follow the periodic table to fill electrons. For Cr^{2+}, remember to fill the (n-1)d orbitals first.
SOLUTION: a) The 2+ cation that is isoelectronic with Ti is Cr^{2+}.
b) The electron configuration of Ti: $[Ar] 4s^2 3d^2$. The electron configuration of Cr^{2+}: $[Ar]3d^4$.

EXERCISE 12: Which illustrate excited state electron configurations? a) Ne: $1s^2 2s^2 2p^5 3s^1$
b) I: $1s^2 2s^2 2p^6 3s^2 3p^6 4s^2 3d^{10} 4p^6 5s^2 4d^{10} 5p^5$ c) Sc: $1s^2 2s^2 2p^6 3s^2 3p^6 4s^2 4p^1$

STRATEGY AND SOLUTION: a) The 2p orbitals are not full, but the 3s orbital contains 1 electron, so this is an excited state. b) This is not an excited state because the orbitals have filled in order of lowest to highest energy. c) This is an excited state because the 3d orbitals should fill before the 4p orbitals.

8.4   PERIODICITY OF ATOMIC PROPERTIES

Several trends in atomic properties occur within the periodic table. It is possible to use these trends in the periodic table to predict properties of an atom or an ion. Keep in mind, though, that all trends are generalities. Be able to use trends to make predictions, but also *understand why* the trends occur so that you can explain the exceptions.

Periodic Trends in Atomic Properties Summary

	Definition	Trend
Atomic Radius	Radius of atom; used to assess size of atom	Increases as you move down a group and from right to left on the periodic table
Ionization Energy	The energy *required* to remove the outermost electron from an atom Example: $Li \rightarrow Li^+ + e^-$	Increases as you move up and to the right
Electron Affinity	The energy *released* when an electron is added to the outermost shell of an atom Example: $F + e^- \rightarrow F^-$	Increases in magnitude (i.e., becomes more negative) as you move up and to the right
Ionic Radius	Radius of ion; used to assess size of ion	- increases down a group - cations decrease left to right - anions decrease left to right - NOTE: as a whole, anions are considerably larger than cations

Helpful Hints

- Values for ionization energies and electron affinities are in Appendix C of the text.
- The main considerations to take into account when analyzing a periodic trend are:
    1. Where are the valence electrons located?
    2. Is this a vertical trend or horizontal trend?
        - As you go down a given column, the Z_{eff} stays the same, but n increases by 1. The principal quantum number is the primary indicator of an atom's size. An increase in principal quantum number significantly increases the distance between the nucleus and the valence electron, thereby decreasing the attractive forces between them.
        - As you move across a given row, n stays the same, but Z_{eff} increases (because the number of protons increases).

EXERCISE 13: Which has a greater electron affinity, a) N or P? b) Se or Br? Explain why.

STRATEGY AND SOLUTION: Electron affinity is the energy released when an atom gains one electron. a) Both N and P contain 5 valence electrons, but N's are in $n=2$ and P's are in $n=3$. An incoming electron is going to feel a much stronger attraction to N's nucleus than P's because the distance between that electron and the nucleus is shorter. N, therefore, has a greater electron affinity than P. b) Se and Br have valence electrons in $n=4$, but Br has one more proton than Se. Their sizes are roughly the same (Br is slightly smaller), but an incoming electron will feel the attraction to Br's nucleus slightly more intensely than to Se's. Br, therefore, has a greater electron affinity than Se.

EXERCISE 14: Which has an a) larger atomic radius, Al or P? b) larger ionic radius, Al^{3+} or P^{3-}?

STRATEGY: a) Al and P are both in the same row, $n = 3$. Their valence shell is $n=3$, indicating that their sizes are roughly similar. The difference between the two atoms is that Al contains 13 protons and P contains 15 protons. Therefore, the valence electrons in P feel the attractive forces of two extra protons in the nucleus (relative to Al). Z_{eff} for P is therefore greater than Al. b) In this situation, Al^{3+} has 3 fewer electrons than atomic Al, making it isoelectronic with Ne. Its valence electrons are in $n=2$. P^{3-} has 3 extra electrons relative to atomic P, making it isoelectronic with Ar. Its valence electrons are in $n=3$. The principal quantum number is the primary factor that dictates an atom's or ion's size.

SOLUTION: a) Al has a larger atomic radius. b) P^{3-} has a larger ionic radius.

EXERCISE 15: Explain the relative values for the first ionization energy of phosphorus and sulfur:

P $=1012$ kJ/mol   S $= 1000$ kJ/mol

STRATEGY AND SOLUTION: Sulfur and phosphorus are both $n=3$ elements and S contains one more proton than P, so we would predict that S would hold on to its valence electrons more tightly than P, resulting in a higher ionization energy. The opposite is true, so there must be some other contributing factor. Both atoms contain a full 3s orbital. Their differences lie in the 3p orbitals: S contains 4 electrons and P contains 3. In P, all of the 3p electrons will be unpaired. Adding that last electron to S, though, requires pairing up two electrons within one orbital. The electron-electron repulsion of the two electrons in that one 3p orbital must be stronger than the stabilizing effect of the extra proton that S has. (This phenomenon actually occurs at regular intervals on the periodic table.)

8.5   ENERGETICS OF IONIC COMPOUNDS

Key Terms:
- **Lattice** – three-dimensional array of alternating positive and negative charges in an ionic compound.
- **Lattice energy** – energy released when ions in the gas phase form a solid crystal lattice.

Key Concepts:
- A release in energy results in a substance moving to a more stable state, or a lower energy state (relative to its initial position). Mathematically, $E_f - E_i$ = a negative number.
- A substance that increases in energy has moved to a relatively higher energy (less stable) state. Mathematically, the change in energy is a positive number.

Useful Relationship:
- $E_{Coulomb} = k \dfrac{q_1 q_2}{r}$ This is called "Coulomb's Law" and is used to calculate the stabilizing energy (attraction) or destabilizing energy (repulsion) that arises between two particles with charges q_1 and q_2 based upon the distance between them, r.

EXERCISE 16: Which should have a more negative lattice energy, $MgCl_2$ or MgF_2? Answer this question qualitatively, then compare your answer to data provided in Table 8-4 of the text.

STRATEGY: Use Coulomb's Law to assess relative lattice energies.

SOLUTION: The charges in both compounds are the same. q_1 and q_2 will be "2+" and "1-" for both compounds.   The distance between the two ions, r, is different in each case. F^- has a smaller ionic radius than Cl^-, so r for MgF_2 will be smaller than r for $MgCl_2$. The lattice energy for MgF_2 will therefore be a larger negative number than the lattice energy for $MgCl_2$. Table 8-4 of the text shows that the lattice energy of MgF_2 is 2913 kJ/mol and the lattice energy of $MgCl_2$ is 2326 kJ/mol.

EXERCISE 17: The second ionization energy for sodium is almost ten times as large as its first ionization energy. The second ionization energy for magnesium is only twice as large as its first ionization energy. Why is this so?

STRATEGY: The second ionization energy is the energy required to remove an electron from a cation with a 1+ charge:

First ionization energy (IE_1): $M (g) \rightarrow M^+ (g) + 1 e^-$
Second ionization energy (IE_2): $M^+ (g) \rightarrow M^{2+} (g) + 1 e^-$

Assess the electronic configurations of these substances for these processes:

Na: $1s^2 \, 2s^2 \, 2p^6 \, 3s^1$ Na$^+$: $1s^2 \, 2s^2 \, 2p^6$

Mg: $1s^2 \, 2s^2 \, 2p^6 \, 3s^2$ Mg$^+$: $1s^2 \, 2s^2 \, 2p^6 \, 3s^1$

SOLUTION: In general, the second ionization energy should be higher than the first ionization energy for any substance, because the M^+ is going to hold its electrons a little more tightly than M would. Thus, IE_2 is greater than IE_1 for both Na and Mg. Of even greater significance in the case of Na is the fact that this electron would have to be removed from the *n*=2 valence shell, rather than the *n*=3 valence shell, as was the case for the first ionization energy. The $n=2$ electron feels a much stronger attraction to the nucleus than the $n=3$ electron, so it will be more difficult to remove.

8.6   IONS AND CHEMICAL PERIODICITY

The material in this section brings together concepts covered in the earlier sections.

EXERCISE 18: Use periodic trends to explain why Sn is classified as a metal, Sb and Te are metalloids, and I is a non-metal.

STRATEGY AND SOLUTION: Sn, Sb, Te, and I are all in the p-block of row 5. The numbers of protons in their nuclei are 50, 51, 52, and 53, respectively. This indicates that although they are *roughly* the same size (all $n=5$), the I holds its valence electrons most tightly while the Sn holds its valence electrons the least tightly. Of the set, iodine has the greatest Z_{eff} and Sn has the lowest Z_{eff}. Appendix C data is not available for these elements, but in looking at similar trends for comparison (i.e., Ga vs. Ge vs. Se), it appears that the first ionization energy spikes approximately 200 kJ/mol as one moves from a metal to metalloid to non-metal. Metals tend to lose electrons to form cations, while non-metals tend to gain electrons to form anions. This correlates to the idea that metals have much lower ionization energies than non-metals. Similar trends are observed with electron affinity: non-metals have higher (negative) electron affinities than metals.

EXERCISE 19: Does O^{2-} have a high electron affinity?

STRATEGY AND SOLUTION: Electron affinity is the energy released when an atom (or ion) gains one electron. In this case, the substance contains a full $n=2$ shell, so the incoming electron, if it were to attach to the O^{2-} ion, would reside in an $n=3$ s orbital. The effective nuclear charge of this incoming electron would be -2 (!), because the 8 protons in the nucleus would be completely screened by the 10 electrons in the $n=1$ and $n=2$ shells. (Remember, electrons in an $n=3$ shell will be completely screened by electrons in any inner shells.) Thus, the incoming electron would not be attracted to the O^{2-} ion. This question could also be assessed using Coulomb's Law.

Chapter 8 Self-Test

You may use a simple periodic table for this test. (i.e., one that has only the symbol, atomic number, and atomic mass for each element)

1. Draw the complete electron configuration *and* the noble gas electron configuration for Sb.

2. a) Draw the orbital diagram for Mn^{2+}. b) Is Mn^{2+} paramagnetic or diamagnetic?

3. Which atom has a greater spin, As or Se? Explain briefly.

4. Which has a smaller atomic radius, O or F? Explain.

5. Consider the element Pb. Within an atom of Pb, which is more stable, an electron in a 4f orbital or an electron in a 5p orbital? Provide an argument in support of each.

6. Here are ionization energies (IE_1) – not listed in any particular order – for Se, S, and Cl. Match each ionization energy to an element and propose reasoning for your explanation.
   1251 kJ/mol   941 kJ/mol   1000 kJ/mol

7. Cation A and Cation B are isoelectronic. Cation A has two more protons than Cation B. Which one has the larger ionic radius?

8. In some forms of the periodic table, zinc, cadmium, and mercury (Zn, Cd, and Hg, respectively) are classified collectively as "Group IIB" metals, while the alkaline earth metals are classified as "Group IIA." In terms of electronic structure, how can this be justified?

Answers to Try Its:

1. $Z_{eff} = 4$. Electron #1 completely screens out 1 proton, resulting in a net 4+ that Electron #2 can feel.
2. The smaller $n=2$ orbital (even though it's a p orbital) will completely screen the larger $n=3$ orbital.
3. Filled: 1s, 2s, 2p, 3s, 3p, 4s, 3d, 4p, 5s. Partially filled: 4d
4. a) 6; b) $3d_{xy}$ & $3d_{xz}$
5. a) 10 valence electrons (4s and 3d); b) 2 valence electrons (5s)
6. a) $1s^2 2s^2 2p^6 3s^2 3p^6 4s^2 3d^{10} 4p^2$; b) $[Ar] 4s^2 3d^{10} 4p^2$; c) see below; d) 1; e) paramagnetic

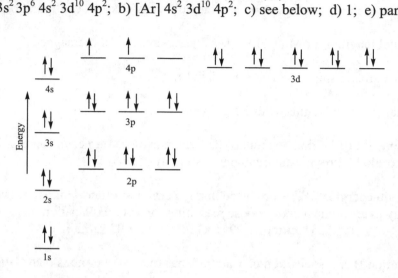

Answers to Self-Test:

1. Sb (Z=51): $1s^2 2s^2 2p^6 3s^2 3p^6 4s^2 3d^{10} 4p^6 5s^2 4d^{10} 5p^3$ and $[Kr] 5s^2 4d^{10} 5p^3$
2. a) see below; b) paramagnetic

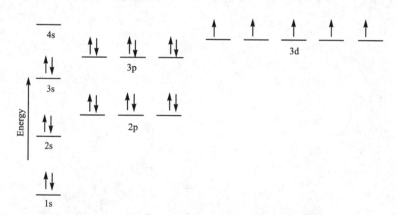

3. Both As and Se have partially filled 4p orbitals, As has 3 electrons in these p orbitals and Se has 4. There are three 4p orbitals that have the same energy, so according to Hund's rule, As's 4p orbitals will have 3 unpaired electrons, and Se's 4p orbitals will have 2 unpaired electrons and 2 paired electrons. As, with a spin of 1.5, has a greater spin than Se, with a spin of 1.0.
4. The valence electrons for both elements are in the same shell ($n=2$). F should have a smaller atomic radius than O because the valence electrons in F can feel a stronger attraction to the more positive nucleus than the electrons in O. The valence electrons in F feel a greater Z_{eff} than the valence electrons in O.

Self-Test Answers, continued

5. The principal quantum number, n, is the primary indicator of the electron's energy state. Based on this concept, an $n=5$ orbital should be less stable than an $n=4$ orbital.
    According to the Aufbau Principle, electrons fill into the lowest-energy orbitals first. Notice from the periodic table that the 5p orbital in Pb will fill before the 4f orbital. This suggests that the 5p orbital – even though it has a higher n value – is lower in energy than the 4f orbital.

6. S and Cl have valence electrons in $n=3$. Cl has 1 more proton than S, so Cl should have a higher first ionization energy than S (due to its greater Z_{eff}). Se is directly below S on the periodic table ($n=4$), so it should be easier to remove an electron from Se than S (same Z_{eff}, but Se's valence electrons aren't as close to the nucleus as S's). The ionization energies should be: Cl = 1251 kJ/mol, S = 1000 kJ/mol, and Se = 941 kJ/mol.

7. If the two cations have the same number of electrons, and Cation A has two more protons than Cation B, then Cation A will be smaller. Cation A, with its greater nuclear charge and smaller radius, will be able to pull electron density towards the nucleus more effectively than Cation B.

8. The alkaline earth metals are representative elements ("s-block" elements) - each has 2 valence electrons. The other metals mentioned (Zn, Cd, and Hg) are all transition metals ("d-block" elements) with filled d orbitals, so they, too, contain 2 valence electrons each. In this periodic table, the "II" may refer to the number of valence electrons and the "A" vs. "B" designation may refer to the different locations in the table.

Chapter 9: Fundamentals of Chemical Bonding

Learning Objectives

In this chapter you will learn how to:

- Assess the polarity of a bond by comparing electronegativities.
- Draw Lewis structures of molecules.
- Evaluate and draw resonance structures for a molecule.
- Predict the electron group geometry and molecular shape of a molecule.
- Approximate bond angles in molecules.
- Predict relative bond lengths and bond strengths.
- Assess the polarity of a molecule by determining its dipole moment.

Practical Aspects

This chapter provides you with the tools to build molecules. Recall that one of the most important themes in chemistry is "structure determines function," which means that the shape or structure of a molecule will dictate its physical and chemical characteristics. It is important to know a given molecule's general shape and symmetry, as well as which orbitals were used for bonding, because these features will influence the characteristics of the molecule. For example, carbon monoxide acts as a poison by strongly binding to hemoglobin, which decreases the number of hemoglobins available to bond with molecular oxygen. The fact that carbon monoxide's structure is flat and compact and contains an orbital that can bond with another atom – features also present in molecular oxygen's structure – is what makes it able to bind to hemoglobin. Most medications and metabolic poisons work in a similar manner to carbon monoxide – they mimic or block a "normal" chemical process inside your body because their structures are similar to the structures of the molecules involved in the normal processes. In this chapter, you will learn the basics for constructing molecules, so that in later chapters you can use this knowledge to predict a given molecule's physical and chemical behavior.

9.1   OVERVIEW OF BONDING

Key Terms:
- **Covalent bond** – sharing of electrons between two atoms as a result of overlapping atomic orbitals.
- **Bond length** – distance between the two nuclei that are involved in the covalent bond.
- **Polar** – term used to describe an unequal distribution of electron density.
- **Polar bond** – a covalent bond which has its two electrons residing nearer to one atom than the other.
- **Electronegativity** – term invented by Linus Pauling to describe the extent to which an atom will pull bonding electrons towards itself. Pauling devised a relative numerical scale to use in evaluating electronegativities. See Figure 9-7 in the text for Pauling's scale.

Key Concepts:
- Bond length is dictated by the balance of attractive forces (between opposite charges) and repulsive forces (between like charges) within the two atoms making up the covalent bond.
- The periodic trend for electronegativity is that it increases up and to the right on the periodic table. As with other periodic trends, it is a balance between Z_{eff} and the size of the atom.

- A polar covalent bond is intermediate between a pure covalent bond (perfect sharing) and a pure ionic bond (complete charge separation).
- The greater the difference in electronegativity between two atoms, the more polar their bond will be.

Helpful Hints

- For all practical purposes, hydrogen atoms have about the same electronegativity as carbon atoms. A carbon-hydrogen bond is considered non-polar (even though $\Delta EN = 0.4$).
- Recall from Chapter 2 that Group I and II metals form ionic bonds with Group 16 and 17 non-metals.

EXERCISE 1: Is electronegativity the same as electron affinity?

SOLUTION: No. Electron affinity is the energy released by an atom when it picks up an electron. Electronegativity describes the extent to which an atom will pull electrons *within a covalent bond* towards itself. The two terms are *not* interchangeable and should not be confused with one another.

EXERCISE 2: Arrange the bonds in order of increasing bond polarity: Se – O, S – O, Ge – O.

STRATEGY: The greater the difference in electronegativity, the more polar the bond.

SOLUTION: S – O, Se – O, Ge – O. On the periodic table, S is directly below O, Se is directly below S, and Ge is to the left of Se. The greatest difference in electronegativity between atoms will therefore be between Ge and O.

EXERCISE 3: Use the periodic table to predict which pair of atoms would form polar bonds. For those that are polar, indicate which atom would pull electron density towards itself more strongly.
    a) H and Cl   b) Fe and C   c) Br and Br

STRATEGY: Look to see relative positions of elements on the periodic table. Electronegativities tend to increase up and to the right on the periodic table. The more electronegative element will pull electron density towards itself, making either a polar bond (difference in electronegativity) or non-polar (no difference in electronegativity).

SOLUTION:
a) H and Cl: Polar. Hydrogen has about the same electronegativity as carbon. Chlorine is more electronegative than carbon (and thus hydrogen), so it will pull electron density towards itself, but not to the extent that a complete charge separation will occur.
b) Fe and C: Polar. Carbon is more electronegative than Fe, so it will pull electron density towards itself.
c) Br and Br: Non-polar. There is no difference in electronegativity between two atoms of the same element, so the electrons in this bond are perfectly shared between the atoms, making the bond non-polar.

EXERCISE 4: Use Figure 9-7 in the text to calculate differences in electronegativities and predict the type of bond that would form for each of the atom pairs in Exercise 3. Compare the results from each method.

STRATEGY: Calculate ΔEN for each pair, categorize, and compare.

SOLUTION:

a) $EN_H = 2.1$; $EN_{Cl} = 3.0$; $\Delta EN = 0.9 =$ Polar.

b) $EN_{Fe} = 1.8$; $EN_C = 2.5$; $\Delta_{EN} = 0.7 =$ Polar.

c) $EN_{Br} = 2.8$; $EN_{Br} = 2.8$; $\Delta_{EN} = 0 =$ Non-polar.

All results correlate with results from Exercise 3.

Try It #1: Use the periodic table to predict which bond between each atom pair has the most unequal sharing of electrons and which has the most equal sharing of electrons.   a) N and Fr; b) C and F; c) Mg and I; d) Br and Te.

9.2   LEWIS STRUCTURES

Key Terms:

- **Lewis structure** – picture that shows how atoms within a molecule are connected to each other.
- **Non-bonding electrons** (or "**lone pair**" electrons) – electrons that are not involved in bonding.
- **Formal charge (F.C.)** – net charge associated with an atom in a Lewis structure.
- **Resonance structures** – Equivalent or near-equivalent Lewis structures that describe the bonding behavior within a molecule or ion.

Key Concepts:

- Orbitals in an atom's valence shell are the orbitals used for making covalent bonds.
- Within a Lewis structure, bonding electrons and non-bonding electrons are distinguished from one another by drawing them differently: bonding electrons are shown as a line and non-bonding electrons as a pair of dots (lone pair).
- Two atoms may form a single bond (2 shared electrons), a double bond (4 shared electrons), or a triple bond (6 shared electrons).
- The optimal Lewis structure will have the minimum formal charges.
- The sum of all formal charges will equal the overall charge on the molecule or ion.

Useful Relationship:

- F.C.$_{given\ atom}$ = (# of valence electrons in atom) – (# of valence electrons assigned to atom in Lewis structure*)

  * Note: Formal charges assume equal sharing of electrons within bonds, so each bonding pair counts as only one electron toward the atom. Both electrons within a lone pair count towards the atom.

EXERCISE 5: Identify the number of lone pairs, bonding pairs, and total electrons in each structure:

$$:C \equiv O:$$

A

B

C

STRATEGY: A lone pair is designated by two dots and a bonding pair is designated by a line. Notice the question asks for lone *pairs*, not for electrons that are in lone pair spaces.

SOLUTION:

Structure	Number of lone pairs	Number of bonding pairs	Total number of electrons
A	2 (1pair on each atom)	3 (triple bond)	10
B	4 (1 pair on N, 3 on Cl)	3 (3 single bonds)	14
C	9 (1 pair on S, 2 on O, 3 on each Cl)	4 (1 double bond & 2 single bonds)	26

EXERCISE 6: Use formal charges to determine the preferred Lewis structure for CH_5N.

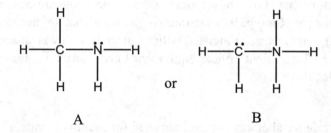

STRATEGY: Calculate formal charges on all atoms in each structure. The structure that contains the smallest individual formal charges is the correct structure.

F.C. = (# of valence electrons in atom) – (# of valence electrons assigned to atom in Lewis structure)

Structure	Atom	# of valence electrons in atom	# of valence electrons assigned to Lewis structure	Formal charge of atom
A	H	1	0 lone pair electrons, 1 electron in 1 bond	$1-(0+1)=0$
A	C	4	0 lone pair electrons, 1 electron in each of 4 bonds	$4-(0+4)=0$
A	N	5	2 electrons from 1 lone pair, 1 electron in each of 3 bonds	$5-(2+3)=0$
B	H	1	0 lone pair electrons, 1 electron in 1 bond	$1-(0+1)=0$
B	C	4	2 electrons from 1 lone pair, 1 electron in each of 3 bonds	$4-(2+3)=-1$
B	N	5	0 lone pair electrons, 1 electron in each of 4 bonds	$5-(0+4)=+1$

SOLUTION: The sum of all formal charges in each structure is zero, which is reasonable because the compound has no charge. Every formal charge in structure A is zero, while B's structure contains a formal charge of –1 and another formal charge of +1. Structure A is the preferred Lewis structure.

Try It #2: Calculate the formal charges on all atoms in Exercise 5.

Writing Lewis Structures

There are many different guidelines for writing Lewis structures. The guidelines shown here are a shortened version of the text's description. This probably seems like a lot to memorize, but after you've drawn a few structures and gotten the pattern down, you'll see it's fairly simple.

1. Add up valence electrons for all atoms in the formula.
2. Determine which atom will be the central atom. This will be the atom that can make the most bonds, and is usually the least electronegative atom in the bunch.
3. Write the formula for the central atom attached with a single bond to each of the other atoms.
4. Take a tally of how many valence electrons remain to be assigned.
5. Place three lone pairs on each of the outer atoms, being sure not to use more electrons than you have available.
6. If there are any extra electrons remaining, place them as lone pairs on the central atom.
7. If there are not enough electrons to give each atom four pairs of electrons (except H), then make multiple bonds as needed.
8. Minimize formal charges by forming extra bonds between atoms, provided the atoms can accommodate extra electrons.

Key Concepts:

*   A Lewis structure can be drawn in different orientations, provided that connectivities don't change.

For Example:

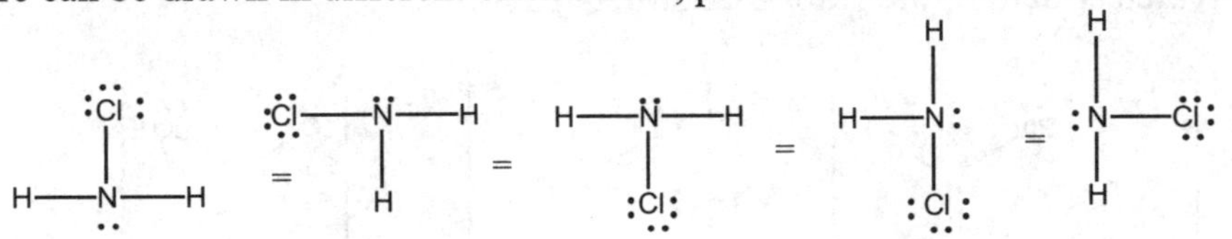

*   Each atom typically will be in its most stable form with an "octet" of valence electrons around it.
*   Only atoms with available "d" orbitals ($n=3$ or greater) can hold more than 8 valence electrons.

Special Considerations:

*   **Ions.** Add one electron for each negative charge; subtract one electron for each positive charge.
*   **More than one central atom.** It is possible to have more than one central atom.
*   **Formulas containing parentheses.** Parentheses indicate how many clusters or groups of a given type are in a formula. For example, $C(CH_2CH_3)_4$ indicates that there are four groups of "$-CH_2CH_3$" attached to a central carbon atom.

EXERCISE 7: We saw three Lewis structures in Exercise 5. Verify that Structure B (NH_2Cl) adheres to the guidelines for drawing Lewis structures.

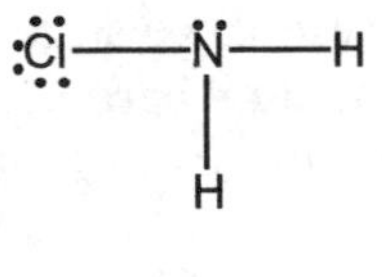

STRATEGY: Walk through the guidelines for drawing Lewis structures and compare the results to the structure from Exercise 5.
*   Total # of valence electrons = 5+1+1+7 = 14
*   N is the central atom: 1) its electronegativity is the same as Cl, and 2) H is never the central atom.

139

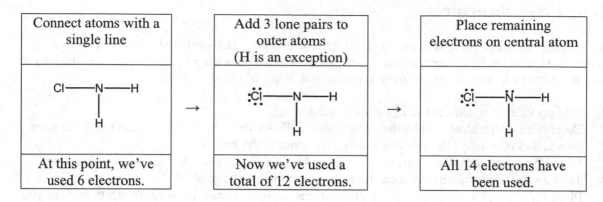

Connect atoms with a single line	Add 3 lone pairs to outer atoms (H is an exception)	Place remaining electrons on central atom
At this point, we've used 6 electrons.	Now we've used a total of 12 electrons.	All 14 electrons have been used.

- The N and the Cl each have 8 valence electrons around them, and the Hs each have 2 valence electrons.
- The formal charge on every atom is zero.

Try It #3: Verify that Structure C in Exercise 5 is the correct Lewis structure.

EXERCISE 8: Which structures are resonance structures for CO_3^{2-}?

STRATEGY: Resonance structures are equivalent or near-equivalent Lewis structures.

SOLUTION: The fifth Lewis structure shown (far right) can immediately be eliminated as a possible resonance structure, because it shows different connectivities of atoms than the others. Resonance structures must contain the same basic framework of atoms. Furthermore, the fifth structure shows only 22 electrons, and CO_3^{2-} contains 24 valence electrons. (The formal charges in this structure do not add up to the overall ion charge either.)

The second structure is not a valid Lewis structure because carbon is an *n*=2 element, and cannot accommodate more than 8 valence electrons. This structure shows carbon with an expanded octet of 10 valence electrons, which is not possible. It is therefore not a resonance structure of CO_3^{2-}.

The first, third, and fourth structures are resonance structures. In each of these structures, the atoms are connected in the same manner (i.e., a central carbon atom is connected to three oxygen atoms), and the rules for drawing Lewis structures are obeyed.

The actual structure of CO_3^{2-} is a blend of these structures.

EXERCISE 9: Draw the Lewis structure for: a) NOF;   b) BH_3;   c) SeF_4;   d) ClO^-

STRATEGY: Follow the guidelines for drawing Lewis structures.

	a) NOF	**b) BH_3**	**c) SeF_4**	**d) ClO^-**
# of valence e-s:	5+6+7 = 18	3+1+1+1 = 6	6+7+7+7+7 = 34	7+6+1 = 14
Central Atom:	N	B	Se	Cl
Make single line between atoms:	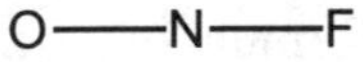 O——N——F	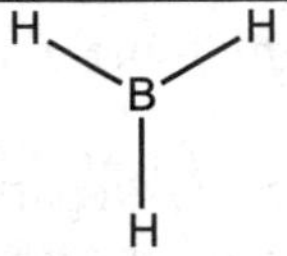 H, H, B, H (H's bonded to B)	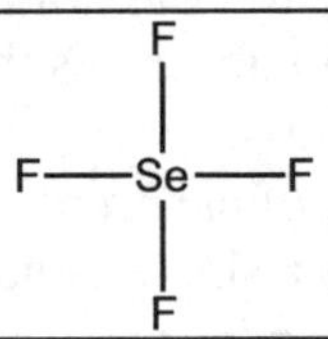 F—Se—F with F above and F below	Cl——O
Tally electrons used:	4	6 (none left)	8	2
Add 3 lone pairs to outer atoms:	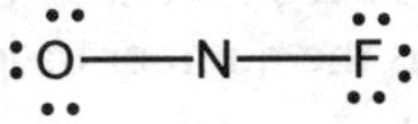 :O——N——F:		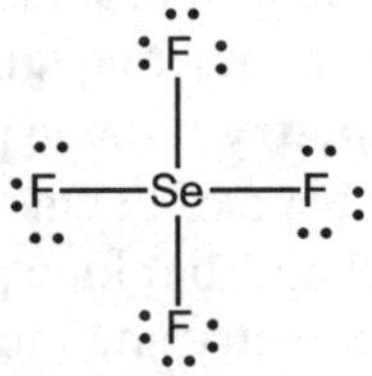 :F: / :F—Se—F: / :F:	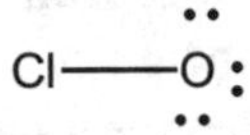 Cl——O:
Tally electrons used:	16 (2 remaining)		32 (2 remaining)	8 (6 remaining)
Place remaining electrons on central atom:	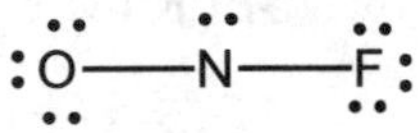 :O——N——F:		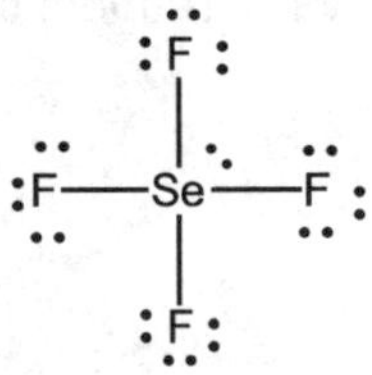 :F: / :F—Se—F: / :F:	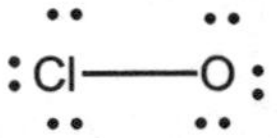 :Cl——O:
Minimize formal charges:	O = -1 / N = +1 / F = 0		F = 0 / Se = 0	Cl = 0 / O = -1
Lewis Structure:	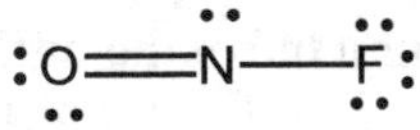 :O══N——F: F.C. on each atom is now zero	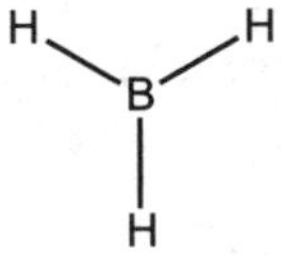 H, B, H, H	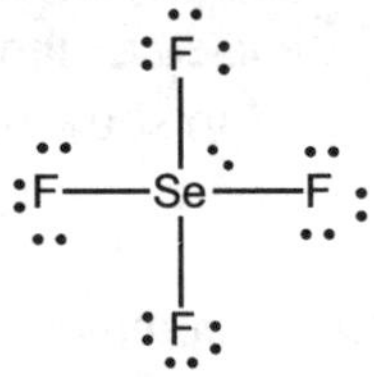 :F: / :F—Se—F: / :F:	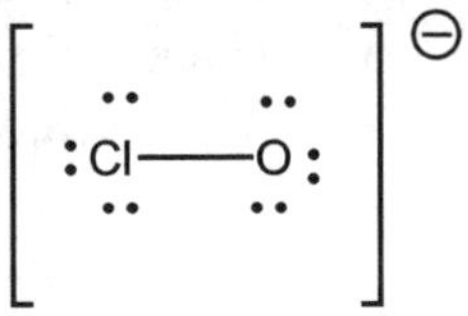 [:Cl——O:]⊖ (anion, so "-" charge)

There are two resonance structures for molecule (d). Remember, the Cl can accommodate more than 8 valence electrons because it is in row 3 of the periodic table and has vacant d orbitals that can be used.

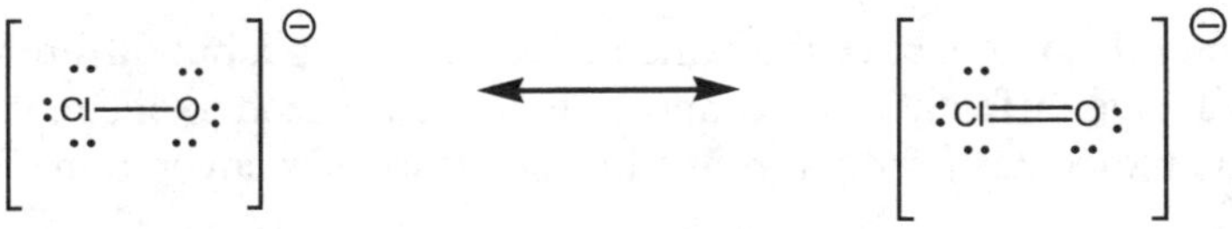

[:Cl——O:]⊖ ⟷ [:Cl══O:]⊖

f.c._Cl = 0; f.c._O = -1   f.c._Cl = -1; f.c._O = 0

Try It #4: Draw Lewis structures for: a) CH_2F_2; b) $XeBr_4$; c) PBr_4^-

9.3 MOLECULAR SHAPES: TETRAHEDRAL SYSTEMS

Key Terms:

- **VSEPR Theory** – "Valence Shell Electron Pair Repulsion Theory" is a theory which states that the orbitals on the central atom of a molecule want to be as far apart from each other as they can, in order to minimize repulsive forces between electrons.
- **Structural isomers** – substances that have the same molecular formula but different structures (connectivities of atoms).
- **Electron group** – set of electrons which occupies a particular space around a central atom. An electron group can be a single bond (2 electrons in a group), a double bond (4 electrons), a triple bond (6 electrons), a non-bonding electron pair (2 electrons), or a single lone electron (1 electron).
- **Ligand** – an atom or group of atoms attached to a central atom.
- **Steric number** – number of orbitals used by the central atom to accommodate all bonded atoms and lone pairs. The steric number equals the number of ligands plus the number of lone pairs.
- **Electron group geometry** – shape predicted by VSEPR theory.
- **Molecular shape** – shape of the molecule that a person actually "sees." The electron group geometry dictates the general shape, but lone pairs can't be seen, so if a central atom contains lone pairs, then the electron group geometry and molecular shape will differ.

Key Concepts:

- The tetrahedral geometry around a central atom X as predicted by VSEPR theory has the 3-D shape:

- The convention for interpreting a 3-D structure is:
  - The central atom is sitting in the plane of the paper.
  - The plain, solid lines are lying on the plane of the paper.
  - Thick solid lines are used to show that the orbital is coming out of the paper towards you.
  - Dashed lines are used to show that the orbital is jutting behind the plane of the paper.

EXERCISE 10: a) Are these compounds structural isomers? b) What is the electron group geometry around each carbon atom?

STRATEGY: Structural isomers have the same molecular formula but different connectivities of atoms. The molecules all have different atom connectivities, so we need to check their molecular formulas. How to determine molecular formula from a line structure was covered in Chapter 3.

SOLUTION: a) The left molecule and the right molecule are structural isomers because they both have the molecular formula C_7H_{16} and they have different structures. The central molecule has the formula C_8H_{18}, so it is not an isomer of the other two. b) All of the carbon atoms shown have a tetrahedral electron group geometry, because each carbon atom is attached to four other atoms (either C or H) and no carbon contains a lone pair.

Key Concepts:
- The steric number determines electron group geometry of a molecule.
- The number of ligands determines the molecular shape that you actually see.

EXERCISE 11: For each molecule, what is the steric number? What is the electron group geometry? What are the approximate bond angles around the central atom? How many ligands does this molecule contain? What is the molecular shape?

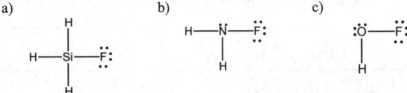

a)   b)   c)

STRATEGY AND SOLUTION:
The steric number is the number of atoms and lone pairs around the central atom. For all three molecules, the steric number is 4: a) has 4 atoms and 0 lone pairs; b) has 3 atoms and 1 lone pair; and c) has 2 atoms and 2 lone pairs.

The steric number dictates the electron group geometry and bond angles. For these compounds, the electron group geometry is tetrahedral and the approximate bond angles are 109.5°.

The number of ligands is based on the number of *atoms* around the central atom. This number differs for each compound shown: a) 4; b) 3; c) 2.

The molecular shape is determined by what is actually visible around the central atom. The lone pairs take up space (so they contribute to the steric number) but they aren't actually seen in the molecule (they aren't counted as ligands).
a) We can see 4 atoms around the central atom – this is a tetrahedral molecular shape.
b) We can see 3 atoms around the central atom. The electron group geometry is tetrahedral, but we don't see one of the orbitals on it; what we see is something that looks like a pyramid with a triangular base. This shape is called "trigonal pyramid."
c) We can see two atoms on this basic electron group geometry. This molecular shape is called "bent."

Here are pictures to correlate with the molecular geometries described. Notice that for simplicity's sake, lone pairs are not drawn on the outer atoms. Notice, too, that there are two options for illustrating a lone pair on the central atom:

a) tetrahedral   b) trigonal pyramidal   c) bent

9.4 OTHER MOLECULAR SHAPES

In Section 9.3 we used the tetrahedral geometry as an example of how VSEPR theory could be used to predict molecular shapes. Of course, it is possible to have other steric numbers and other shapes. The table below summarizes the possibilities as predicted by VSEPR theory.

Summary Table of Information Derived From Steric Number

Steric Number	Electron group geometry	Hybridization (to be covered in 10)	Bond Angle	3-D Shape
2	Linear	sp	180°	——X——
3	Trigonal Planar	sp^2	120°	
4	Tetrahedral	sp^3	109.5°	
5	Trigonal Bipyramidal	sp^3d	90° – axial-equatorial 120° – equatorial-equatorial	
6	Octahedral	sp^3d^2	90°	

Key Concepts:
- The steric number predicts electron group geometry, hybridization, bond angle, and 3-D shape. (Hybridization will not be covered until Section 10.2, but is included in the above summary table because it also depends upon steric number.)
- If there are no lone pairs on the central atom, then the molecule's molecular shape and electron group geometry are the same.
- The central atom's number of ligands plus the number of lone pairs predicts molecular shape and slight distortions to bond angles. (Angle distortions will be covered in Section 9.5.)

Helpful Hint
- Table 9-3 in the text provides an overall summary of electron group geometry, molecular shape, and bond angles.

EXERCISE 12: Determine the molecular shape and approximate bond angles for:
a) NOF;   b) BH_3;   c) ICl_3;   d) IBr_4^-

STRATEGY: Draw the Lewis structure of each molecule, and then use VSEPR theory.

SOLUTION:

	Lewis Structure Steric # Number of ligands	Electron group geometry Bond Angles (approx)	Molecular Shape And 3-D Picture
a)	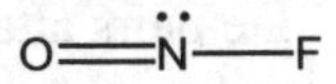 O=N—F Steric # = 3 Ligands = 2	Trigonal Planar sp^2 120°	Bent 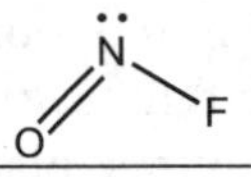
b)	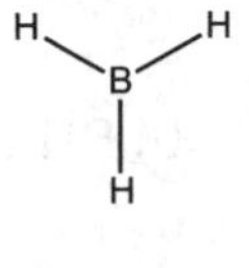 H—B(—H)—H Steric # = 3 Ligands = 3	Trigonal Planar sp^2 120°	Trigonal Planar 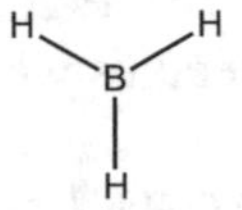
c)	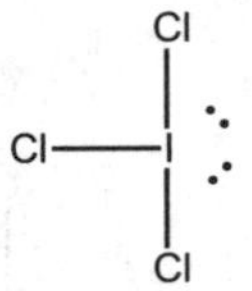 Cl—I(Cl)(Cl) Steric # = 5 Ligands = 3	Trigonal Bipyramidal sp^3d Cl_a-I-Cl_e = 90° Lone pair-I-Cl_e = 120°	T-Shaped 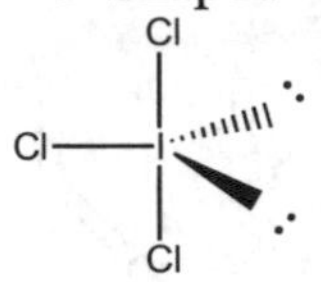
d)	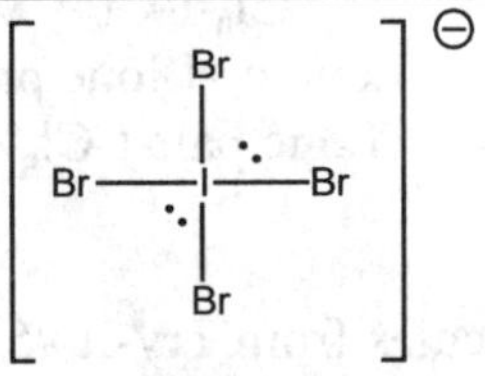 [Br—I(Br)(Br)—Br]$^{\ominus}$ Steric # = 6 Ligands = 4	Octahedral sp^3d^2 90°	Square Planar 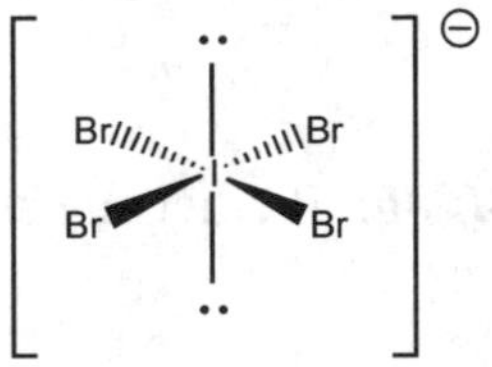

Try It #5: Determine the 3-D structure, molecular shape, and approximate bond angles for: a) CBr_4; b) PBr_4^-; and c) $XeBr_4$.

9.5   PROPERTIES OF COVALENT BONDS

Four main properties of covalent bonds are covered in this section: bond angles, dipole moments, bond lengths and bond strengths.

Bond Angles

Key Concepts:
- VSEPR predictions for bond angles correlate with experimental results.

- Experimental results do show slight deviations from VSEPR predictions when lone pairs are on the central atom: lone pairs repel more strongly than bonding pairs, altering the bond angles slightly.

EXERCISE 13: Use VSEPR theory to verify that a steric number of 3 will result in trigonal planar electron group geometry and bond angles of 120°.

STRATEGY AND SOLUTION: According to VSEPR theory, atoms and lone pairs attached to a central atom want to be as far apart as they can, in order to minimize repulsive forces between electrons. A steric number of 3 indicates that there are a total of three atoms and/or lone pairs on the central atom. Three things arranged as far as possible around a central point will be located 120° apart from each other in one plane, hence the bond angle and the name "trigonal planar."

EXERCISE 14: Predict any slight deviations in bond angles for: a) NOF;   b) BH_3;   c) ICl_3 (Note: these molecules were studied in Exercise 12.)

STRATEGY: Lone pairs take up more space than bonding pairs.

SOLUTION:

a)

b)

c)

O-N-F angle < 120°

No lone pairs. No deviation from anticipated 120°

Cl_a-I-Cl_e < 90°
Lone pr-I-lone pr > 120°
Lone pair-I-Cl_e < 120°

Try-It #6: Predict any slight deviations in bond angles for the molecules from Try-It #5.

Dipole Moments

Key Terms:
- **Dipole Moment** – overall dipole in a molecule; the sum of all individual dipoles within the molecule.
- **Polar** – unequal distribution of electron density.
- **Non-polar** – equal distribution of electron density.

Key Concepts:
- A molecule that does not have a dipole moment contains an overall even distribution of electron density, and is classified as a non-polar molecule.
- A molecule that has a dipole moment contains an overall *uneven* distribution of electron density, and is classified as a polar molecule.
- If a molecule containing polar bonds is so symmetrical that all of its polar bonds cancel each other out, then the overall molecule will have no dipole moment (i.e. nonpolar molecule).

- Note that the terms "polar" and "non-polar" can refer to a bond within a molecule or to the overall molecule. When you use these terms, be sure to specify what you're referring to: a specific bond's polarity or a molecule's polarity.

EXERCISE 15: Classify each molecule as polar or non-polar: a) CF_4;   b) CH_2F_2;   c) $COCl_2$

STRATEGY: If the molecule has a dipole moment, then it is polar. If not, then it is non-polar. The only way to figure this out is to determine the molecule's structure.

SOLUTION:

a)	b)	c)
The C-F bonds are polar, but the overall molecule is perfectly symmetrical, so all the equivalently polar bonds cancel each other out. The overall molecule has no dipole moment, so it is non-polar.	The C-F bonds are polar, but the C-H bonds are not. The net result is that there is an electron pull towards the F atoms. The molecule has a dipole moment, so it is polar.	O is more electronegative than Cl, so the molecule has an overall dipole moment. It is a polar molecule.

Bond Length

Key Terms:
- **Bond length** – distance between the two nuclei that make up a chemical bond.
- **Bond multiplicity** – number of bonds between two atoms. For example, a double bond has a bond multiplicity of 2.

Key Concepts:
- Factors that affect a bond length include: principal quantum number, bond multiplicity, bond polarity, and Z_{eff}.

Helpful Hint
- Table 9-2 in the text lists common bond lengths.

EXERCISE 16: Rank the bond lengths in this compound from shortest to longest. Explain your reasoning.

STRATEGY: Tally up each type of bond and compare the bond lengths.

SOLUTION: This compound contains six different types of bonds, which fall into two groups:
   Group A (hydrogen attached to O, N, or C): H-O, H-N, H-C
   Group B (carbon attached to O, N, or C): C-O, C-N, C-C

Group A's bonds are shorter than Group B's bonds because the hydrogen's $n=1$ valence shell is significantly smaller than the carbon's $n=2$ valence shell. Within each group, the Z_{eff} increases as you go from C to N to O, which shortens the atomic radius, thereby shortening the bond length.
The overall ranking from shortest to longest bond length is: H-O < H-N < H-C < C-O < C-N < C-C.

Bond Strength

Key Concept:

- The strength of a given bond depends upon the degree of orbital overlap between the atoms that make up the bond: the greater the overlap the stronger the bond. Three factors contribute to the degree of orbital overlap:
   1. Bond length – shorter bond has better overlap than longer bond.
   2. Bond multiplicity – triple bonds have greater orbital overlap than double bonds, and double bonds have greater orbital overlap than single bonds.
   3. Difference in electronegativity – a greater difference in electronegativity between atoms will cause the atoms to pull together closer (better overlap).

EXERCISE 17: Predict which bond – O-Cl vs. O-Br – will have the higher average bond energy. Provide a brief explanation for your choice.

STRATEGY: Determine which factor must be influencing the bond's strength.

SOLUTION: Cl has a smaller atomic radius than Br because it has a smaller valence shell. The O-Cl bond will be shorter; therefore it will have a better orbital overlap than O-Br. Furthermore, the O-Cl bond is more polar than the O-Br bond, so in order for the bond to cleave into equal fragments (one electron to each atom), the bond must first be made symmetrical. O-Cl will have a higher bond energy than O-Br.

Try It #7: Use Table 6-2 in the text to look up actual average bond energy values for the O-Cl vs. O-Br bond. Compare these values to the predictions made in Exercise 17.

EXERCISE 18: As we saw in Chapter 7, high-energy electromagnetic radiation from the sun can remove an electron from a nitrogen molecule to form N_2^+ and can fragment an O_2 molecule into two O atoms.

(These reactions occur in the outermost region of our atmosphere, the thermosphere.)   Explain in terms of relative bond strengths why the oxygen fragments into atoms, while the nitrogen does not.

STRATEGY AND SOLUTION: Nitrogen molecules contain a triple bond, and oxygen molecules contain a double bond. Both N and O have valence shells of $n=2$, so the bond lengths are *roughly* similar. Nitrogen molecules have one extra bond relative to the oxygen molecules. This extra bond makes the overall N-N triple bond stronger than the O-O double bond. The energy provided by the high energy electromagnetic radiation must be stronger than the bond energy of the O_2, but not strong enough to overcome the bond energy of the N_2. (Calculations can be performed to support this.)

Chapter 9 Self-Test

You may use a periodic table for this test.

1. Answer the following questions for: NH_3F^+

   a) Draw the Lewis structure.
   b) Are the bonds between N and the hydrogen atoms ionic, polar covalent, or non-polar covalent?
   c) What is the electron group geometry of this ion?
   d) What are the bond angles in this ion?
   e) How many ligands does N have in this ion?
   f) Which do you predict to be the strongest bond? Provide an explanation for your choice.
   g) Which is the longest bond?

2. A compound containing two lone pairs and four atoms attached to a central atom will have what molecular shape?

3. Draw the structure of $SeCl_4$. Approximate all bond angles. What is the molecular shape of $SeCl_4$?

4. Draw all resonance structures of the SCN^- ion. Indicate all formal charges.

5. Explain this apparent contradiction: if a carbon-fluorine bond is more polar than a nitrogen-fluorine bond, then why is NF_3 more polar than CF_4?

6. Which structure contains the more distorted bond angle? Explain. (Distorted means "deviating from normal.")

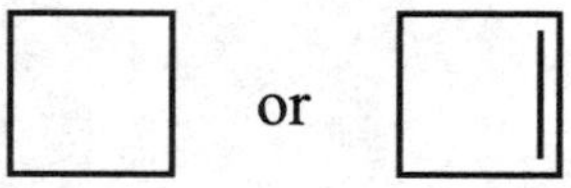

      A   B

7. a) Are the molecules shown in question 6 structural isomers?
   b) Draw the line structure for a different molecule that is a structural isomer of Structure A.

Answers to Try Its:

1. A bond between N and Fr would have the greatest unequal sharing of electrons because (from the choices provided) those atoms are the farthest apart from each other on the periodic table. The bond between Br and Te would be expected to have the most equal sharing of electrons because these atoms are diagonally adjacent to each other on the periodic table.

2. a) C = -1 and O = +1; b) & c) every atom has a formal charge of 0

3. S is the central atom because it is the least electronegative and it can accommodate the most electrons. All 26 valence electrons are accounted for, all formal charges are zero, S can accommodate more than 8 valence electrons because it is in $n=3$ and has available d orbitals.

4.

a)   b)   c)

5.

a) tetrahedral, 109.5°   b) see-saw, 120° (e-e), 90° (a-e)   c) square planar, 90°

6. CBr_4 – no deviations because no lone pairs are present (and all ligands are identical). PBr_4^- - the lone pair in the equatorial position takes up slightly more space than the bonding pairs, so the Br_e-P-Br_e bond angle will be slightly less than 120° and the Br_e-P-Br_a bond angle will be slightly less than 90°.

7. O-Cl = 220 kJ/mol and O-Br = 200 kJ/mol. These values correlate with our predictions.

Answers to Self-Test:

1.

a)

b) polar covalent
c) tetrahedral
d) 109.5°
e) 4
f) All bonds are single bonds, so multiplicity is not a factor. The H-N bond is much shorter than the N-F bond because it contains H, which uses its $n=1$ valence shell. All other atoms are using their larger $n=2$ valence shell. The bond between N-F is probably slightly more polar than the bond between N-H, since F is the most electronegative element of all (this comparison is actually hard to make, though, without a table of electronegativities.) The atoms' sizes probably are the greatest factor, so the H-N bond is probably the strongest bond.
g) The bond between N and F will be the longest bond. ($n=2$, $n=2$)

2. square planar molecular shape

3. The structure is shown below. Its electron group geometry is trigonal bipyramidal because its steric number is 5. It contains one lone pair, so its molecular shape is "see-saw." The lone pair will occupy an equatorial position to minimize electron repulsions. This lone pair will distort bond angles slightly: the $Cl_{equatorial}$-Se-$Cl_{equatorial}$ bond angle will be slightly less than the predicted 120°, and the $Cl_{equatorial}$-Se-Cl_{axial} bond angle will be slightly less than the predicted 90°.

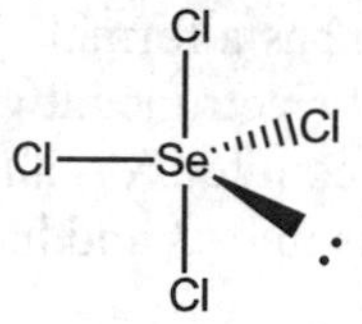

4.

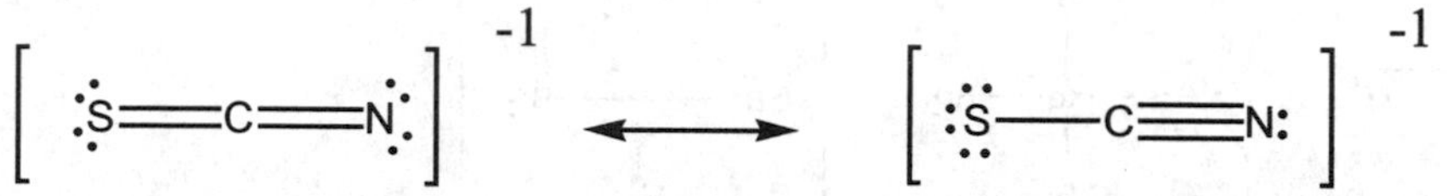

Left structure: $FC_S = 0$; $FC_C = 0$; $FC_N = -1$.   Right structure: $FC_S = -1$; $FC_C = 0$; $FC_N = 0$.

5. CF_4 is perfectly symmetrical so all of the dipoles (or polar bonds) cancel each other out. The net result is that this compound has no dipole moment. It is non-polar.
    NF_3 has a lone pair on the nitrogen atom, making its molecular shape trigonal pyramidal. The dipoles of the polar N-F bonds can't all cancel out, so the compound has a net dipole moment. It is a polar molecule.

6. These pictures are line structures for:

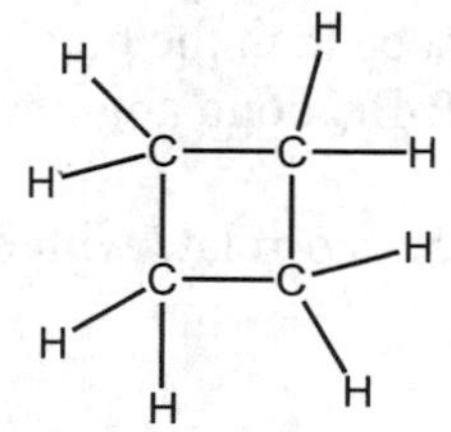

    and

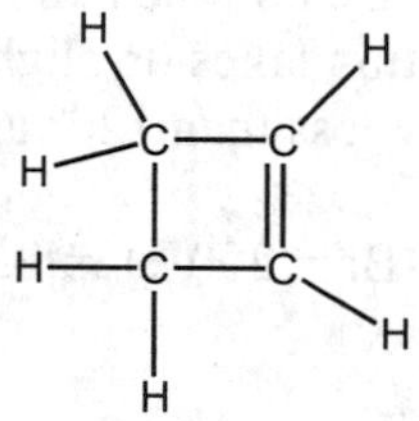

    In A, all of the carbons are steric number 4, which indicates their ideal bond angle is 109.5°. B contains two steric number 4 carbons and 2 steric number 3 carbons (on the right). The normal bond angle for steric number 3 atoms is 120°. Although both molecules contain bonds that deviate from normal, B's bond angles are more distorted, because the apparent 90° angles are closer to 109.5° than to 120°.

7. a) No, Structure A has the chemical formula C_4H_8, and Structure B has the chemical formula C_4H_6.
    b) This is one isomer of A:

Chapter 10: Theories of Chemical Bonding

Learning Objectives

In this chapter, you will learn how to:

- Determine which orbitals are involved in making covalent bonds.
- Create hybrid orbitals from atomic orbitals to use in making equivalent covalent bonds.
- Draw orbital pictures to illustrate σ and π bond formation in molecules.
- Describe covalent bonds in terms of molecular orbitals.
- Recognize and draw delocalized π systems.
- Use Band Theory to assess electrical conductivity of elements.

Practical Aspects

At the end of Chapter 9, you learned how to look at a molecule's structure and assess relative bond strengths and lengths, dipole moments and bond angles. In this chapter, you will learn how to look in more detail at a molecule's structure to develop an idea of how its bonds are constructed. In the two chapters collectively, you will learn how to look at a molecule from a chemist's perspective. Which bond is the strongest bond? Which is the weakest bond? Is the bond of interest a sigma bond or a pi bond? Is this compound stabilized by delocalization of its electrons? The concepts in Chapters 9 and 10 are used extensively in organic chemistry, where you will learn how to predict products for chemical reactions by assessing bond strengths, bond lengths, hybridization, and molecular shapes.

10.1 LOCALIZED BONDS

Key Terms:
- **Bonding orbital** –atomic orbital on one atom that is used to make a covalent bond with another atom.
- **Orbital overlap** – the extent to which the bonding orbitals on two atoms overlap with one another.

Key Concepts:
- Orbitals that contain valence electrons are used to make bonds. (Recall that valence electrons are used to make bonds.)
- Only unfilled orbitals will participate in bond formation.
- Electrons in molecules obey the Aufbau Principle and the Pauli Exclusion Principle.
- The orbital overlap model has its limitations because it cannot accurately describe many observed molecular shapes.

Helpful Hint
- Conventions of the Orbital Overlap Model are highlighted on page 389 of the text.

EXERCISE 1: NaH exists as a diatomic molecule. Which picture (A, B, or C) best illustrates the covalent bond between Na and H?

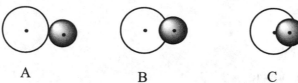

A   B   C

STRATEGY: The covalent bond is an overlap of atomic orbitals between two atoms. The sodium atom will use its 3s orbital in making the bond, while the hydrogen atom will use its 1s orbital. Picture A shows *no* orbital overlap, so Picture A is definitely not the best representation of the bond. As we learned in Chapter 9, the length of a covalent bond is determined by the balance between proton-proton repulsions, electron-electron repulsions, and electron-proton attractions. Picture C looks like the orbitals have overlapped to the degree that the hydrogen nucleus is blending with the sodium orbital. This overlap is a little too close – repulsive forces would predominate.

SOLUTION: Picture B best illustrates the covalent bond between Na and H because it shows an overlap of orbitals and a reasonable distance between the nuclei.

EXERCISE 2: Which orbital(s) can each element use to make a covalent bond? a) Si; b) Br; c) Fe; d) H

STRATEGY: Orbitals used for bonding are the unfilled orbitals in the valence shell.

SOLUTION: a) Silicon's only unfilled orbitals are two of the 3p orbitals.
b) Bromine's only unfilled orbital is one of the 4p orbitals.
c) Fe (the atom) will use some of its 3d orbitals.
d) Hydrogen's 1s orbital is not full, so it can form a bond using its 1s orbital.

Try It #1: Which orbital(s) can each element use to make a covalent bond? a) Ni; b) F; c) Xe

10.2   HYBRIDIZATION OF ATOMIC ORBITALS

Key Terms:
* **Hybridization** – the blending of different atomic orbitals to make equivalent hybrid orbitals.

Key Concepts:
* We'll focus exclusively on hybridization of the central atom(s).
* Only valence orbitals are used to make hybrid orbitals. (Recall that only valence electrons are used in bonding.)
* The number of atomic orbitals mixed together equals the number of hybrid orbitals produced. For example, if four hybrid orbitals are needed, then one s orbital and three p orbitals from a valence shell will be blended to produce four equivalent sp^3 hybrid orbitals. "sp^3" is pronounced "s p three" and indicates the hybrid is one part s and three parts p (or 25% s and 75% p).
* The lowest energy valence shell atomic orbitals are used to make hybrid orbitals.
* A central atom's steric number determines its hybridization.
* The concept of hybrid orbitals explains two main problems with the simple orbital overlap model: 1) many examples of molecules exist that contain more bonds on the central atom than would be

predicted by the orbital overlap model; and 2) non-hybridized orbital overlaps can only create 90° and 180° bond angles, and we know of many other bond angles that exist.

Helpful Hint
- Table 10-1 in the text provides a summary of hybridization information.

EXERCISE 3: CF_4 contains hybrid orbitals. What is carbon's motivation to create hybrid orbitals in this molecule?

STRATEGY: Consider what we know about CF_4:
1. If we drew its Lewis structure, we'd see that it has four F atoms attached to a central C.
2. The carbon in this molecule has a steric number of 4, which we learned in Chapter 9 predicts a tetrahedral geometry with bond angles of 109.5°.
3. Atomic orbitals have specific orientations. Using simple overlap orbital models, we would look at carbon's electron configuration: $1s^2 2s^2 2p^4$, and would say that carbon would use the two unfilled 2 p orbitals for bonding. This doesn't seem reasonable, because CF_4 contains 4 fluorine atoms, not two – we need to make four bonds.
4. If the fluorine atoms bond directly to carbon's two available atomic orbitals, they would be 90° to each other. (Recall: the p orbitals within a given n value are all 90° to each other.) Again, this does not seem reasonable, because VSEPR predicts the bond angles to be 109.5°.

SOLUTION: Carbon's motivation to create hybrid orbitals stems from two factors: 1) it needs to make four bonds, not two, and 2) the F atoms need to be as far apart as possible from each other. By making four equivalent sp^3 hybrid orbitals, carbon can accommodate four ligands and they will be at angles of 109.5°.

EXERCISE 4: Use VSEPR theory to verify that a steric number of 3 will result in sp^2 hybridization.

STRATEGY AND SOLUTION: According to VSEPR theory, atoms and lone pairs attached to a central atom want to be as far apart as they can, in order to minimize repulsive forces between electrons. A steric number of 3 indicates that there are a total of three atoms and/or lone pairs on the central atom. Three things arranged as far as possible around a central point will be located 120° apart from each other in one plane, hence the bond angle and the name "trigonal planar." To accommodate these three orientations, we need three equivalent (hybrid) orbitals. We therefore need to blend three atomic orbitals: an s orbital (always use the lowest energy valence atomic orbital first) and two of the p orbitals (we need three atomic orbitals total, so we only need two of the p orbitals). Blend the s orbital and the p orbitals and we end up with three sp^2 hybrid orbitals, each of which is one part s and two parts p (33.3% s and 66.6% p).

EXERCISE 5: Which hybrid orbital do you expect to be more spherical, sp or sp^3?

STRATEGY AND SOLUTION: The designation "sp" indicates a hybrid orbital that is one part s and one part p; it has a blend of 50% s and 50% p characteristics. An sp^3 hybrid orbital has 25% s and 75% p character. A pure s orbital is spherical and a pure p orbital is dumbbell shaped, so the hybrid orbital that is more spherical will be the one with the greater amount of s character to it: the sp hybrid orbital.

EXERCISE 6: Determine the hybridization for: a) NOF;   b) BH_3;   c) ICl_3;   d) IBr_4^-
Note: These molecules were also shown in Chapter 9's Exercise 12.

STRATEGY: Hybridization depends upon steric number.

SOLUTION:
   a)   NOF had a steric number of 3, so the central N atom will contain sp^2 hybridized orbitals.
   b) BH_3 had a steric number of 3, so the central B atom will contain sp^2 hybridized orbitals.
   c) ICl_3 had a steric number of 5, so the central I atom will contain sp^3d hybridized orbitals.
   d) IBr_4^- had a steric number of 6, so the central I atom will contain sp^3d^2 hybridized orbitals.

Try It #2: Determine the hybridization on the central atom for: a) CBr_4; b) PBr_4^-; and c) $XeBr_4$.
Note: these molecules were studied in Chapter 9's *Try-It #5.*

10.3 MULTIPLE BONDS

Key Terms:
- **Sigma bond (σ bond)** – a bond formed from the head-on overlap of atomic orbitals.
- **Pi bond (π bond)** – a bond formed from the sideways overlap of atomic orbitals.
- **Bonding picture (orbital representation)** – picture of the molecule that shows its structure in terms of sigma and pi bonds.

Key Concepts:
- A single bond is made up of a σ bond. A double bond contains 1 σ bond and 1 π bond. A triple bond contains 1 σ bond and 2 π bonds.
- The electron density in a σ bond is concentrated along the bond axis.
- The electron density in a π bond is concentrated above and below the bond axis.
- To make a good π bond, you need a good sideways overlap of orbitals. This can only occur if the original σ bond between the atoms is short. For this reason, π bonds usually involve $n=2$ atoms.

EXERCISE 7: Classify each as a sigma bond or a pi bond.

a)   b)   c)   d)

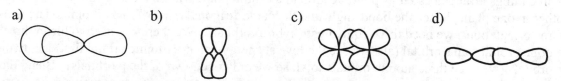

STRATEGY: Use the definitions of sigma and pi bonds.
SOLUTION: A sigma bond is a head-on overlap of atomic orbitals. Both a) and d) are sigma bonds. A pi bond is a sideways overlap of atomic orbitals; b) and c) are pi bonds.

EXERCISE 8: Consider a π bond between two atoms of nitrogen versus a π bond between two atoms of phosphorus. Which is a stronger pi bond? Why?

STRATEGY: Compare overlap of p orbitals that make up the π bond. The greater the overlap, the stronger the bond.

SOLUTION: A π bond is made by the sideways overlap of two atomic orbitals. The strength of the bond is dependent upon the extent of the overlap of these orbitals. Nitrogen is an *n*=2 element, while phosphorus is *n*=3. The *n*=2 valence electrons are closer to the nucleus, which results in a shorter distance between nuclei, and thus a better σ bond. A shorter σ bond will allow for better sideways overlap of p orbitals, resulting in a stronger π bond. The N-N π bond will be stronger than the P-P π bond.

EXERCISE 9: a) How many total π bonds are in each molecule? b) How many carbon-carbon σ bonds are in each molecule?

A   B   C

STRATEGY: a) A single bond = 1 σ bond. A double bond = 1 σ bond and 1 π bond. A triple bond = 1 σ bond and 2 π bonds. b) Count only bonds between carbon atoms.

SOLUTION: a) Molecule A contains 2 double bonds, so it has 2 π bonds. Molecule B contains 5 π bonds (5 double bonds). Molecule C contains 6 π bonds (4 double bonds and 1 triple bond).
b) Molecule A contains 3 C-C σ bonds. Molecule B contains 8 C-C σ bonds. Molecule C contains 9 C-C σ bonds.

EXERCISE 10: Describe the bonding framework in this molecule.

STRATEGY: Draw the complete structure of the molecule and use VSEPR theory to determine the hybridization for each inner atom.

SOLUTION:

The inner atoms have been labeled for quick reference. All hydrogen atoms will use 1s orbitals to make bonds. The oxygen atom and each carbon atom are using $n=2$ valence electrons in hybrid orbitals. Atoms #1 and #2 each have a steric number of 3, so they are each sp^2 hybridized. Atoms #3 and #4 each have a steric number of 4, so they are each sp^3 hybridized.

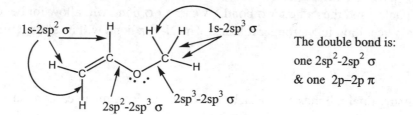

The double bond is:
one $2sp^2$-$2sp^2$ σ
& one $2p$–$2p$ π

Helpful Hint

- There are two conventions used for describing a bonding framework. The more thorough convention is shown in the above molecule. A shorter version involves excluding the n value. For example, a "$1s$-$2sp^2$ σ" bond would be written "s-sp^2 σ."

EXERCISE 11: Draw a bonding picture for: a) the bond indicated in Molecule A; b) The entire Molecule B.

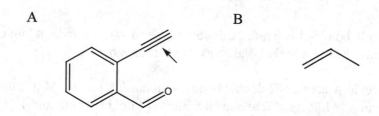

A   B

STRATEGY: Follow the guidelines from the text for drawing bonding pictures of compounds.

SOLUTION: Molecule A. (triple bond only)

1. Draw the Lewis structure.

2. Determine steric # & hybridization.   Each C has a steric # = 2 = sp hybridization.

3. Construct a σ framework.

The p_x orbital was used in hybridization, so pure p_y and p_z orbitals are left unused at this point.

4. Add in the π bonds.

The p_z orbitals are perpendicular to the plane of the paper. We'll indicate the p_z orbitals in this structure by drawing them at an angle.

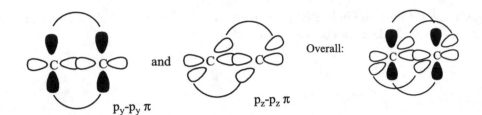

p_y-p_y π and p_z-p_z π Overall:

Molecule B.

1. Draw the Lewis structure.

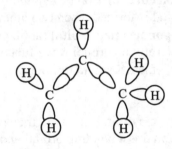

2. Determine steric # & hybridization. C_1 & C_2: steric # = 3 = sp². C_3 has steric # = 4 = sp³.

3. Construct a σ framework.

If the sp² hybrid orbitals are lying in the plane of the paper, then we have pure p_z orbitals left unused at this point.

4. Add in the π bonds.

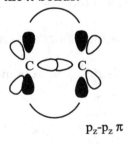

p_z-p_z π

The $2p_z$-$2p_z$ π bond is perpendicular to the plane of the paper.

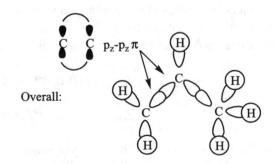

Overall:

For clarity, the overall picture for this molecule is shown in two parts.

159

Try It #3: Answer the questions for the bond indicated with an arrow in the molecule below. a) Describe the bonding framework in this bond. b) Draw an orbital picture for this bond.

10.4 MOLECULAR ORBITAL THEORY: DIATOMIC MOLECULES

Molecular Orbitals

Key Terms:
- **Molecular orbital** – 3-dimensional wave function made from the combination of two atomic orbitals.
- **Bonding molecular orbital** (bonding orbital) – constructive interference of two wave functions; a bonding orbital is formed when two atomic orbital wave functions add together.
- **Antibonding molecular orbital** (antibonding orbital) – destructive interference of two wave functions; two atomic orbital wave functions subtract each other out, resulting in a node.
- **Bond order** – the net amount of bonding between two atoms. (Note: it can be a non-integer.)

Key Concepts:
- When two atomic orbitals interact, they generate two new orbitals, called molecular orbitals. One molecular orbital is a bonding orbital and the other is an antibonding orbital. (Figure 10-27 in the text illustrates this for an s orbital, and Figure 10-30 illustrates this for p orbitals.)
- A molecular orbital is designated by its bond type (σ or π) and the orbital used to generate it (s or p or d). An asterisk is used to designate an antibonding orbital.
- Electron configurations and energy level diagrams can be generated for molecules. The same rules apply to molecules as atoms (Aufbau Principle, Pauli Exclusion Principle, Hund's Rule).
- Molecular orbital energy level diagrams are used to determine: 1) bond order; and 2) paramagnetic vs. diamagnetic properties.
- Only valence electrons are used in bonding, so molecular electron configurations and energy level diagrams show only valence electrons.
- The order of filling molecular orbitals depends upon the atomic numbers (Z) of the atoms within the molecule:

$$\text{If } Z_{average} \leq 7: \sigma_s \; \sigma^*_s \; \pi_p \; \sigma_p \; \pi^*_p \; \sigma^*_p$$

$$\text{If } Z_{average} > 7: \sigma_s \; \sigma^*_s \; \sigma_p \; \pi_p \; \pi^*_p \; \sigma^*_p$$

- There are two degenerate π orbitals. (If you made a σ bond with two p_x orbitals, then the p_y and p_z orbitals on each atom would still be available to make two degenerate π bonds: π_{py} and π_{pz}.)

Useful Relationship:
- Bond order = ½ (# of electrons in bonding orbitals – # of electrons in antibonding orbitals)

EXERCISE 12: Determine the bond order for molecule "XY" given that its electron configuration is:

$$(\sigma_s)^2 (\sigma^*_s)^2 (\sigma_p)^2 (\pi_p)^4 (\pi^*_p)^3$$

STRATEGY: Calculate bond order from the equation:

Bond order = ½ (# of electrons in bonding orbitals – # of electrons in antibonding orbitals)
# of electrons in bonding orbitals = 8;   # of electrons in antibonding orbitals = 5
Bond order = ½ (8 – 5) = 1.5

SOLUTION: The bond order is 1.5, which means the bond can be described as halfway between a single bond and double bond.

EXERCISE 13: What is the valence electron configuration for B_2?

STRATEGY: Each boron has three valence electrons, so the total number of electrons to use is six. Fill the orbitals with electrons, starting with the lowest orbital. Be sure to use the correct order (The atomic number of each boron is 5, so $Z_{average} \leq 7$).

SOLUTION: The electron configuration for B_2 is: $(\sigma_s)^2 (\sigma^*_s)^2 (\pi_p)^2$

EXERCISE 14: Consider the ion N_2^+, which was mentioned in Exercise 8. a) Construct an energy level diagram for N_2^+. b) What useful information can be obtained from this diagram?

STRATEGY: Determine the total number of valence electrons present in N_2^+: (5+5) – 1 = 9. Then choose the correct order of filling electrons, based on N's atomic number (7): $\sigma_s \ \sigma^*_s \ \pi_p \ \sigma_p \ \pi^*_p \ \sigma^*_p$. Construct the diagram.

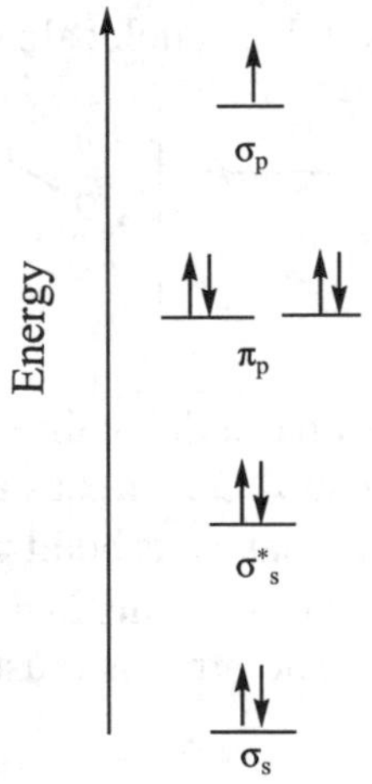

b) Molecular orbital energy level diagrams are useful for determining bond order and magnetic properties of molecules. This particular diagram shows that N_2^+ contains one unpaired electron, indicating that the substance is paramagnetic (it has magnetic properties). The diagram also shows that the bond order of N_2^+ is: ½ (7 – 2) = 2.5. The electron removed from a molecule of N_2 is removed from a bonding orbital. The resulting ion is less stable than the original N_2 (which has a bond order of 3).

Try It #4: What is the electron configuration for N_2^+?

Try It #5: Draw an energy level diagram for Cl_2. Determine the bond order of Cl_2.

10.5 THREE-CENTER π ORBITALS

Key Terms:
- **Delocalized orbitals** – orbitals that are spread out over several atoms; not localized to one atom.

Key Concepts:
- A delocalized π system is present whenever p orbitals on three or more adjacent atoms are in position for sideways overlap.
- All p orbitals must be aligned along the same axis in order to delocalize.
- A delocalized pi system that spans three atoms will contain three molecular orbitals: bonding, antibonding, and nonbonding. See Figure 10-37 in the text for pictures of nodes in these types of orbitals.

EXERCISE 15: Which molecules contain delocalized π systems?

a) NO_2^- b)

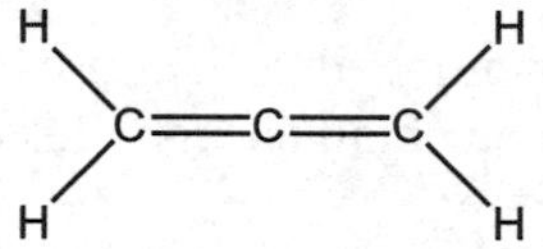

STRATEGY AND SOLUTION:

a) NO_2^- has two resonance structures, so it has a delocalized π system.

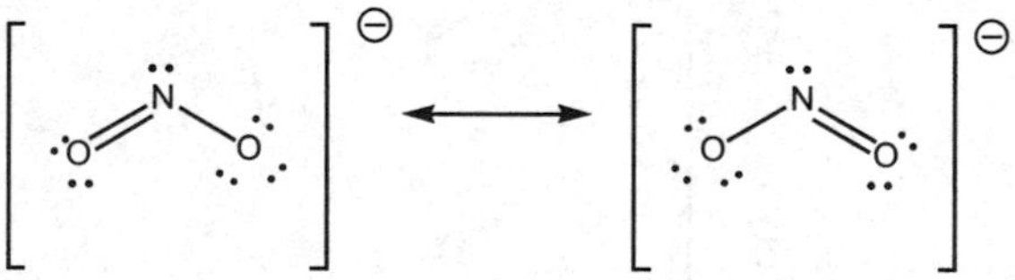

b) This compound may at first look like it has a delocalized π system, but if you analyze the orbitals used to construct it, you'll see that the central C atom has a pure p_y and a pure p_z orbital available for making π bonds. If it uses its p_z orbital to make a π bond with the left carbon, then the p_y orbital remains to make a π bond with the right carbon. This makes the two π bonds 90° to each other, so the electrons within them cannot delocalize. The orbitals must be aligned for the electrons to delocalize.

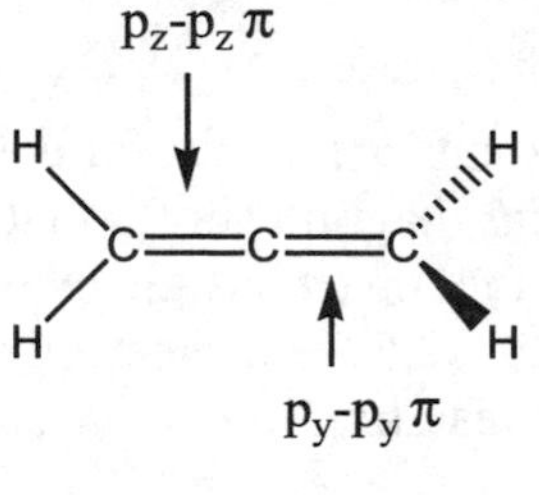

EXERCISE 16: We saw that NO_2^- in Exercise 15 contained a delocalized pi system. Describe the nonbonding molecular orbital in this ion.

STRATEGY AND SOLUTION:

The nonbonding molecular orbital will contain the delocalized electrons not involved in bonding. These electrons are located on either oxygen, but not on the nitrogen. Thus a node for the nonbonding molecular orbital exists which bisects the molecule through the nitrogen, perpendicular to the plane defined by the atoms. The node indicates that there is zero probability of finding these nonbonding electrons on the nitrogen, but that they could be found on the two oxygen atoms.

10.6 EXTENDED π SYSTEMS

Key Concepts:
- Delocalized π systems are not limited to just three carbon atoms. They can extend to an infinite number of carbons, as long as the conditions are right for delocalization.
- The more delocalized an orbital, the more stable its electrons are.
- Chemicals with large delocalized π systems can absorb certain wavelengths of visible light, and therefore are usually colored. This occurs because of a decrease in the separation between filled bonding orbitals and unfilled antibonding orbitals. Visible light is of sufficient energy to bump an electron up to an antibonding orbital (thereby absorbing a color of light, and reflecting other colors).

EXERCISE 17: Describe the orbitals of the delocalized electrons in the resonance structures of ClO^-.

$$\left[:\overset{\cdot\cdot}{\underset{\cdot\cdot}{Cl}}\!-\!\overset{\cdot\cdot}{\underset{\cdot\cdot}{O}}: \right]^{\ominus} \quad\longleftrightarrow\quad \left[:\overset{\cdot\cdot}{\underset{\cdot\cdot}{Cl}}\!=\!\overset{\cdot\cdot}{\underset{\cdot\cdot}{O}}: \right]^{\ominus}$$

STRATEGY AND SOLUTION: Cl is considered the "central" atom in this ion. Its steric number is 4 in both structures, indicating sp^3 hybridization. Cl's valence shell is $n=3$, so it does have vacant d orbitals available. An electron pair that is residing in a $2p_y$ orbital on oxygen (left structure) is able to move to form a $2p_y\text{-}3d_{xy}$ π bond with the Cl (right structure). Here is a picture of the new bond that is formed in the right structure:

The above-mentioned electron pair is capable of being in two different locations; it is delocalized.

EXERCISE 18: The anthocyanins are a group of chemicals that give plants their color. Delphinidin is an anthocyanin; it is responsible for the deep blue/purple in eggplants and blue larkspur. Indicate how many carbons are involved in the delocalized π system in this compound.

Delphinidin

STRATEGY: Look for alternating single and double bonds that are be arranged within one plane, which will leave pure p orbitals aligned along one axis for delocalized π bonding.

SOLUTION: All of the carbon atoms are sp^2 hybridized. If all of the sp^2-sp^2 σ bonds between carbon atoms are lying in the plane of the paper (the x-y plane), then all of the pure p orbitals on the carbons in this compound can be aligned along the z axis (perpendicular to the paper). The π bonds that these p orbitals form can therefore be delocalized. All 15 carbons in delphinidin are therefore involved in the delocalized π system.

Try It #6: One of these compounds is a vibrantly colored dye used to make lipstick and the other is not. Which is the lipstick dye? How can you tell?

10.7 BAND THEORY OF SOLIDS

Key Terms:
- **Band Theory** – theory which states that for a given solid, all of the molecular orbitals of the same type (i.e. σ_s or σ_s *) cluster together to form an energy band. The ability of electrons to delocalize within one band or between two bands will determine the substance's ability to conduct an electrical current.
- **Band gap (E_g)** –difference in energy between a bonding orbital band and an antibonding orbital band.

- **Insulator** – a substance that cannot conduct an electrical current. Non-metals are typically good insulators.
- **Conductor** – a substance that can conduct an electrical current. Metals are typically good conductors.
- **Semiconductor** – a substance that can conduct an electrical current under certain conditions. Metalloids are typically good semiconductors.

Key Concepts:

- For a substance to be able to conduct an electrical current, it must have space for the electrons to move (i.e., electrons must be delocalized).
- There are two situations in which electrons can be delocalized:
    1. If the occupied bonding orbital band is only partially filled, then there is space for electrons to move around in it, or
    2. If the bonding orbital band is filled, but the band gap is small, then electrons can easily move from the full bonding orbital to the unoccupied antibonding orbital.
- Insulators have large band gaps, conductors have essentially no band gaps, and semiconductors have intermediate band gaps. This picture illustrates the relative band gaps between the occupied bonding orbital bands (black) and the vacant antibonding orbital bands (white).

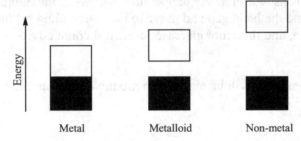

- Temperature and light energy both influence a semiconductor's ability to conduct electricity.

EXERCISE 19: The longest wavelength of light that can cause silicon to conduct an electrical current at room temperature is 1020 nm. Calculate silicon's band gap.

STRATEGY: In order for an electron to move up to the unoccupied antibonding orbital band, it must be supplied with enough energy to overcome the band gap. This minimum energy for Si corresponds to a wavelength of 1020 nm. We can use the relationship: $E = hc/\lambda$ from Chapter 7 to calculate the energy.

SOLUTION:

$$E = \frac{hc}{\lambda} = \frac{6.626 \times 10^{-34} \text{ J sec}}{1020 \times 10^{-9} \text{ m}} \times \frac{2.9979 \times 10^8 \text{ m}}{\text{sec}} = 1.95 \times 10^{-19} \text{ J}$$

Key Term:

- **Doped semiconductor** – a semiconductor whose ability to conduct electricity has been enhanced by the addition of a small amount of an impurity. There are two types of doped semiconductors:
    - *n*-type – contains extra electrons (extra *n*egative charges) in its high energy bands.
    - *p*-type –contains gaps or holes where electrons are missing from its low energy bands.

EXERCISE 20: Draw a band picture to show Ge that has been doped with a small amount of Ga.

STRATEGY: Germanium has four valence electrons, so it has a filled bonding orbital band. Ga has three valence electrons, so addition of Ga to Ge will result in holes in the bonding orbtial band. This is a *p*-type semiconductor.

SOLUTION:

EXERCISE 21: Why does the electrical conductivity of a semiconductor increase with increasing temperature?

SOLUTION: A semiconductor's capability to conduct an electrical current is dependent upon the number of delocalized electrons within it. As temperature increases, the number of electrons with sufficient energy to overcome the band gap and move to the antibonding orbital band also increases. This increases delocalization, and therefore increases electrical conductivity.

Try It #7: What type of semiconductor will be made from mixing a small amount of As in with Si?

Chapter 10 Self-Test

1. Determine the hybridization on the central atom and describe what atomic orbitals were used to create the hybrids in: a) NH_3F^+, and b) SF_6.

2. It requires 345 kJ/mol to break a C-C bond. Estimate how much energy is required to break a C=C bond. Explain your reasoning.

3. Lycopene is the chemical that gives tomatoes and watermelons their intense, red color. Decreased occurrences of certain types of cancer have been observed in people who regularly consume lycopene-rich foods in their diets.

   Lycopene

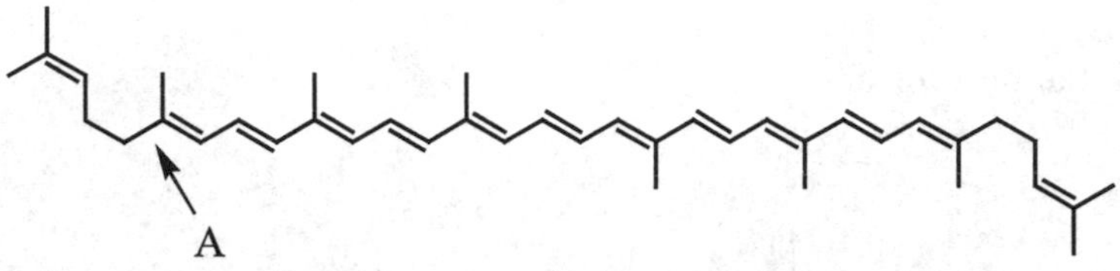

   A

   a) How many carbon atoms make up the delocalized π system in lycopene?
   b) Describe the bond marked "A" using VSEPR theory terms.
   c) Describe the bond marked "A" using molecular orbital terms.

4. Draw a bonding picture for methyl isocyanate. You may draw a set of bonding pictures if you wish. (Note: in 1984, one of the worst industrial disasters in history occurred at the Union Carbide plant in Bhopal, India, when over thirty tons of highly toxic methyl isocyanate leaked out into the environment, killing thousands of people.)

   Methyl Isocyanate

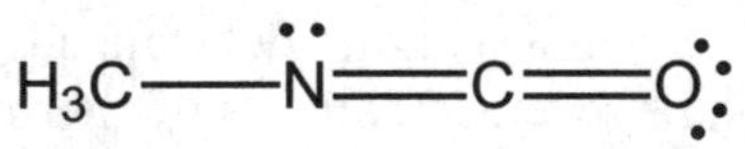

5. Superoxide (O_2^-) is a very reactive substance that may play a role in the aging process. Use molecular orbital diagrams to illustrate why superoxide ion is more reactive than molecular oxygen. Be sure to include a brief interpretation of your diagrams.

6. Silicon is a semiconductor. List two different elemental impurities that could be added to a sample of silicon to make a p-type semiconductor.

7. Carbon is a non-metal and therefore is considered to be a good insulator. Is it possible to create a carbon-based molecule that could conduct electricity? If so, what features would it need to have?

Answers to Try Its:
1. a) 3d; b) 2p; c) None. All are full. (Note: technically, this is not true because Xe can make bonds. This will be explained in Section 9.2.)
2. sp^3, sp^3d, sp^3d^2.

3. a) $2sp^2$-$2sp^3$ σ or sp^2-sp^3 σ   b)

4. $(\sigma_s)^2 (\sigma^*_s)^2 (\pi_p)^4 (\sigma_p)^1$
5. MO energy level diagram for Cl_2

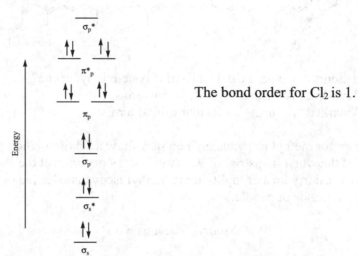

The bond order for Cl_2 is 1.

6. Molecule B is the lipstick dye. It contains a highly delocalized π system (alternating single and double bonds). Molecule A contains several small sets of alternating single and double bonds, but the delocalization is not nearly as vast as in Molecule B.
7. As contains five valence electrons, one more electron than silicon. As mixed with Si would therefore create an *n*-type semiconductor. It has extra electrons which will occupy spaces in the antibonding orbital band.

Answers to Self-Test:

1. a) hybridization of N = sp^3 (N needed 4 hybrid orbitals). The orbitals that N used to make the hybrid would be: 2s, $2p_x$, $2p_y$, $2p_z$. b) hybridization of S = sp^3d^2 (S needed 6 hybrid orbitals). S used its 3s, $3p_x$, $3p_y$, $3p_z$, and two of its 3d orbitals to create the hybrids.

2. The bond energy (energy required to break the bond) will be greater than 345 kJ/mol but less than 690 kJ/mol, maybe somewhere around 600 kJ/mol. The single bond is a sigma bond (head-on overlap of atomic orbitals) and the double bond contains a sigma bond plus a pi bond. It is easier to overcome the attractive forces in the pi than the sigma bond because a pi bond is made of sideways overlap of atomic orbitals. Therefore, the bond energy will be greater than one sigma bond but less than twice a sigma bond (2 x 345 kJ/mol = 690 kJ/mol).

3. a) 22;   b) This bond is made from a $2sp^3$ hybridized atomic orbital of the left carbon overlapping with a $2sp^2$ hybridized atomic orbital of the right carbon;   c) This is a single bond, which indicates a bond order of 1. This means there are (net) two electrons present in a bonding molecular orbital between the two atoms.   This bonding molecular orbital was formed from the constructive interference of an atomic orbital from each atom surrounding the bond.

4.

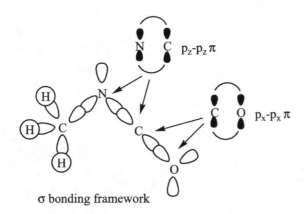

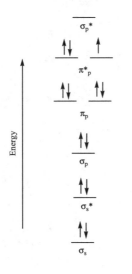

σ bonding framework

If the sp^3-sp^2 σ bond between C and N is in the plane of the paper (xy plane), then the pure p orbital left on N will be a p_z orbital. The sp hybridized C will therefore need to use its p_z orbital to make a π bond with the N. This would leave a p_x orbital available on that C to make a π bond with O. The sp^2 hybridized O, then, must be perpendicular to the plane of the paper.

5.

MO energy level diagram for superoxide   MO energy level diagram for molecular oxygen

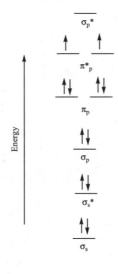

Superoxide has an extra electron in an antibonding orbital (relative to molecular oxygen). This is a destabilizing factor. The bond order of superoxide is 1.5 and the bond order of oxygen is 2.0. Therefore, superoxide is less stable than molecular oxygen.

6. To make a *p*-type semiconductor, you must add an elemental impurity that will create positive holes in the bonding orbital band. Two possibilities are B or Al; each of these elements contains one less valence electron than Si.

7. Yes, it would be possible. A very highly delocalized π system would be required, in order to stabilize the molecule to the extent that its π and π* orbitals would be separated by a very small energy gap. For example, a solid composed of molecules containing 1,000 carbon atoms in a row connected by alternating single and double (or triple) bonds might work.

Chapter 11: Effects of Intermolecular Forces

Learning Objectives

In this chapter, you will learn how to:

- Approximate physical properties such as boiling point, melting point, surface tension, and viscosity of a substance based on its structure.
- Categorize solids and assess their properties.
- Assess the different types of attractive forces in a given solid.
- Differentiate between different types of crystal structures.
- Assess energy changes for phase changes.
- Interpret and construct a phase diagram for a substance.

Practical Aspects

An understanding of intermolecular forces is essential to a proper understanding of chemistry, physics, biology and engineering. You have learned how atoms combine to form compounds, and the shapes that these compounds have. A chemist will use intermolecular forces in every subsequent course he/she takes. With an understanding of intermolecular forces, you can answer questions such as: Why and how does DNA hold together in a double helix formation? Why are some solids brittle and others malleable? Why does ice float on water, when most solids are denser than their respective liquid phase? Why do we have to worry more about toxicity of chlorine at room temperature than that of iodine?

11.1 REAL GASES AND INTERMOLECULAR FORCES

Key Terms:
- **Intermolecular forces** – attractive forces between molecules.
- **Real gas** – gas deviates from Ideal Gas Law behavior because the pressure is so high or the temperature is so low that 1) interactions between molecules occur; and 2) the space the molecules occupy is no longer insignificant relative to the size of the container.

Key Concept:
- The physical properties of a substance depend upon a given substance's intermolecular forces.

Useful Relationship:
- $\left(P + \dfrac{n^2 a}{V^2}\right)(V - nb) = nRT$

  This is van der Waal's equation for real gases. Table 11-1 in the text provides values for "a" and "b," which account for attractions between molecules and the space that the molecules occupy, respectively.

EXERCISE 1: 15.0 moles of argon occupy 5.00 L at 311 K. Compare the pressure of the argon gas using the ideal gas law versus van der Waal's equation.

STRATEGY: Calculate P using the ideal gas law and using van der Waal's equation.

$$P_{ideal} = \frac{nRT}{V} = \frac{15.0 \text{ mol}}{5.00 \text{ L}} \times \frac{0.08206 \text{ L atm}}{\text{mol K}} \times 311 \text{ K} = 76.6 \text{ atm}$$

$$\left(P_{real} + \frac{n^2 a}{V^2}\right)(V - nb) = nRT \text{ so } P_{real} = \frac{nRT}{(V - nb)} - \frac{n^2 a}{V^2}$$

$$P_{real} = \left[\frac{15.0 \text{ mol}}{(5.00 \text{ L} - (15.0 \text{ mol} \times 0.0322 \text{ L/mol}))} \times \frac{0.08206 \text{ L atm}}{\text{mol K}} \times 311 \text{ K}\right] - \left[\frac{(15.0 \text{ mol})^2}{(5.00 \text{ L})^2} \times \frac{1.345 \text{ L}^2 \text{ atm}}{\text{mol}^2}\right]$$

$$= 72.6 \text{ atm}$$

SOLUTION: The pressure using the ideal gas law is 76.6 atm versus 72.6 atm for the van der Waal's equation. The real gas experiences a lower pressure because attractive forces between argon atoms are present under these conditions.

Try It #1: Refer back to Exercise 1. Would you expect helium's deviation from ideal behavior to be greater or less than argon's under these conditions?

Vaporization and Condensation

Key Terms:
- **Boiling point** – temperature at which the vapor pressure of a substance equals the atmospheric pressure. At this temperature, the gas phase and liquid phase are in dynamic equilibrium.
- **Freezing point** – temperature at which the solid and liquid phases are in dynamic equilibrium.
- **Normal boiling or freezing point** – the boiling or freezing point of a substance under "normal" atmospheric conditions at sea level (where atmospheric pressure = 1 atm).

Key Concepts:
- A substance melts and freezes at the same temperature (freezing point = melting point).
- A substance evaporates and condenses at the same temperature (boiling point = condensation point).
- In order to melt a substance, enough energy must be put into it to overcome at least some of its intermolecular forces, so that the molecules can flow over one another.
- In order to evaporate a substance, enough energy must be put into it to completely overcome all intermolecular forces (the molecules must be completely separated from one another).
- The stronger the intermolecular forces, the higher the melting point or boiling point.

EXERCISE 2: Does water boil at a higher or lower temperature at high altitude than at sea level?

STRATEGY AND SOLUTION: The boiling point is the temperature at which the vapor pressure of the liquid equals atmospheric pressure. Atmospheric pressure decreases as altitude increases, which means that the boiling point of water is lower at high altitudes than at sea level.

EXERCISE 3: Compound A has a boiling point of 28°C and Compound B has a boiling point of 229°C. Which substance has stronger intermolecular forces?

SOLUTION: Compound B, with a higher boiling point, has stronger intermolecular forces than Compound A. The relative boiling points indicate that it requires less thermal energy to completely separate molecules of Compound A from each other than it would to separate molecules of Compound B from each other.

11.2 TYPES OF INTERMOLECULAR FORCES

Key term:
- **Polarizability** – ability of an electron cloud to distort or shift position around the nucleus. *n* value and Z_{eff} influence polarizability.

There are three main types of attractive forces that exist between molecules.

Type of Force:	How it works:	Factor to assess:
Dispersion Forces	Attractions between molecules as a result of temporary or "instantaneous" dipolar attractions.	Polarizability
Dipolar Forces (dipole-dipole forces)	Attractions between the δ+ and δ- ends of two polar molecules.	Dipole Moment
Hydrogen Bonding	Attractions between a hydrogen that is attached to N, O, or F on one molecule to a lone pair of N, O, or F on a second molecule.	# of hydrogen bonding sites

Several Key Concepts can be derived from this table.

Key Concepts:
- To determine relative properties of two molecules, determine the types of intermolecular forces that predominate in each and compare the relative strengths of these forces.
- In general, the strongest type of intermolecular force is hydrogen bonding and the weakest type is dispersion forces. (*Remember that this is just a trend and that exceptions are possible.*)
- To compare two molecules that have the same type of intermolecular forces, assess the specific factor that influences the force's strength. For example, in comparing two polar molecules, the one with the larger dipole moment will most likely be the one with the stronger intermolecular forces.
- Chemists usually consider only the strongest type (predominant) of intermolecular force for a given chemical. Keep in mind, though, that *all* molecules have dispersion forces. Similarly, all molecules that can hydrogen bond also have dipolar forces.
- The polarizability of a molecule is influenced by its overall size and shape.
- The general trend for relative strengths of intermolecular forces is:
    hydrogen bonding attractions > dipolar forces > dispersion forces
  Remember that this is just a trend, so exceptions can occur.

EXERCISE 4: Draw a picture to illustrate two molecules attracted to each other by: a) dispersion forces; b) dipolar forces; c) hydrogen bonding attractions.

SOLUTION:

a) dispersion forces	b) dipolar forces	c) hydrogen bonding attractions
temporary δ⁻ ... temporary δ⁺	Dipoles line up:	X = N, O, or F:

EXERCISE 5: What is the strongest type of intermolecular force that each chemical has? a) Kr; b) C_6H_{10}; c) CH_3CH_2Cl; d) C_3H_5OH

STRATEGY: Determine if the compound is polar or non-polar and whether or not it can hydrogen bond.

SOLUTION:
a) Kr is non-polar, so dispersion forces will predominate between Kr atoms.
b) C_6H_{10} is non-polar (recall that the bond between carbon and hydrogen is considered non-polar) because all bonds within the molecule are non-polar. Dispersion forces will predominate between C_6H_{10} molecules.
c) CH_3CH_2Cl has a dipole moment because the polar C-Cl bond is not canceled out by any opposing polar bond. The molecule does not contain an N, O, or F attached to an H, so hydrogen bonding cannot occur between these molecules. Dipolar forces will predominate between CH_3CH_2Cl molecules.
d) C_3H_5OH has a dipole moment due to the highly polar bonds between C & O and O & H. Furthermore, this molecule can hydrogen bond because it contains O attached to H. Hydrogen bonding will be the predominant intermolecular force between C_3H_5OH molecules.

EXERCISE 6: Answer each question and provide a brief explanation.

a) Which isomer of C_3H_9N should have the higher boiling point?

b) Which isomer should have the higher melting point?

c) Which liquid do you expect will be more viscous?
(viscosity is the resistance of a liquid to flow)

STRATEGY: Assess relative strengths of intermolecular forces.

SOLUTION:
a) A is a polar covalent compound (don't forget that the central nitrogen atom has a lone pair on it!) so dipolar forces predominate in it. B is also polar AND molecules of B can hydrogen bond to each other. Hydrogen bonding attractions are typically stronger than dipolar attractions, so B should have a higher boiling point than A.

b) Both of these compounds are flat (all central atoms have sp^2 hybridization). A has a dipole moment, so dipolar forces predominate. The polar C-Cl bonds in B are directly opposite each other, so they cancel each other out. B is non-polar, so dispersion forces predominate in it. A should have a higher melting point than B, due to greater attractions between molecules.

c) Both of these compounds are polar covalent and both can hydrogen bond because they contain H attached to O. A has three –OH groups and B has one –OH group. A has more hydrogen bonding sites than B, so molecules of A will be more attracted to each other than molecules of B. A will be more viscous than B.

Try It #2: a) Why is Ar a gas at room temperature, while C_3H_7OH is a liquid? b) Element X and element Y are both non-metals that exist mono-atomically in nature. Element X has a lower boiling point than element Y. Propose an explanation for why this is so.

11.3 LIQUIDS

Liquid Properties

Key Terms:
- **Surface tension** – resistance of a liquid to increase its surface area.
- **Viscosity** – resistance of a liquid to flow.
- **Meniscus** – the curvature of a liquid's surface inside a container.
- **Capillary action** – upward movement of a liquid against the force of gravity.

Key Concepts:
- A substance's surface tension and viscosity are directly related to the strength of that substance's intermolecular forces.
- A substance's meniscus and capillary action are a result of the balance between cohesive forces and adhesive forces.
  - **Cohesive forces** – attractive forces between molecules of the same substance.
  - **Adhesive forces** – attractive forces between two different substances.

- In Chapter 5, section 7, you learned that vapor pressure (*vp*) is the pressure that vapor exerts above a solid or liquid inside a closed system at equilibrium and that vapor pressure depends on temperature and the identity of the substance. At a given temperature, a chemical with strong intermolecular forces will have a lower vapor pressure than a chemical with weak intermolecular forces. (Think about it – strong intermolecular forces mean the molecules are attracted to each other, so it will be difficult for them to escape to the vapor phase.)

EXERCISE 7: A liquid is inside a container. The liquid's cohesive forces are stronger than the adhesive forces it feels. Draw a picture of a meniscus to illustrate this.

STRATEGY AND SOLUTION: The attractions between the liquid and the container (adhesive forces) are weaker than the attraction of liquid molecules to each other (cohesive forces), so the liquid will occupy space in a manner to minimize its interaction with the walls of the container. The liquid will rise up in the center, as shown here:

Try-It # 3: Which substance should have a higher vapor pressure at room temperature, Cl_2 or I_2?

11.4 FORCES IN SOLIDS

This table summarizes the types of forces present in different types of solids.

Type of Solid:	Example:	Exist as:	Forces that must be overcome to melt it:	Strength of Attractive Forces: (kJ/mol)
Molecular	H_2O	Individual molecules	Attractions between molecules	0.05 – 40
Network	Diamond	Network of atoms that are covalently bonded to each other	Huge 3-D network of covalent bonds	150 – 500
Metallic	Cu	Atoms held together by a "net" or "sea" of valence electrons	Delocalized bonding	75 -1000
Ionic	NaCl	3-D array of opposite charges	Attractions between ions (Coulombic forces)	400 - 4000

EXERCISE 8: What type of solid will each form? a) CsCl;   b) Si;   c) brass;   d) $CHCl_3$

STRATEGY AND SOLUTION:
a) CsCl consists of a Group I metal and a halogen, so it will form an ionic solid;
b) Si is in the same group as carbon, with four valence electrons. It will exist as a network solid;
c) Brass is a mixture of two metals, so it will form a metallic solid;

d) $CHCl_3$ is composed of all non-metals that are covalently bonded to each other to make a molecule, so it will form a molecular solid.

Try It #4: What type of solid will each form? a) Ge; b) CH_3OH; c) $MgSO_4$

Try It #5: Of the choices in Try It #4, which will have the lowest melting point?

EXERCISE 9: Tungsten, W, has a melting point of 3410°C. Sodium, Na, has a melting point of 98°C. Propose an explanation for their huge differences.

STRATEGY AND SOLUTION: Both substances are metallic solids. The forces that must be overcome to melt a metallic solid are the stabilizing forces of delocalized electrons. Each atom of Na has 1 valence electron, while each atom of W has 6 valence electrons. There are more delocalized electrons on W, so W will have a higher melting point than Na. (Also worth noting is the fact that the delocalized electrons are in bonding orbitals. W feels a greater stabilizing effect from this delocalization than Na.)

EXERCISE 10: When NaCl melts, the NaCl solid is broken down into Na^+ ions and Cl^- ions. When ice melts, is H_2O converted into individual H atoms and O atoms?

STRATEGY: Compare the types of solids.

SOLUTION: No. NaCl is an ionic solid. In order for it to melt, enough energy must be added to it to disrupt the attractive forces in the 3-D network of opposite charges. Ice is a molecular solid, and molecular solids exist as individual, small molecules. In the solid phase, these molecules are packed together. For ice to melt, the attractions *between* the molecules – hydrogen bonding – must be overcome, so that the molecules can move around. The H_2Os stay as molecules, only they are able to move around each other in the liquid phase.

EXERCISE 11: Which has the lower melting point: $CaCl_2$ or OCl_2?

STRATEGY: Compare the types of solids.

SOLUTION: $CaCl_2$ is an ionic solid; it exists as an enormous 3-D array of opposite charges. OCl_2 is a covalent compound. It exists as individual small molecules, so it will require much less energy to get this substance to melt than the $CaCl_2$. OCl_2, therefore, has a lower melting point than $CaCl_2$.

11.5 ORDER IN SOLIDS

Crystal Packing

Key Terms:
- **Simple cubic structure** – simplest crystal lattice to visualize, with all atoms stacking directly on top of each other. See Figure 11-26 in the text for an illustration.

- **Close-packed structure** – packing within a crystalline substance that results in the minimum amount of empty space between particles. Figure 11-29 in the text illustrates particle arrangements within a close-packed structure. Each particle has 12 nearest neighbors.
- **Unit cell** – the smallest repeating portion of a solid from which a crystal can be constructed. Two common types of unit cells are:
  - **Face-centered cubic structure** – packing of particles in a cubic close-packed arrangement. See Figure 11-28 in the text for an illustration.
  - **Body-centered cubic structure** – packing of particles in which each particle sees 8 nearest neighbors. This is not as tightly packed as a close-packed structure. See Figure 11-26 in the text for an illustration.

Key Concepts:
- Solids can be categorized as either crystalline or amorphous. Amorphous solids have no specific arrangement of particles, while crystalline solids have definite packing structures.
- There are two types of close-packed structures, which differ by how the particle layers stack within the structure:
  - **Hexagonal close-packed** – layers stack in an "ABAB" or alternating arrangement; the particle arrangement repeats every other layer.
  - **Cubic close-packed** – layers stack in an "ABCABC" arrangement; every third layer repeats. The face-centered cubic cell has a cubic close-packed arrangement.
- The colors of many gemstones are a result of slight imperfections within a crystal lattice.

Helpful Hint
- Figure 11-30 in the text illustrates hexagonal close-packed vs. cubic close-packed structures.

EXERCISE 12: The atomic radius of Fe is close to that of Pb. One of these solids packs in a body-centered cubic arrangement, and the other packs in a face-centered cubic arrangement. Given that Pb is denser than Fe, match the metal to the packing type.

STRATEGY: The body-centered cubic arrangement is not a close-packed structure. An atom within a body-centered cubic structure will have only eight nearest neighbors, as opposed to the twelve for the face-centered cubic (which is one type of close-packed arrangement).

SOLUTION: Pb must be face-centered cubic and Fe must be body-centered cubic.

EXERCISE 13: Carbon dioxide molecules in dry ice arrange themselves so that each carbon atom occupies one spot of a face-centered cubic arrangement. How many total carbon atoms are in one unit cell of carbon dioxide?

STRATEGY: Visualize the unit cell that has a face-centered cubic arrangement:
- Each of the six faces of the unit cell contains one carbon atom, which is shared with one neighboring unit cell. (6 faces x ½ C per face) = 3 carbons total on the faces
- Every corner of the unit cell also contains carbon. Each corner of a unit cell touches 7 other unit cells; 8 unit cells meet at the corners. This means that 1/8 of the carbon atom in a corner is within our unit cell. There are 8 corners total. (8 corners x 1/8 C per corner) = 1 carbon total on the corners

SOLUTION: There are four carbon atoms within one unit cell of dry ice.

11.6 PHASE CHANGES

Heats of Phase Changes

Key Terms:
- **Molar Heat of Fusion (ΔH_{fus})** – energy required to melt one mole of a given substance.
- **Molar Heat of Vaporization (ΔH_{vap})** – energy required to evaporate one mole of a given substance.
- **Sublimation** – phase change from a solid directly to a gas.

Key Concepts:
- It requires energy to overcome intermolecular forces.
- Energy is released when intermolecular forces are established.
- The energy required or released during a phase change is dependent upon the identity and the quantity of the substance undergoing the change.
- There is no change in temperature during a phase change.

Helpful Hint
- By convention, only endothermic values for phase changes are tabulated. To find $\Delta H_{freezing}$, simply take the negative of ΔH_{fusion}. Similarly, $\Delta H_{condensation} = -\Delta H_{evaporation}$.

EXERCISE 14: Which has a larger heat of vaporization, water or methane, CH_4?

STRATEGY AND SOLUTION: Since heat of vaporization is the energy required to evaporate a mole of substance, we must assess which chemical is more difficult to convert from the liquid to gas phase. Water is a polar covalent compound that can hydrogen bond; so a group of water molecules will be held together by hydrogen bonding attractions. Methane is a non-polar covalent compound with weak dispersion forces holding the molecules together. It will therefore require much more energy to evaporate a mole of water than methane.

EXERCISE 15: 10.00 g aluminum is at 500.0°C. How much energy needs to be put into the aluminum to melt it? Given this data for aluminum: melting point = 660.0°C, boiling point = 1800.°C, ΔH_{fus} = 10.7 kJ/mol, ΔH_{vap} = 225 kJ/mol, C_{solid} = 24.35 J/mol K.

STRATEGY: Initially, we have Al (s) at 500.0°C. We want to heat it until it melts at 660.0°C. The Al is going through a temperature change *and* a phase change, and heat is required for both processes. We need to break the problem down:

$$q_{process} = q_{to\ heat\ the\ Al\ solid} + q_{to\ melt\ the\ solid}$$

$$q_{to\ heat\ the\ Al\ solid} = nC\Delta T = 10.00\ g \times \frac{1\ mol}{26.982\ g} \times \frac{24.35\ J}{mol\ K} \times 160.0\ K = 1444\ J$$

$$q_{to\ melt\ the\ solid} = \Delta H_{fus} = \frac{10.7\ kJ}{mol} \times \frac{1\ mol}{26.982\ g} \times 10.00\ g = 3.9645\ kJ = 3964.5\ J$$

$$q_{process} = 1444\ J + 3964.5\ J = 5409\ J$$

SOLUTION: The heat energy required for this process is 5410 J. (3 sig figs from data) The answer seems reasonable because it is a positive value, which indicates energy is required.

Try It #6: How much heat energy is released when 5.00 g of aluminum metal condenses at 1800°C? Refer to Exercise 15 for aluminum constants.

Phase Diagrams

Key Terms:
- **Phase diagram** – plot of temperature (x-axis) vs. pressure (y-axis) for a given substance, with lines drawn to indicate conditions at which the substance is in equilibrium between phases. Phase diagrams are described thoroughly in 11.6 of the text.
- **Triple point** – the one pressure/temperature combination in which a given substance simultaneously coexists in all three phases.

Key Concepts:
- The main pieces of information used to construct a phase diagram are the substance's normal melting point, normal boiling point, triple point, and the relative density of the solid vs. the liquid phases.
- A phase diagram can be used to predict the phase of a given substance at any temperature or pressure, as well as any phase changes that may occur if conditions are altered.
- Any point that is on a line in a phase diagram will be a temperature/pressure combination for which that substance is in equilibrium between two phases.
- The densest phase for a given substance can be determined by observing what phase exists above the triple point. (i.e., if the substance is at its triple point pressure and temperature conditions, and pressure is increased, it will spontaneously convert to the phase with the closest molecular packing.)

EXERCISE 16: Here is the phase diagram for carbon dioxide. Answer the questions below.

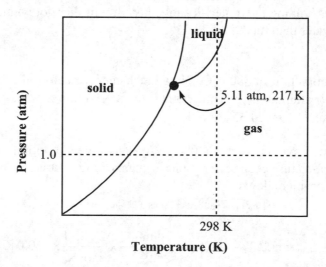

a) Under what conditions do all three phases of carbon dioxide exist simultaneously?
b) What is the phase of carbon dioxide when its temperature is 200 K and its pressure is 0.50 atm?
c) What is the normal boiling point of carbon dioxide?
d) Consider carbon dioxide at 298 K and 1 atm. What would happen if the pressure were increased without changing the temperature?

STRATEGY: Interpret information from the phase diagram. First determine which phase corresponds to which area on the diagram. Solids normally exist at low temperatures, gases at high temperatures.

Gases exist at low pressures, but as pressure is increased at a given temperature, the gas molecules may get packed so tightly that they change to a liquid or solid. On the diagram then, the solid phase will be the left-most region because it corresponds to low temperature and a huge range of pressures. The right-most region will be the gas phase because it shows situations of high temperature and (relatively) low pressure. The one remaining phase to be assigned is the liquid phase. All phase diagrams have this basic construction.

SOLUTION:
a) P = 5.11 atm, T = 217 K (This is the triple point.)
b) gas
c) CO_2 does not have a normal boiling point. The phase diagram shows that CO_2 converts from a solid directly to a gas (sublimes) as temperature is increased at a constant pressure of 1 atm. (The line that intersects 1 atm is the solid-gas equilibrium line.)
d) It would condense. At 298 K and 1 atm, CO_2 is in the gas phase. If the pressure is increased at 298 K, we will eventually cross the gas-liquid equilibrium line. The CO_2 will convert from a gas to a liquid.

Chapter 11 Self-Test

You may use a periodic table and a calculator for this test.

1. For each property, circle the appropriate chemical choice and briefly explain your reasoning.

Property	Choices
a) higher boiling point	Ar vs He
b) a better amino acid side chain within a protein for a hydrophobic environment	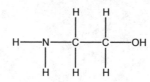

2. Draw a picture to illustrate what aqueous $MgBr_2$ looks like at the molecular level. Illustrate the intermolecular forces between the solute and solvent in the picture.

3. Rank these from lowest to highest boiling point and explain your reasoning:
   I. H_2O II. KBr (aq)   III. CH_2Cl_2 IV. KBr

4. Categorize each solid by type based on its characteristics:
   a) It is a smelly solid that melts at 115°C.
   b) It is a hard solid that melts above 2000°C. It doesn't conduct electricity in the liquid phase.

5. Circle the atoms in the molecule that can hydrogen bond with a lone pair on water.

$$H-N-\overset{\overset{\displaystyle H}{|}}{\underset{\underset{\displaystyle H}{|}}{C}}-\overset{\overset{\displaystyle H}{|}}{\underset{\underset{\displaystyle H}{|}}{C}}-OH$$

6. List the energy transformations that occur when water freezes to ice cubes in a freezer.

7. Consider this process: 4.00 moles of water condenses at 100°C at a constant pressure of 1.00 atm. ΔH_{vap} for water is 40.79 kJ/mol and ΔH_{fus} for water is 6.01 kJ/mol. Calculate ΔH for the process.

8. Molecular nitrogen has a normal boiling point of 77.35 K and a normal freezing point of 63.29 K. Liquid nitrogen is less dense than solid nitrogen. Construct a rough phase diagram for nitrogen.

Answers to Try-Its:

1. Helium's deviation from ideal behavior should be less than argon's. Helium is smaller than argon, so it 1) takes up less space, and 2) has weaker attractive forces between atoms.

2. a) The strongest attractive force between non-polar Ar atoms is dispersion forces. C_3H_7OH molecules can hydrogen bond with each other, which is a stronger attraction than Ar's dispersion forces.
   b) The elements are mono-atomic, so they are non-polar (atoms are symmetrical). The predominant intermolecular forces are dispersion forces. Y has greater dispersion forces than X, so it must be larger and more polarizable than X.

3. Cl_2 will have a higher vapor pressure, because it has weaker intermolecular forces than I_2. Both are nonpolar molecules, but I_2 is more polarizable than Cl_2 is.

4. a) network; b) molecular; c) ionic

5. CH_3OH will have the lowest melting point because it takes less energy to disrupt its individual molecules than to disrupt a 3-D array of atoms (network) or ions (ionic).

6. -41.7 kJ (heat released so negative)

Answers to Self-Test:

1. a) Ar – it has greater dispersion forces; b) the structure on the right because it is non-polar and would therefore be attracted to a hydrophobic environment.

2.

1 Mg^{2+} for every 2 Br^-. Dipoles in water are aligned to maximize attractions to ions: ion-dipole attractions.

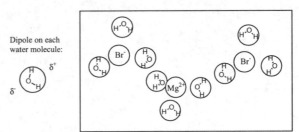

3. Lowest = CH_2Cl_2 w/ dipolar forces; 2nd = water with hydrogen bonding capabilities; 3rd = KBr (aq), which is an aqueous solution, whose boiling point is elevated relative to the pure solvent, water; 4th (highest) = KBr which is an ionic compound– it will take a tremendous amount of energy to get KBr to boil because every ion-ion attraction will have to be overcome.

4. a) molecular; b) network

5.

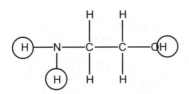

6. The water molecules transfer thermal energy to the freezer.
7. −163 kJ

8.   Notice that the solid is denser than the liquid (increased P from triple point forms solid).

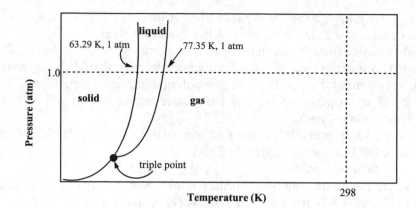

Chapter 12: Properties of Solutions

Learning Objectives

In this chapter, you will learn how to:

- Predict whether or not two substances should be able to dissolve in each other.
- Convert between different types of concentration units.
- Quantitatively and qualitatively assess energy changes for solutions.
- Use colligative properties to calculate boiling points, freezing points, and osmotic pressures of solutions.
- Identify colloidial suspensions.
- Identify dual-nature molecules and predict their behavior in various chemical environments.

Practical Aspects

This chapter completes the basic concepts of the theme "structure determines function." You have learned how atoms combine to form compounds, and the shapes that these compounds have. In this chapter, you will learn how to assess properties of solutions qualitatively (based on the structures of the solution components) and quantitatively (by performing a wide variety of calculations). Give special attention to the calculations involving unit conversions for solutions and the intermolecular forces explanations behind why solutions gain or lose heat energy when they are prepared as these topics are necessary for every chemistry course you ever take, and will surely help you in other science courses as well.

12.1 THE NATURE OF SOLUTIONS

Key Terms:
- **Solvent** – the major component of the solution. (Water is considered the "universal solvent.")
- **Solute** – substance present in minor amounts within a solution.

Key Concepts:
- The words "dissolve" and "melt" are not interchangeable. Melting is a phase change of a single substance from the solid phase to the liquid phase. Dissolving is the even distribution of one substance in another to make a solution.

Helpful Hints
- See the summary table on the next page for common units used to describe solution concentrations.
- To convert between concentration units, you may need the density of the solution. (Recall, D=m/V). This will allow you to convert between volume of solution and mass of solution.
- Concentration is independent of the amount of solution you have, so it is fine to assume you have a given quantity of solution if it will make your conversion calculations easier. If you

make the following assumptions, you will always know how many moles of solute you're starting with:

- o If you are converting from molarity units, then assume you have one liter of solution.
- o If you are converting from molality units, then assume you have one kilogram of solvent.
- o If you are converting from mole fraction units to other units, then assume you have 1 mole total.

Common Units for Solution Concentrations

Units	Expressed in	Notes
Molarity, M	$M = \dfrac{\text{moles of solute}}{\text{volume of solution}}$	Volume in denominator **must** be in liters.
Molality, C_m	$C_m = \dfrac{\text{moles of solute}}{\text{kg of solvent}}$	Useful for solutions that are changing temperature, because it doesn't depend upon volume.
Mole fraction, X	$X_A = \dfrac{\text{moles of component A}}{\text{total moles}}$	The sum of the mole fractions of all components in the solution will equal 1.

EXERCISE 1: What is the mole fraction of ethanol in an aqueous solution of ethanol that has a molality of 3.88mol/kg?

STRATEGY: Molality is moles of solute (ethanol) per kilogram of solvent (water, because the problem specifies that it is aqueous). We're starting with molality, so assume we have one kilogram of solvent. Thus, we have 3.88 moles of ethanol. We now need to find total moles of solution.

$$\text{Total moles} = \text{moles of ethanol} + \text{moles of water}$$

$$\text{moles of water} = 1000\,\text{g}\,H_2O \times \frac{1\,\text{mol}}{18.015\,\text{g}} = 55.51\,\text{moles; total moles} = 3.88 + 55.51 = 59.39\,\text{moles}$$

$$\text{mol fraction of ethanol} = \frac{\text{moles of ethanol}}{\text{total moles}} = \frac{3.88\,\text{mol}}{59.39\,\text{mol}} = 0.0653$$

SOLUTION: The mole fraction of ethanol in the solution is 0.0653 (3 sig figs from molality data). This number may sound small, but it makes sense when you consider how many moles of water are in 1 kg. Note that we did not need density of the solution for this exercise.

EXERCISE 2: What is the molality (m) of solute particles for a solution that is 2.165 M Na_2SO_4 (aq)? (Given: density of solution = 1.145 g/mL, density of water = 1.00 g/mL)

STRATEGY: We're given molarity (M) and we're asked to find molality. Molarity is moles of solute (Na_2SO_4) per liter of solution (water, because the problem specifies that it is aqueous). We should assume that we are starting with 1L of solution, which will mean we have 2.165

moles of Na_2SO_4 solute. We now need to find how many kilograms of solvent are present in the one liter of solution, so we can divide that into our moles of solute to get molality.

We have one liter of solution, and we know its density, so we can find the mass of solution.

$$1 \text{ L of solution x } \frac{1000 \text{ mL}}{1 \text{ L}} \text{ x } \frac{1.145 \text{ g of solution}}{1 \text{ mL of solution}} = 1145 \text{ g of solution}$$

Notice that we needed density of *solution*, not density of the *solvent*, water. We now know mass of solution and we know that the solution is made up of solvent and solute:

Mass of solution = mass of solvent + mass of solute = mass of water + mass of Na_2SO_4

Calculate mass of solute, Na_2SO_4, and then find mass of solvent.

$$\text{mass of } Na_2SO_4 = 2.165 \text{ moles } Na_2SO_4 \text{ x } \frac{142.04 \text{ g}}{1 \text{ mol}} = 307.5 \text{ g } Na_2SO_4$$

So, mass of water (solvent) = 1145 g – 307.5 g = 838 g

$$C_m \text{ of } Na_2SO_4 = \frac{2.165 \text{ moles } Na_2SO_4}{0.838 \text{ kg water}} = 2.59 \text{ mol } Na_2SO_4 \text{ / kg water}$$

SOLUTION: The molality of the solution is 2.59 mol Na_2SO_4/kg water (3 sig figs because of subtraction step). Note that density of solution had to be used to answer this exercise.

Try It #1: Calculate the mole fraction of sodium chloride in a solution that is 2.69 M NaCl (aq). Given: the density of the solution is 1.142 g/mL and the density of water is 1.00 g/mL.

12.2 DETERMINANTS OF SOLUBILITY

Key Terms:
- **Miscible** – ability of two substances to dissolve in each other in any proportion.
- **Saturated** – describes a solution that contains the maximum amount of solute possible for the amount of solvent present.
- **Alloy** – mixture of two or more metals dissolved in each other.

Key Concepts:
- "Like dissolves like" – two substances that have similar intermolecular forces will typically dissolve in each other. Substances with different types of intermolecular forces typically won't dissolve in each other.
- The rule "like dissolves like" is a general rule; keep in mind that it won't apply to every situation.
- Molecules with dipole moments contain a permanent partial charge separation and can therefore interact with ionic compounds.

- The ability of an ionic compound to dissolve in water depends upon the relative strengths of attractive forces between ions within the crystal lattice versus attractive forces between individual ions and the polar water molecules.
- Covalent network solids (such as diamond) are insoluble in all solvents, because to dissolve them would mean breaking covalent bonds.
- Metals are insoluble in water because they contain delocalized bonding networks that would have to be broken down (in a reaction) to disperse in a solvent.

EXERCISE 3: Both KBr and NH₃ dissolve in water. Describe why these chemicals dissolve in water, and then draw molecular pictures to illustrate the intermolecular forces in aqueous solutions of these chemicals.

STRATEGY AND SOLUTION:

KBr is an ionic compound. Ionic compounds can typically dissolve in solvents that contain a permanent dipole. In water, the KBr dissociates and yields one K^+ for every one Br^-; water dipoles are arranged around ions to maximize attractions.

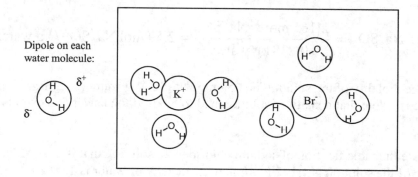

NH₃ is a polar covalent compound that is capable of hydrogen bonding. Water also has these structural features. NH₃ molecules, because they are covalently bonded, will not dissociate into ions like the KBr does. Instead, the NH₃ molecules will align themselves so that they can form hydrogen bonds with water molecules. In the picture here, hydrogen bonds are shown with dotted lines.

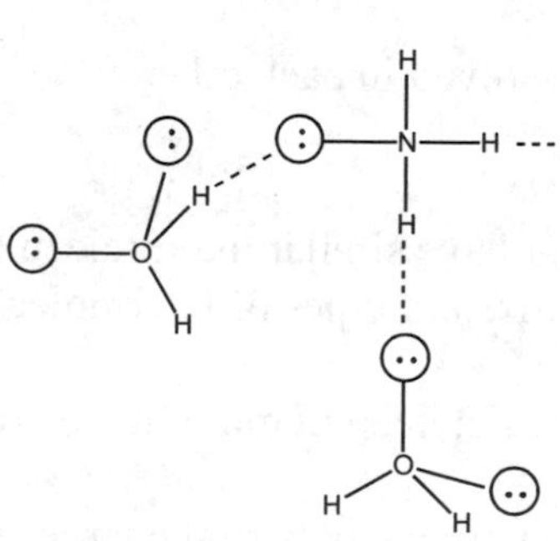

EXERCISE 4: Which pairs of substances should dissolve in each other? a) Sn and Fe (molten); b) H₂S and CCl₄; d) CCl₄ and oil; e) CH₃CH₂OH and NH₃.

STRATEGY: "Like dissolves like." Look to see if the intermolecular forces are similar.

SOLUTION:
a) Both Sn and Fe are metallic solids. When molten, they should dissolve in each other due to their similar structures – both contain delocalized electrons. (Note: in the solid phase, they wouldn't be able to dissolve in each other.)
b) H_2S is a polar covalent molecule, so dipolar forces predominate among H_2S molecules. CCl_4 is non-polar covalent; although each bond is polar, the net result is a zero dipole moment, because the molecule is symmetrical and the dipoles cancel out. CCl_4 molecules are attracted to each other by dispersion forces. H_2S and CCl_4 should not dissolve in each other because their intermolecular forces are different.
c) CCl_4 and oil – these two substances should dissolve in each other. Oil is a carbon-based compound, and we know that oil and water don't mix, which indicates that oil must be a non-polar molecule. CCl_4 is also a non-polar molecule. Both substances have dispersion forces as their dominant intermolecular force, so they should dissolve in each other.
d) CH_3CH_2OH and NH_3 – both of these substances are polar covalent molecules that can hydrogen bond. They should dissolve in each other because they have similar intermolecular forces.

Try It #2: Which pairs of substances should be able to dissolve in each other? a) N_2 and O_2; b) H_2O and CsCl; c) H_2O and Fe.

12.3 CHARACTERISTICS OF AQUEOUS SOLUTIONS

Key Terms:
- **Heat of Solution (ΔH_{soln})** – the energy released or required to dissolve a specific quantity of a given substance in water. (Molar heat of solution refers to 1 mole of substance.)
- **Heat of Dilution** – energy released or required to dilute an existing solution.

Key Concepts:
- Water is an excellent solvent. The fact that it has a strong dipole moment and can hydrogen bond allows it to interact with many chemicals.
- Solubility of solids in water usually increases as temperature increases, but some exceptions do exist.
- Solubility of gases in water decreases as temperature increases. This can be quantitatively assessed using Henry's Law.
- The solute molecules in a saturated solution are constantly moving between the aqueous and pure solute phases. The molecules are said to be in dynamic equilibrium.
- Energy changes occur when a solute is dissolved or diluted in a solvent because a net overall change in intermolecular forces occurs. For example, when salt is dissolved in water, ion-ion attractions are broken (energy is required) and new ion-dipole attractions are formed (energy is released). Figure 12-7 in the text details the various factors that influence the overall change.

Useful Relationship:
- $(P_{gas})_{eq}K_H = [gas(aq)]_{eq}$ This is Henry's Law. It shows the relationship between the amount of gas that can be dissolved in water and the pressure

of the gas at equilibrium. K_H is Henry's Law constant, and it depends upon the identity of the gas. Table 12-2 in the text provides constants for common gases. To use this equation, gases must have pressure units of "atm" and aqueous solutions must have concentration units of "mol/L".

EXERCISE 5: The molar heat of solution of calcium chloride is –81.8 kJ/mol. a) How many grams of $CaCl_2$ must be added to 65.0 mL of water at 19.0°C to get the water's temperature to change by 5.0°C? b) Explain the value of the molar heat of solution for calcium chloride.

STRATEGY: a) We're asked to find the mass of calcium chloride. When the solid calcium chloride comes in contact with the water, it will dissolve, releasing heat energy in the process (The heat of solution of calcium chloride is a negative value, indicating the heat is released when $CaCl_2$ dissolves). As we learned in Chapter 6, the heat released by the $CaCl_2$ will be absorbed by the water. This means that the water temperature will rise. Here is what we know:

- $q_{soln} = -q_{water}$
- $\Delta T_{water} = +5.0°C$, $m_{water} = 65.0$ g (recall that density of water is 1.00 g/mL), $C_{water} = 75.291$ J/molK
- $\Delta H_{soln} = -81.8$ kJ/mol

Working backwards from the goal, we see that we need moles of $CaCl_2$ to find mass of $CaCl_2$. We can find moles of $CaCl_2$ from comparing the q_{soln} to the heat released by one mole of $CaCl_2$.

$$q_{soln} = -q_{water} = -[n_{water}\, C_{water}\, \Delta T_{water}]$$

$$q_{soln} = -65.0\,g\, x\, \frac{1\,mol}{18.015\,g}\, x\, \frac{75.291\,J}{mol\,K}\, x\, 5.0\,K = -1358\,J$$

Set up a ratio to determine moles of $CaCl_2$:

$$\frac{-81.8 x 10^3\,J}{1\,mol} = \frac{-1358\,J}{x\,mole};\ x = 0.01660\ \text{moles of}\ CaCl_2$$

$$0.01660\ \text{moles}\ CaCl_2\ x\ \frac{110.98\,g}{mol} = 1.8\,g\ CaCl_2\ \text{(2 sig figs based on temperature data)}$$

Part b) strategy: Determine the types of intermolecular forces being broken and formed in the process. Consider the fact that the negative value for calcium chloride's ΔH_{soln} indicates that the net change in these forces involves an overall *release* in energy. (Keep in mind that positive value for ΔH_{soln} would indicate the opposite: it would require more energy to break apart attractions than the amount of energy released forming new attractions.)

SOLUTION: a) 1.8 g of $CaCl_2$ must be added to the water. b) The net attractive forces that are formed (ion-dipole between calcium ions/water and chloride ions/water) must be stronger than the net attractive forces being broken (ion-ion between calcium ions and chloride ions, hydrogen bonding between some water molecules) to make the value for molar heat of solution of calcium chloride to be a negative value.

EXERCISE 6: a) Draw a molecular picture to illustrate a saturated solution of oxygen dissolved in water. b) The earth's atmosphere is about 21% oxygen. The concentration of dissolved oxygen in water changes with temperature. Different types of fish are biologically-adapted with varying capabilities to absorb dissolved oxygen in the ocean, thus different types of fish are found in different water climates. Calculate the concentration of oxygen that can dissolve in water at 0°C and at 25 °C.

STRATEGY: A saturated aqueous solution containing a gaseous solute means that the maximum amount of solute (in this case oxygen) is dissolved in a given amount of water. This can be calculated using Henry's Law.

SOLUTION: a) No quantitative data was provided for part "a", so we'll show a qualitative view. Oxygen atoms within molecular oxygen are represented by black dots.

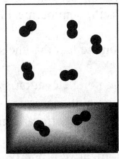

b) Atmospheric pressure at sea level is 1.00 atm. As we learned in Chapter 5, the partial pressure of oxygen will be: (1.00 atm)(0.21) = 0.21 atm. Table 12-2 in the text provides us with Henry's Law constants for oxygen at various temperatures. Use Henry's Law equation to solve the problem, remembering that pressure units are "atm" and concentration units are "mol/L":

$$(P_{gas})_{eq}K_H = [gas(aq)]_{eq}$$

At 0°C, oxygen's $K_H = 2.5 \times 10^{-3}$, so $(2.5 \times 10^{-3})(0.21) = [O_2] = 5.3 \times 10^{-4}$ M at 0°C

At 25°C, oxygen's $K_H = 1.3 \times 10^{-3}$, so $(1.3 \times 10^{-3})(0.21) = [O_2] = 2.7 \times 10^{-4}$ M at 25°C

These numbers seem reasonable because gas solubility decreases with increasing temperature.

Try It #3: Given that the heat of solution of magnesium sulfate, $MgSO_4$, is –91.2 kJ/mol, how many grams of magnesium sulfate must be dissolved in 44.0 mL of water for the water to gain 500.0 J of heat energy?

12.4 COLLIGATIVE PROPERTIES

Key Terms:
- **Colligative property** – property that depends upon the *number* of solute particles in solution but not upon the identity of the solute.
- **Osmosis** – the movement of small (solvent) molecules across a semipermeable membrane.

- **Molality (m)** – concentration unit expressed as "moles of solute per 1 kg of solvent." The units are "molal," expressed as "m."

Key Concepts:
The text covers four colligative properties:

- **Vapor pressure reduction** – the vapor pressure of a solution is lower than the corresponding vapor pressure of the pure solvent.
- **Boiling point elevation** – a solution boils at a higher temperature than the pure solvent from which it was made.
- **Freezing point depression** – a solution freezes at a lower temperature than the pure solvent from which it was made.
- **Osmotic pressure** – pressure that must be applied to a solution in an osmosis apparatus to stop osmosis from occurring. This pressure re-establishes the dynamic equilibrium of solvent molecules moving across the semipermeable membrane between the solution and solvent.
- Colligative properties are all based on the fact that the solute particles disrupt an equilibrium process:
  - vapor-liquid equilibrium (boiling point elevation, vapor pressure),
  - solid-liquid equilibrium (freezing point depression), and
  - movement back and forth across a semipermeable membrane (osmotic pressure).
- Colligative properties are often used for determining the molar mass of a compound. A given mass of solute is analyzed using a colligative property and the number of moles of solute particles can be obtained from the experiment, providing data for a mass to mol ratio (molar mass).

Useful Relationships:

- $vp_{solution} = X_{solvent} vp_{pure\ solvent}$ This is Raoult's Law, and it relates the vapor pressure (vp) of a solution to that of the corresponding pure solvent. X is mole fraction of solute.

- $\Delta T_b = k_b \underline{m}$ This equation relates the boiling point elevation of a solution to the *molal* concentration of solute particles (m). ΔT_b = difference between the boiling point of the solution and the boiling point of the pure solvent and k_b = boiling point elevation constant for the *solvent*.

- $\Delta T_f = - k_f \underline{m}$ This equation relates the freezing point depression of a solution to the *molal* concentration of solute particles (m). ΔT_f = difference between the freezing point of the solution and the freezing point of the pure solvent and k_f = freezing point depression constant for the *solvent*. Note: the "-" is shown here to emphasize that the freezing point is depressed rather than elevated.

- $\Pi = MRT$ This equation relates osmotic pressure (Π) to *Molar* of solute particles (M).

EXERCISE 7: Rank these aqueous solutions from highest to lowest concentration of solute particles.

         A: 0.2 M C_2H_6O B: 0.1 M Na_2CO_3 C: 0.12 M NaCl

STRATEGY: Colligative properties depend only upon the number of solute particles present. Determine and compare the concentration of solute particles.

Solute	Type of compound	# of particles	Particle Concentration (M)
A	Covalent compound (all non-metals)	1 C_2H_6O	0.2 M x 1 = 0.2 M
B	Ionic compound (Group I metal & a non-metal)	3 Na^+, Na^+, and CO_3^{2-}	0.1 M x 3 = 0.3 M
C	Ionic compound (Group I metal & a non-metal)	2 Na^+ and Cl^-	0.12 M x 2 = 0.24 M

SOLUTION: Solution B has the highest concentration of solute particles, Solution C has the next highest, and Solution A has the lowest concentration of solute particles.

EXERCISE 8: Which solution in Exercise 16 will have   a) the highest boiling point? b) the lowest freezing point?

STRATEGY AND SOLUTION: All solutions are aqueous, so the only difference between the solutions is the concentration of solute particles. The solution with the greatest concentration of solute particles will have the greatest boiling point elevation and the greatest freezing point depression relative to pure water. Solution B, therefore, will have the highest boiling point *and* the lowest freezing point.

EXERCISE 9: An aqueous solution is 0.222 m glucose. What is the boiling point of this solution? What is its freezing point? ($k_b = 0.512$ °C/m and $k_f = 1.86$ °C/m for water)

STRATEGY: Calculate the change in boiling point temperature using: $\Delta T_b = k_b \underline{m}$

Glucose is a covalent compound, so it will not break into smaller particles when dissolved in water. The molality of solute particles is therefore 0.222 m.

$\Delta T_b = k_b \underline{m} = (0.512$ °C/m$)(0.222$ m$) = 0.114$°C; $\Delta T_f = k_f \underline{m} = (1.86$ °C/m$)(0.222$ m$) = 0.413$°C

SOLUTION: a) The boiling point of the solution will be 100°C + 0.114°C = 100.114°C. The freezing point will be 0°C - 0.413°C = -0.413°C. The values seem reasonable because the boiling point is higher than pure water and the freezing point is lower than pure water.

Try It #4: Repeat the calculations in the previous exercise for 0.222 m $MgSO_4$ in place of glucose.

EXERCISE 10: Consider a solution made from two chemicals, X and Y. At 65°C, the vapor pressure of pure Y is 450.0 torr. If you have a 38.0 mol% solution of X in Y and the total pressure of the system is 520.0 torr, then what is the vapor pressure of pure X at this temperature?

STRATEGY: We're asked to find the vapor pressure of pure X at this temperature. If we visualize this problem, we notice that the vapor pressure of the solution is actually greater than the vapor pressure of component Y. This can only be true if component X is also volatile, so that both components X and Y are contributing to the total pressure present. As we learned in Chapter 5, Dalton's Law says that the total pressure of the system equals the sum of the partial pressures present. Therefore,

$$P_T = P_X + P_Y$$

According to Raoult's Law, the vapor pressure of a given chemical in a solution is found by:

$$P_X = X_X P°_X$$

So, for this problem we have:

$$P_T = P_X + P_Y = X_X P°_X + X_Y P°_Y$$

We know that a 38.0 mol% of X means a mole fraction of .38 for X. Since Y is the only other component present in the mixture, the mole fraction of Y must be 1 – 0.38 or 0.62. Now, we can simply plug all of our known values in and solve for $P°_X$:

$$P°_X = \frac{[P_T - X_Y P°_Y]}{X_X} = \frac{[520.0 \text{ torr} - (0.620)(450.0 \text{ torr})]}{0.380} = 634 \text{ torr}$$

SOLUTION: The vapor pressure of pure X at this temperature is 634 torr (3 sig figs from data). This answer makes sense because X's vapor pressure would have to be large in order to increase the total pressure of the system by such a great amount.

EXERCISE 11: Calculate the osmotic pressure of a solution at 25°C that contains 4.50 g of $CaCl_2$ in 500.00 mL of solution.

STRATEGY: Determine the *molar* concentration of solute particles present and calculate osmotic pressure using $\Pi = MRT$.

$$\text{Molarity of solute particles} = 4.50 \text{ g } CaCl_2 \times \frac{1 \text{ mol}}{110.984 \text{ g}} \times \frac{3 \text{ mol solute particles}}{1 \text{ mol } CaCl_2} \times \frac{1}{0.500 \text{ L}} = 0.2433 \text{ M}$$

$$\Pi = MRT = \frac{0.2433 \text{ mol of solute particles}}{\text{L of solution}} \times \frac{0.08206 \text{ L atm}}{\text{mol K}} \times 298 \text{ K} = 5.95 \text{ atm}$$

SOLUTION: The osmotic pressure of this solution is 5.95 atm (3 sig figs from data – the temperature data used was in Kelvins).

EXERCISE 12: Consider two beakers. Beaker A contains 100 mL of water and Beaker B contains 100 mL of aqueous 0.25 M $MgBr_2$. We have to make the boiling points of the contents of each beaker be the same by adding table sugar ($C_{12}H_{22}O_{11}$) to one of the containers. a) To which beaker should the sugar be added?   b) Approximately how much sugar should be added?

STRATEGY AND SOLUTION:
a) The boiling point of a solution will be higher than the boiling point of the pure solvent used to make the solution, so 0.25 M $MgBr_2$ (aq) will have a higher boiling point than pure water. The sugar should therefore be added to Beaker A.

STRATEGY:
b) Consider the information provided.
- The boiling point elevation is based upon the *molal* concentration of solute particles. Beaker B contains 3 x 0.25 moles (or 0.75 moles) of solute particles in 1 L of solution because the solute breaks into three pieces in solution (Mg^{2+}, Br^- and Br^-).
- We have 100 mL of dilute solution, which roughly equals 0.1 kg of water, so we have 0.075 moles of solute particles in Beaker B.
- Sugar is covalently bonded, so it won't break into ions in solution. Therefore the moles of solute particles equal the moles of sugar.

SOLUTION: We need to add 0.075 mol of sugar to Beaker A.

EXERCISE 13: A solution is prepared by dissolving 3.90 g of an unknown in 50.0 g of water. The molality was determined to be 0.672 $\underline{m}$ by freezing point depression experiments. What is the molar mass of this unknown? Assume the compound does not form ions.

STRATEGY: The molar mass of a compound is its mass to mole ratio: "grams/mol." We need to find grams of unknown and moles of unknown and take a ratio of these numbers.

We've been given mass of unknown. We need to find moles of unknown from molality (0.672 moles of solute per 1 kg of solvent) and 50.0 g of water (solvent).

$$\frac{0.672 \text{ moles of solute}}{1 \text{ kg of solvent}} \text{ x } 50.00\text{x}10^{-3} \text{ kg of solvent} = 0.0336 \text{ moles of solute}$$

SOLUTION:

$$\text{MM of unknown} = \frac{3.90 \text{ g of unknown}}{0.0336 \text{ moles of unknown}} = 116 \text{ g/mol}$$

12.5 BETWEEN SOLUTIONS AND MIXTURES

Key Terms:
- **Colloidial suspension** – a mixture of two or more substances that has properties intermediate between a solution and a heterogeneous mixture. The phase of each substance in the suspension will dictate how the suspension is named. Table 12-3 in the text provides the names of the different types of colloidial suspensions.
- **Hydrophilic** – "water loving;" term used to describe a polar molecule or polyatomic ion (or region of a molecule).

- **Hydrophobic** – "water fearing;" term used to describe a non-polar molecule (or region of a molecule).
- **Dual nature molecule** – molecule that can interact with both polar and non-polar molecules because its structure contains a hydrophilic "head" and a hydrophobic "tail."

Schematic drawing of a dual-nature molecule:

tail   head

- **Surfactant** – a type of dual-nature molecule specifically designed to decrease the surface tension of water (for example, soap or shampoo.)

Key Concepts:
- Dual-nature molecules will arrange themselves in one of three ways in water: micelles, vesicles, or mono-layers. Schematic pictures of these are shown in Figure 12-17 of the text.
  - **Micelles** – spherical aggregates with non-polar tails facing inward.
  - **Vesicles** – double-layered aggregates in which the hydrocarbon tails are clustered.
  - **Mono-layers** – single layer on the water's surface, with hydrophilic heads facing the water.
- Soaps and detergents are dual-nature molecules that lift non-polar dirt and oil from clothing and skin by forming micelles around the dirt/oil.
- Cell membranes are dual-nature molecules arranged in vesicles.

EXERCISE 14: Many foods we eat are colloidial suspensions. Determine what type of colloidial suspension each of these is: a) mayonnaise, which is made by pouring vegetable oil into liquid egg yolks in a blender; b) Taffy: when the candy is prepared it is quite stiff. It is made edible by "pulling" – the solid candy is stretched over and over again to incorporate air into it until it is soft enough to eat. c) Pudding, which is made by creating a suspension of milk and various dissolved flavors in insoluble cornstarch (a solid).

STRATEGY: Refer to Table 12-3 in the text for a list of how colloidial suspensions are classified. (You may need to memorize the names of the types.)

SOLUTION: a) Mayonnaise is an emulsion because it is contains a liquid suspended in a liquid. b) Taffy is a foam because it contains a gas suspended in a solid. c) Pudding is a gel because it contains liquids suspended in a solid.

EXERCISE 15: Some of the compounds shown are dual-nature molecules and some are not. For the compounds that are dual-nature molecules, identify which end is hydrophobic and which end is hydrophilic.

a)

c)

b)

d)

SOLUTION: Molecules A and D are dual-nature molecules because they contain a long non-polar carbon chain with an ionic end to it. Molecule B has polar groups evenly distributed along the molecule, so it is not a dual-nature molecule. Molecule C is non-polar with a hydrogen-bonding site, but the molecule is too compact to be able to act in a dual-nature fashion.

a)

Na⁺

hydrophilic end
(or "head")

hydrophobic end
(or "tail")

d)

NH₄⁺

Cl⁻

hydrophobic end
(or "tail")

hydrophilic end
(or "head")

EXERCISE 16: Draw a schematic drawing of surfactant molecules "dissolving" a droplet of water in a container of oil.

STRATEGY: The water is polar and the oil is non-polar, so the surfactant molecules will show the polar "heads" facing towards the water droplet. The non-polar "tails" will face outward towards the oil. This will occur in a spherical shape around the water droplet; the drawing shows only two dimensions.

SOLUTION:

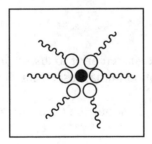

Chapter 12 Self-Test

You may use a periodic table and a calculator for this self-test.

1. The vapor pressure of toluene at some temperature is 20.0 mmHg. Consider a solution made by dissolving 22.65 g of naphthalene, $C_{10}H_8$, in 75.0 g of toluene, C_7H_8.
   a) What is the vapor pressure of this solution? Assume that naphthalene's vapor pressure is negligible.
   b) Calculate the concentration of naphthalene in units of molality.

2. Given this information: NH_4^+ (aq) + NO_3^- (aq) → NH_4NO_3 (s)   ΔH = -21.1 kJ/mol
   a) Calculate q_{water} when 45.0 g of solid ammonium nitrate dissolves in 325.0 mL of water.
   b) Explain the value of $ΔH_{soln}$ of ammonium nitrate in terms of changes in intermolecular forces.

3. "Adding salt to boiling water makes pasta cook faster":…myth or fact? To answer this question, determine the boiling point of salted water according to the typical proportions used in cooking: 1.00 teaspoon of NaCl per 2.00 quarts of water. [Note: 1 teaspoon of salt weighs 3.33 g; 1 quart ≅ 1 L.] k_b = 0.512°C/$\underline{m}$ for water & k_f = 1.86°C/$\underline{m}$ for water.

4. Iron (III) sulfate is the chemical present in dietary iron supplements. It is not very soluble in water. Explain, from an intermolecular forces perspective, why iron (III) sulfate does not easily dissolve in water.

5. Both silver and sodium chloride arrange in a cubic close-packing structure. Silver can be pounded into different shapes to make jewelry, but sodium chloride shatters when it is hit with a hammer. Why?

6. Canned whipping cream must be shaken prior to dispensing it, or it will be runny when it exits the container. What type of colloidial suspension is whipping cream?

7. What is a dual-nature molecule? Describe the necessary structural features of a dual-nature molecule.

Answers to Try Its

1. 0.0469 (Assume 1L of solution. Use density of solution to find mass total, then find mass of water, then moles of water, then add moles water to moles of NaCl for total moles.)
2. a) They should be able to dissolve in each other because they have the same type of intermolecular forces (dispersion forces). b) The CsCl might be able to dissolve in water. It will if the ion-dipole attractions between the ions and water are strong enough to compete with the ion-ion attractions in the crystal lattice. (CsCl should be able to dissolve in water, because salts made from Group I metals and halogens are typically water-soluble.) c) Water is a polar covalent molecule that can hydrogen bond. Iron is a metal. The two will not dissolve in each other.
3. 0.660 g (Notice that you didn't need the volume of water to solve this problem.)
4. Boiling point = 100.227°C; freezing point = -0.826°C

Answers to Self-Test:

1. a) 16.4 mm Hg; b) 2.36 mol/kg
2. a) −11.9 kJ (Note: did you notice that the process being analyzed is the reverse of the reaction shown? ΔH_{soln} of ammonium nitrate is therefore 21.1 kJ/mol. Also, note that you don't even need the volume of water to answer this question. Just multiply ΔH_{soln} of ammonium nitrate by the number of moles of ammonium nitrate provided. The negative of this value is q for water for this process.)

   b) ΔH_{soln} of ammonium nitrate is positive, so energy is required overall for this process to occur. This means that the collective ion-ion attractions between ammonium ions and nitrate ions to start along with the hydrogen bonds in water that are broken during the process must be stronger than the ion-dipole interactions between ammonium ions/water and nitrate ions/water that are established.
3. The solution's boiling point temperature will be 100.029°C, which is hardly more than the boiling point of pure water. This is a myth!
4. The ion-ion attractions between Fe^{3+} and SO_4^{2-} must be significantly stronger than the ion-dipole attractions of the ions with water. (Recall the effect of ion charge in Coulomb's Law.)
5. The packing describes how the atoms or ions are arranged within the crystalline solid. Hitting a metallic substance, like Ag, with a hammer will shift the *atoms* inside the crystal. Hitting an ionic substance, like NaCl, will shift *ions*, which may result in a series of nearest neighbors having the same charge. Repulsive forces would be exerted all along the line of the impact, and the crystal would break apart.
6. Whipping cream is a foam because it contains a gas (air, or other gaseous chemicals inside the container) suspended in a liquid (cream).
7. A dual-nature molecule is one that is capable of interacting with both polar and non-polar molecules. It can do this because its structure contains a long carbon backbone with one end of its chain being non-polar and the other end of its chain being polar or ionic. The non-polar "tail" is attracted to other non-polar molecules and may be called hydrophobic, water fearing, or oil-soluble. The polar or ionic "head" is attracted to polar molecules and can be referred to as hydrophilic, water loving, or water-soluble. Soaps, detergents, and cell walls all contain dual-nature molecules.

Chapter 13: Macromolecules

Learning Objectives

In this chapter, you will learn how to:
- Identify functional groups in molecules.
- Link reactant molecules together by reacting functional groups.
- Identify monomers for free radical and condensation polymerization processes.
- Draw structures of polymers from their monomers.
- Recognize and draw structures of three types of biological polymers: carbohydrates, nucleic acids, and proteins.

Practical Aspects

This chapter is devoted to the study of polymers, huge molecules composed of tens to thousands of repeating monomer units. Nature makes a wide variety of polymers. For example, DNA, starch, fiber, enzymes, hair, skin, fingernails, cotton, wool, and silk, are all naturally-occurring polymers. Some examples of synthetic polymers are: polyethylene, PVC, PET, nylon, rayon, Dacron, Teflon, and Melmac.

In nature and in the chemistry lab, the structure of the monomer and the overall structure of the resulting polymer will determine for what purpose the polymer is used. In the lab, chemists use this concept to design polymers with specific needs in mind. A given polymer may need to be inert to air or water or temperature. Its structure may need to be either rigid or soft, sticky or slippery. In this chapter, you will learn how polymers are constructed, and how their overall shapes determine their properties.

13.1 STARTING MATERIALS FOR POLYMERS

Key Terms:
- **Polymer** – giant organic molecule consisting of many repeating parts (monomers) that are covalently bonded in a long chain.
- **Monomer** – A small molecule that is capable of undergoing polymerization.
- **Functional group** – a reactive portion of an organic molecule that participates in a characteristic set of reactions.
- **R-group (-R)** –the portion of a molecule that is not chemically significant in terms of the reaction taking place (aka: the **R**est of the molecule).

Key Concepts:
- Functional groups are the locations in molecules where reactions take place.
- A molecule can be broken down into the functional group and the R group. For example, benzoic acid contains a carboxylic acid functional group. The entire structure can be drawn (left structure) or we can focus on just the carboxylic acid group and designate the rest of the compound as "R" (right structure):

Quick Reference of Functional Groups

Table 13-1 in the text provides a detailed list of functional groups for polymerization. The groups shown in the table below, along with the linking groups shown later in this section (ester, amide and ether), make up the functional groups involved in most of the polymerizations you will encounter in this chapter.

Alkene	Alcohol	Carboxylic acid	Amine	Phosphate
C=C *this is shown as a functional group because the π bond is a reactive site.	R–OH *covalently-bonded to a carbon atom (*no ionic bonding*)	$\overset{O}{\underset{R}{\overset{\|}{C}}}\text{—OH}$	R—NH₂ R—NH $\overset{}{\underset{R}{R—N—R}}$ *N with 3 single bonds to R and/or H.	$\overset{O}{\underset{OH}{R—O—P—OH}}$

These functional groups are also useful to know:

Aldehyde	Thiol	Ketone
$\overset{O}{\underset{R}{\overset{\|}{C}}}\text{—H}$	R –SH	$R\overset{O}{\overset{\|}{C}}R$

Helpful Hints

- One visual analogy for polymers is to think of individual boxcars of a train as monomers, and the entire train is the polymer that they form when they link together.
- This structure is called a benzene ring:

  It is a particularly stable structure, so it typically won't contribute one of its double bonds in making a polymer.

EXERCISE 1: Identify the functional group(s) in each compound.

a) Acetic acid – the main solute in vinegar

b) Cysteine: an amino acid present in hair.

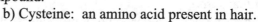

c) Adrenalin: neurotransmitter responsible for fight-or-flight response.

d) Testosterone: hormone responsible for male sexual characteristics.

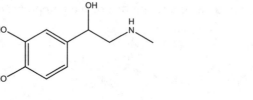

STRATEGY: Use the tables above as a guide. Use guidelines from Chapter 3 to interpret line structures.

SOLUTION:

a)

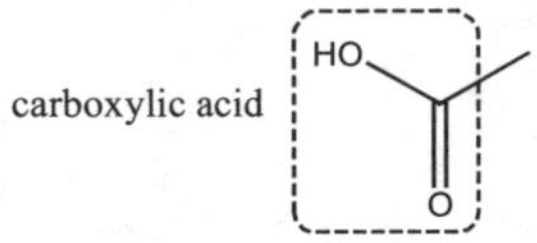

b)

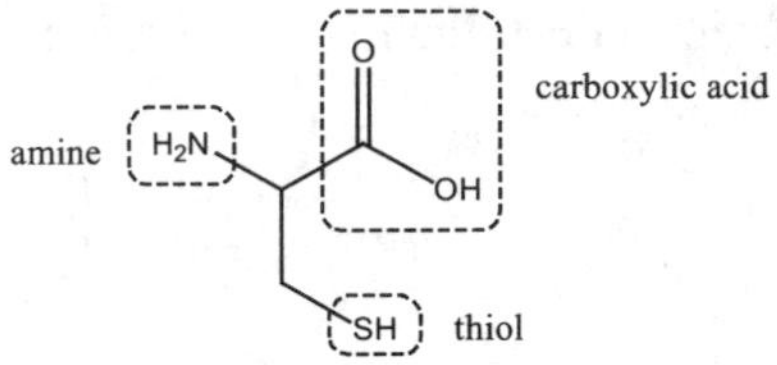

c)

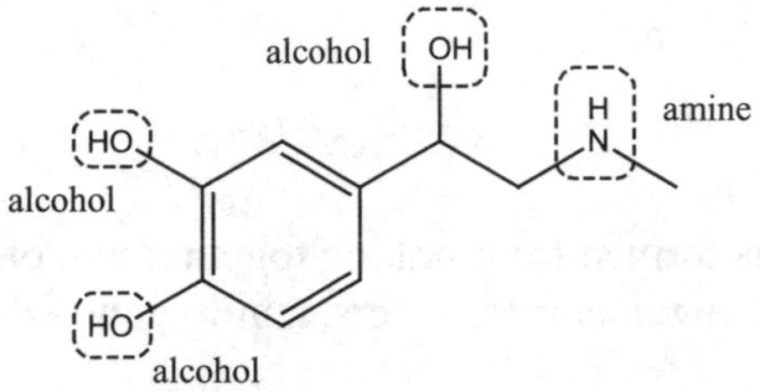

d)

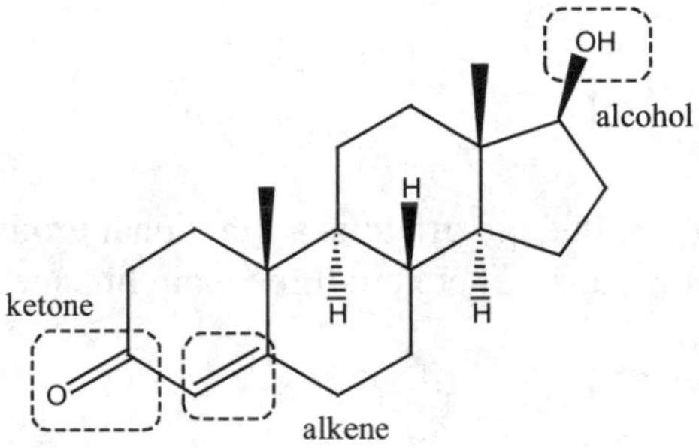

Linkage Groups

Key Term:

- **Condensation reaction** – reaction in which two functional groups react to form a new functional group. A small molecule (often water) is produced as a side product. There are three main condensation reactions to remember for this chapter:

    1. Carboxylic acid + alcohol → ester
    2. Carboxylic acid + amine → amide
    3. Alcohol + alcohol → ether

Quick Reference of Important Linkage Functional Groups

Ester	Amide	Ether
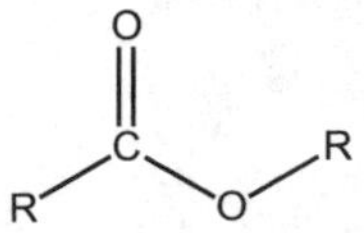	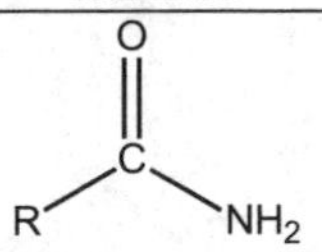 *the H's on N can be R groups.*	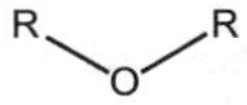

Key Concepts:

- The new functional group formed in a condensation reaction is sometimes called a linkage group, because it links the two reactant molecules together.
- Two other useful condensation reactions to know are:
    - Phosphoric acid + phosphoric acid form P-O-P linkages.
    - Phosphoric acid or phosphate + alcohol form P-O-C linkages.

EXERCISE 2: a) Identify the functional group that is considered a linkage group. b) What two functional groups had to react to form this linkage group? c) Determine the structures of the reactant molecules that would have to be used to form these compounds.

i. Acetaminophen – an over-the-counter analgesic (pain-killer).

ii. Benzyl acetate – one of the fragrant chemicals in jasmine.

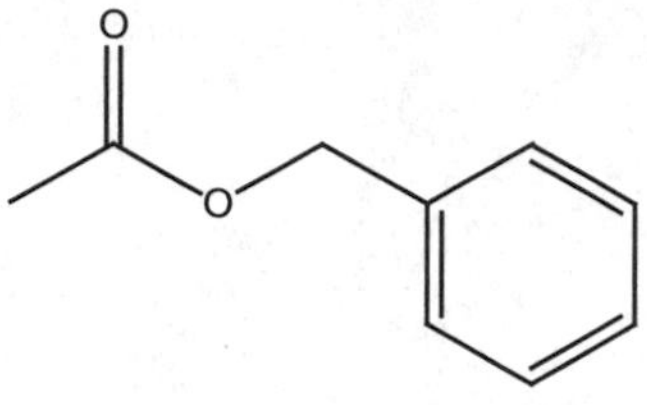

iii. Diethyl ether – organic solvent and anesthetic.

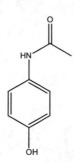

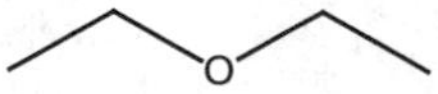

STRATEGY: A linkage group is a functional group that is formed from linking together two other functional groups. There are three main linkage groups to remember: esters, amides, and ethers.

SOLUTION:

i)

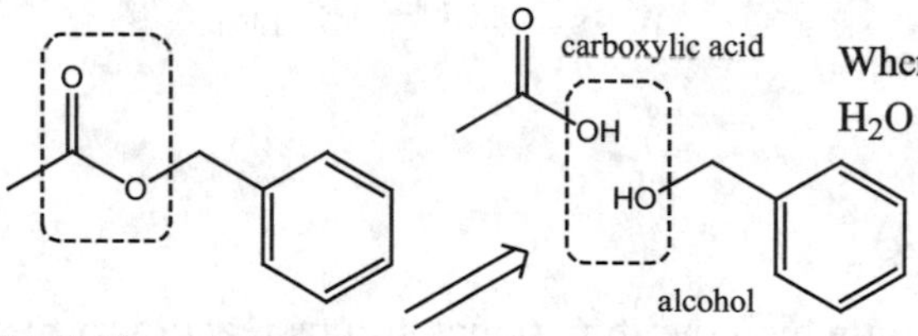

carboxylic acid

When the ester group forms, H_2O is formed as a side-product.

This is the ester group. It is made from reacting a carboxylic acid and an alcohol.

alcohol

ii)

amine

carboxylic acid

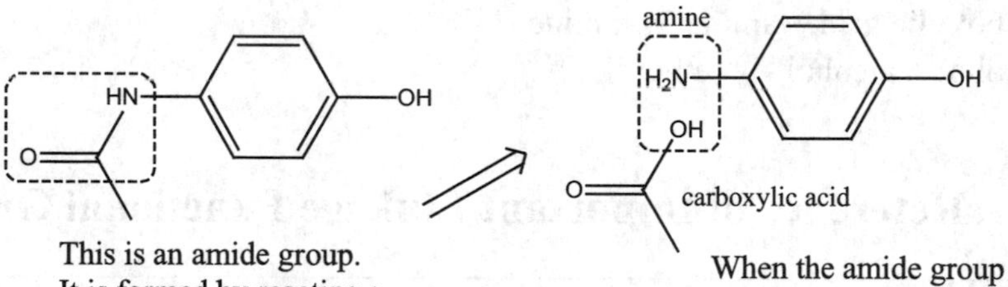

This is an amide group. It is formed by reacting a carboxylic acid with an amine.

When the amide group forms, H_2O is formed as a side-product.

iii)

This is the ether group. It can be made by reacting two alcohols with each other.

When the ether group forms, H_2O is formed as a side product.

EXERCISE 3: Draw the structure of the compound that will form when the two starting materials react. Name the new functional group.

STRATEGY: Identify the functional groups in the reactant molecules, and line them up, if necessary, so that the two reacting functional groups are near each other. Form the new functional group, and break off a water molecule.

SOLUTION: Here are the structures. The functional groups are highlighted with dashed lines.

These groups link together to form an ester, and release H_2O as a side product.

Interesting side note: Both reactant molecules in Exercise 3 have unpleasant odors – one reactant is acetic acid, which is in vinegar. The ester product is called isopentyl acetate, which is banana oil. Low molecular-weight (molar mass) esters often have fruity or floral aromas, and are therefore quite interesting molecules to perfume chemists.

Try It #1: Identify the functional groups in capsaicin, the compound that makes serranos and other peppers hot.

13.2   FREE RADICAL POLYMERIZATION

Key Terms:
- **Free radical** – substance that contains one unpaired electron in its Lewis structure. Free radicals are very unstable and highly reactive.
- **Free radical initiator (R-O-O-R)** – compound with a very weak bond that will easily break upon heating. A tiny amount of this material in the reaction mixture initiates the polymerization process.
- **Cross-linking** – covalent bond linkages between polymer chains.
- **Co-polymer** – polymer made from two or more different monomers.

Key Concepts:

- A compound that contains a carbon-carbon double (or triple) bond can form a polymer via free radical polymerization.
- There are typically at least 250 repeating monomer units in a free radical polymer.
- The free radical polymerization process occurs in three steps:

Step	General Process (Only C atoms on the alkene are shown, in order to focus on the carbon backbone of the polymer.)
- **Initiation** – a free radical is generated and reacts with one monomer unit.	RO—OR $\longrightarrow$ RO··OR RO· C⚌C $\longrightarrow$ RO—C—C·
- **Propagation** –the polymer grows, step by step, as additional monomers react with the free radical.	RO—C—C· C⚌C $\longrightarrow$ RO—C—C—C—C·
- **Termination** – two free radicals combine to form a stable compound.	

- *The structure of the monomer will affect the properties of the resulting polymer.* For example:

This monomer:	This monomer:	This monomer:
══		
makes polyethylene, which is used to make plastic bags and bottles.	makes polytetrafluroethylene, (aka Teflon), which is used as a non-stick coating for pans.	makes polyvinylchloride (aka PVC), which is used to make plastic pipes and credit cards.

- A conjugated monomer can use all of its conjugated bonds in the polymerization process. This is illustrated in Exercise 4.

EXERCISE 4: Draw the structure of the polymer that will form from each monomer. Show at least three monomer units.

a)   b)

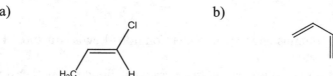

STRATEGY: Draw three monomers in a row, so that their C=C bonds are aligned. We'll use arrows to show the movement of the electrons during the polymer formation. We can either show the carbon atoms as "C"s, or leave them as line structures.

SOLUTION:

a)

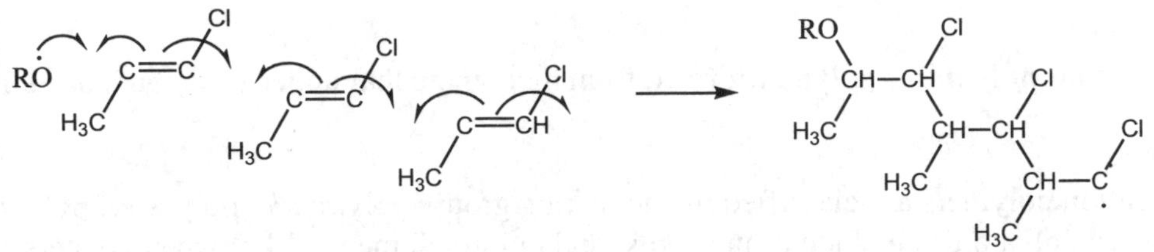

b)

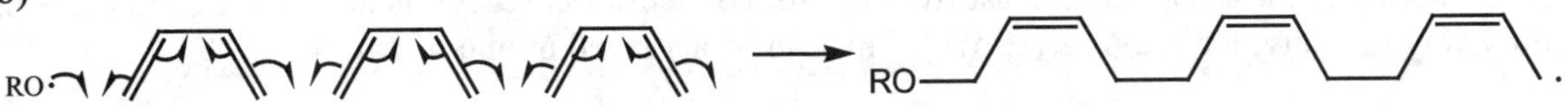

Helpful Hint
- If the polymer chain contains a double bond, then the monomer must have contained a conjugated π system, just like in Exercise 4b. (Or, it could have contained a triple bond.)

Try It #2: Draw the structure of three repeating monomer units in the free radical polymerization of:

EXERCISE 5: Here is a portion of a co-polymer chain. Identify the monomers used to construct this polymer. Determine the proportions of each monomer used.

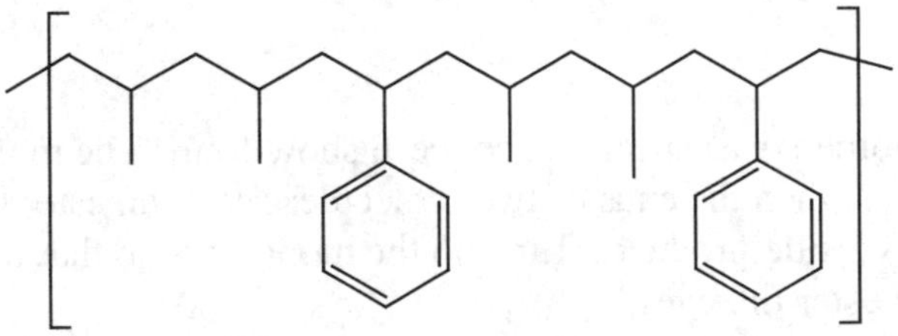

STRATEGY: The polymer chain contains all single bonds, so all of the monomers contained only one double bond that participated in the polymerization reaction. "Reading" the structure from left to right, divide the polymer into two-carbon bundles. Construct the monomers from these bundles:

Polymer, segmented:

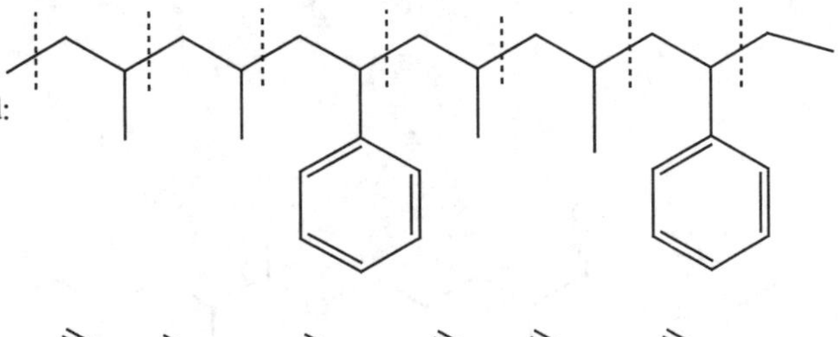

Monomers:

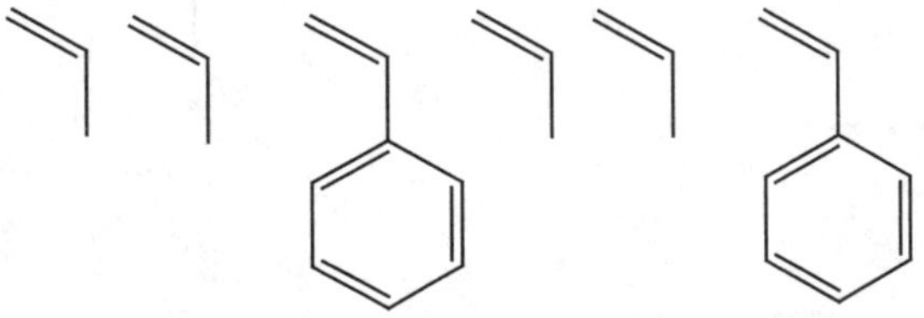

SOLUTION:

The polymer was made from:

Two parts propylene,

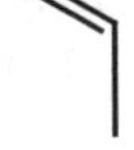

and one part styrene,

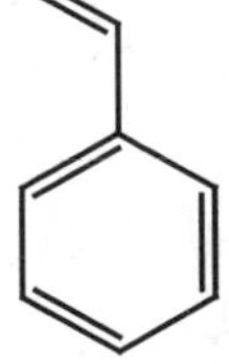

13.3 CONDENSATION POLYMERIZATION

Key Term:
- **Condensation polymer** – polymer formed from monomers that underwent condensation reactions.

Key Concepts:
- Condensation polymers are classified by the linking group: poly*amide*, poly*ester*, poly*ether*. Some well-known polyamides include protein, silk, and nylon. Some well-known polyesters are PET, polyester, Dacron, and Mylar. Carbohydrates are polyethers.
- Each monomer must contain at least two sites for condensation reactions to take place. (This is the only way the polymer chain can grow.) This can be achieved in one of two ways:

Use one monomer that can react with itself:	Use two *different* monomers that can react with each other:

*This example shows carboxylic acids and alcohols which can form esters; the same concept applies to the formation of amides and ethers as well. Figure 13-6 in the text provides a general conceptual scheme.

EXERCISE 6: Draw the structure of the polymer that would form from:

STRATEGY: No number of monomers was specified, so we'll show four. The monomer contains an alcohol and a carboxylic acid, so when the ends of two molecules react, an ester linkage will form (along with a water molecule as a side-product). Line up the monomers so that their functional groups can react, then form the ester linkages.

SOLUTION:

EXERCISE 7: What monomer(s) was(were) used to construct this polymer?

STRATEGY: Determine the type of linkage group in the polymer chain, divide the polymer into monomer segments, and break the polymer down into monomers.

The linkage groups are all amides, which are made from reacting amines with carboxylic acids.

SOLUTION: There are two monomers. They are:

and

Try It #3 – Construct a polymer from these monomers. Show two of each monomer unit.

13.4   TYPES OF POLYMERS

Polymers can be classified into three categories: plastics, fibers, and elastomers. Examples from these respective categories are: plastic bottles, nylon, and rubber tires.

Key Terms:
- **Plastic** – type of polymer that hardens when cooled or when solvent is evaporated from it. This allows it to be molded into definite shapes, like blocks or sheets. There are two categories of plastics:
  - **Thermoplastic** – plastic that melts or distorts when heated, due to weak interactions between polymer chains.
  - **Thermosetting** – plastic that maintains its shape when heated, due to its highly-cross-linked polymer structure.

- **Plasticizer** – chemical added to a plastic that will increase its flexibility and decrease brittleness.
- **Fiber** –thin thread of a polymer that is made by forcing a thermoplastic polymer through tiny pores.
- **Elastomers** – polymer that is flexible, and can be stretched without breaking. Elastomers are amorphous solids in which the polymer chains are randomly tangled together.

Key Concepts:
- Factors that influence a given polymer's properties:
  - **Rigidity** – the greater the interaction between polymer chains, the more rigid the structure. Interactions ranked strongest to weakest are: crosslinking > hydrogen bonding > dispersion forces.
  - **Density** –polymer chains that can pack together tightly will be denser than those that cannot. The size and number of side-groups and the number of kinks in each polymer chain will influence how tightly a given set of polymer chains can pack.
- The overall rigidity of a given polymer is a balance between the desired strength and flexibility of the polymer.
- Polymers do not easily degrade in nature. This makes them durable, but a potential waste problem.

Helpful Hint
- See Figure 13-9 in the text for structures of high-density vs low-density polyethylene to illustrate how polymer packing will influence its physical properties.

EXERCISE 8: Categorize these polymers: a) astro-turf; b) weather-stripping; c) shower curtains; d) Lycra; e) trash cans; f) o-rings; g) lipstick containers.

STRATEGY: The three categories of polymers are: plastics, fibers, and elastomers.

SOLUTION:
Plastics: shower curtains, trash cans, lipstick containers.
Fibers: astro-turf, Lycra.
Elastomers: weather-stripping, Lycra, and o-rings. (Lycra is a fiber and an elastomer.)

EXERCISE 9: Here are the structures of the two isomers of polyisoprene. Predict the physical properties of these two polymers. One is natural rubber, which is used to make tires, and the other is gutta-percha, which is used to make golf-ball casings.

STRATEGY: The physical properties of the polymers depend upon how closely their chains can pack together, and how much interaction exists between the chains.

SOLUTION: The top structure is quite linear, so several polymer chains should be able to pack together tightly. The bottom structure has lots of kinks in it, so its polymer chains will pack less tightly and tangle more than the top structure's will. The top structure is gutta-percha and the bottom structure is natural rubber.

Try It #4: a) Predict the properties of a polymer made of polyethylene chains that were crosslinked to each other. b) How would this polymer be classified?

13.5   CARBOHYDRATES

Key Terms:
- **Carbohydrate** – class of chemicals that contain the general structure "$C_x(H_2O)_y$." Most carbohydrates end with the suffix "-ose." Categories of carbohydrates include:
  - **Monosaccharide** – an individual carbohydrate unit. Most monosaccharides contain five or six carbon atoms. (Example: glucose.)
  - **Oligosaccharide** – small chain of 2 to 10 monosaccharide units. (Example: sucrose.)
  - **Polysaccharide** – a polymer of monosaccharides. (Example: starch.)

Monosaccharides

Key Concepts:
- Monosaccharides exist in many structural forms, but their ring form is the one used to build polysaccharides. The general ring structure of a monosaccharide is a five- or six-membered ring of carbons that contains an ether linkage. α-Glucose is shown below, as an illustration:

All atoms that form the ring have tetrahedral geometry, so the ring is puckered, like a lawn chair:

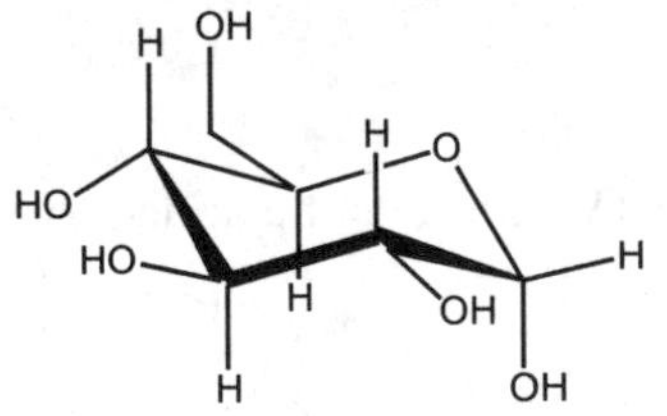

The 3-D structure on the left is often translated to a 2-D "flat" form:

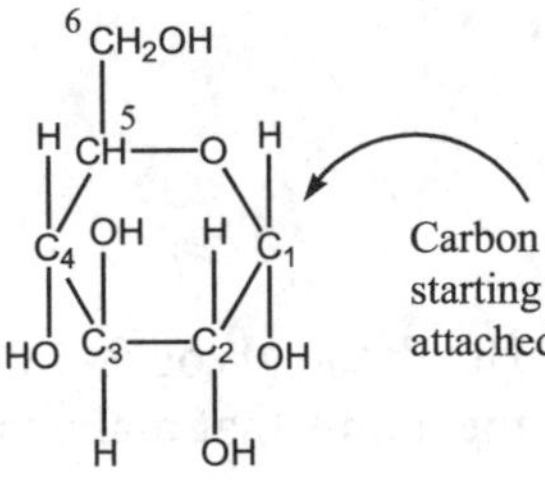

Carbon atoms are numbered, starting with the "HCOH" attached to the ether linkage.

- β-glucose differs from α-glucose in the positioning of the "H" and the "OH" on C_1. This tiny difference has a huge impact on the polymers that they form.

EXERCISE 10: Re-draw the structure of α-allose in the flat form. Label all of the carbon atoms.

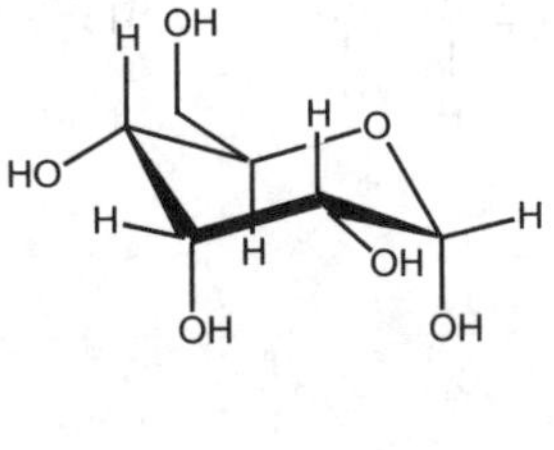

α-allose

STRATEGY AND SOLUTION:
Use the scheme shown above for translating sugar structures to two dimensions and numbering them.

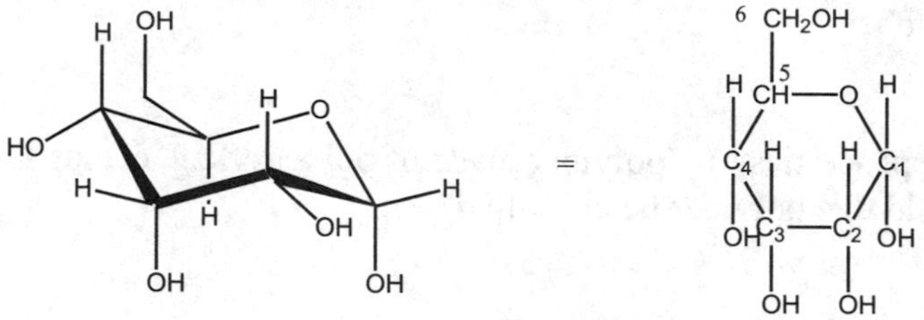

Disaccharides

Key Term:
• **Glycosidic bond** – ether linkage between two sugar units.

Key Concept:
• In all naturally-occurring glycosidic bonds, at least one sugar uses the hydroxyl group on a carbon atom next to a ring oxygen.

EXERCISE 11: Here is the structure of lactose, showing flattened views of the rings. Draw the structures of the two monosaccharides that make up this structure.

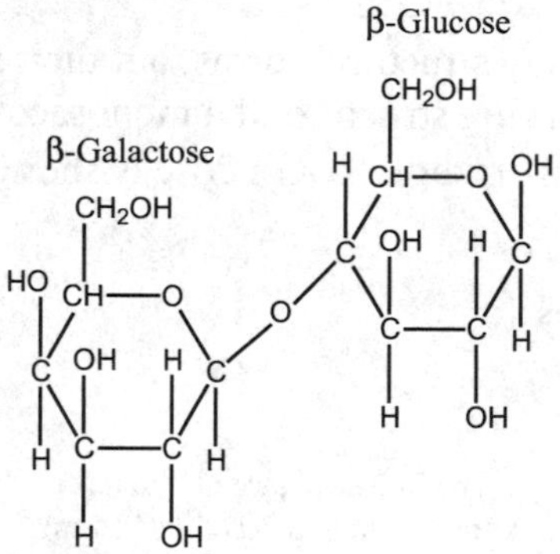

STRATEGY: Break the glyocosidic bond between the two sugar units. (Water must be added to account for the atoms that make up the alcohols.)

SOLUTION:

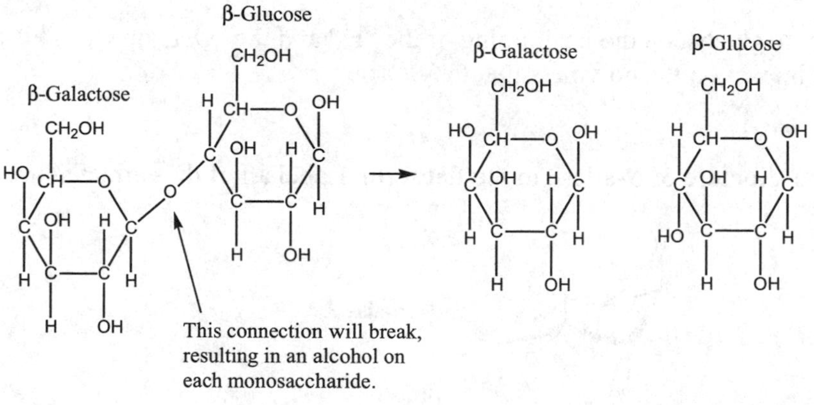

This connection will break, resulting in an alcohol on each monosaccharide.

Polysaccharides

Key Terms:

- **Starches** – polymers of *α-glucose*, ranging from about 200-1000 glucose monomer units. Some have branches of glucose molecules extending from the polymer chain. The linkages of the monomer units cause starch molecules to form coils. Some important starches are:
    - **Amylose and Amylopectin** – produced by plants during photosynthesis.
    - **Glycogen** – produced by animals in the liver.
- **Cellulose** – straight-chain polymer of *β-glucose* (2000-3000 monomer units). The cellulose polymer forms a long ribbon whose hydroxyl groups can hydrogen-bond to neighboring celluloses, resulting in a sheet-like structure. See Figure 13-21 in the text for an illustration.

Key Concepts:

- Starches store sugar until it is needed for energy production.
- The overall polymer structures of starch and cellulose are quite different from one another. One tiny variation in the monomer structure results in two completely different polymers. See Figure 13-20 in the text to see an illustration of the structures.
- Enzymes break starch down into smaller fragments, eventually breaking it down into maltose (a disaccharide of α-glucose), then glucose.
- The enzymes used to break down starch cannot break down cellulose, because the polysaccharide structures are different. For this reason, we cannot digest cellulose.

13.6 NUCLEIC ACIDS

Key Terms:

- **Nucleotide** – molecule composed of a pentose sugar, a nitrogen-containing organic base, and a phosphate linkage.
- **Nucleic acid** – macromolecule made from nucleotides. There are two types of nucleic acids:
    - **Deoxyribonucleic acid (DNA)** – nucleic acid that *stores* genetic information. DNA has these basic structural features:
        - molar mass of up to a trillion g/mol;
        - deoxyribose is the pentose sugar;
        - uses these organic bases: Cytosine, Guanine, Adenine, and *Thymine,* which are abbreviated C, G, A, and T, respectively.
        - is usually double-stranded.
    - **Ribonucleic acid (RNA)** – nucleic acid that *transmits* genetic information. There are several types of RNA (for example, messenger RNA, transfer RNA). RNA has these basic structural features:
        - molar mass ranging from 20,000 – 40,000 g/mol;
        - ribose is the pentose sugar;
        - uses these organic bases: Cytosine, Guanine, Adenine, and *Uracil*, which are abbreviated C, G, A, and U, respectively.
        - is usually double-stranded.

Key Concepts:

- The components of the nucleotide link together via condensation reactions:
    - a C–N bond forms from reaction of the –OH on the 1'-position of the sugar and N–H from a base.

- a P –O –C link forms from the reaction of the –OH on the 5'-position of the sugar and a phosphoric acid molecule.
- The convention is to show the components in this order: phosphate, sugar, base.

Helpful Hints
- Figure 13-22 in the text shows the structures of the five organic bases found in nucleic acids. These structures indicate the hydrogen atoms that are eliminated during the condensation reaction.
- Figure 13-23 in the text shows the structures of ribose and deoxyribose.
- In nucleic acids, the sugars are numbered with primes (1', 2', etc) to distinguish them from the numbering on the organic bases.

EXERCISE 12: Draw the structure of the RNA nucleotide that contains cytosine.

STRATEGY: A nucleotide contains a pentose sugar, an organic base, and a phosphate, which are attached in a specific manner via condensation reactions. In RNA, the sugar is ribose. The structures of ribose and the organic base, cytosine, can be found in the text. Line up the individual component molecules from left to right: phosphoric acid, then ribose, then cytosine. Link them together like this:
- the –OH on the 1'-position of the sugar with the indicated N–H from the base.
- the –OH on the 5'-position of the sugar with the phosphoric acid molecule.

The Primary Structure of Nucleic Acids

Key Term:
- **Primary structure** – the order of bases along a nucleic acid polymer chain.

Key Concepts:
- The polymer backbone in a nucleic acid is constructed of alternating sugar and phosphate units.
- The phosphates attach to the sugars in the 3' and 5' positions of the sugar.
- By convention, the chain starts on the left with a phosphate attached to the 5' position of a sugar, and ends with an unreacted hydroxyl group in the 3' position.
- The nucleotides in a chain are listed by their one-letter designation. For example, "CCGT" is a chain consisting of four nucleotide units attached in the order of cytosine, cytosine, guanine, thymine.

EXERCISE 13: Draw the structure of a G-C portion of a DNA sequence.

STRATEGY: Draw each nucleotide following the guidelines we used in Exercise 12. Then, link the two nucleotides together at the 3' position of the G nucleotide and the phosphate of the C nucleotide. The sequence is significant: G must be to the left of C.

SOLUTION:

Note: The 3' OH is available to react with the next nucleotide.

The Secondary Structure of Nucleic Acids

Key Terms:
- **Secondary structure** – the three-dimensional shape that the nucleic acid polymer takes.
- **Complementary base pairs** – the pairs of bases that hydrogen bond to each other within a DNA molecule:
    - guanine pairs with cytosine
    - adenine pairs with thymine

Key Concepts:
- The secondary structure of DNA contains two polymer strands wound together in a double helix, with these notable features:
    - the hydrophilic sugars and phosphates are on the surface of the double helix;
    - the hydrophobic bases are tucked inside the structure;
    - the structure is held in place by hydrogen bonds between complementary base pairs.
- RNA is usually single-stranded. However, it can contain regions where complementary base pairs along one given strand will hydrogen bond. In these cases, remember, *uracil* pairs with adenine.

Helpful Hints
- See Figure 13-27 in the text for an illustration of the double helix.
- Figure 13-28 in the text shows the hydrogen bonding that occurs between complementary base pairs.

EXERCISE 14: If part of one strand of DNA contains the sequence: A-G-C-C-T-G-A-T, what is its complementary sequence?
 STRATEGY: In DNA, A pairs with T and C pairs with G.
 SOLUTION: The complementary sequence will be: T-C-G-G-A-C-T-A.

13.7 PROTEINS

Key Terms:
- **Amino acid** – the building blocks of proteins. Every amino acid has this general structure:

Tetrahedral carbon atom bonded to:
- an amine,
- a carboxylic acid,
- a hydrogen atom,
- and an R-group.

- **Protein** – a polyamide made from condensation reactions of amino acids. Proteins vary in length from about 8 amino acids to over 1750 amino acids.
- **Peptide linkage** –amide link that is formed between two amino acids in a condensation reaction.
- **Primary structure (proteins)** – the sequence of amino acids within the polypeptide chain.

Key Concepts:
- There are twenty different R-groups in naturally-occurring amino acids. See Figure 13-32 in the text for illustrations of the R-groups. Notice that each amino acid:
  - is given a three-letter name; and
  - is categorized by its R-group as polar or non-polar.
- The structure of a given protein will be arranged so that its non-polar R-groups interact with hydrophobic environments and its polar R-groups interact with hydrophilic environments.
- By convention, the amine end of the peptide is written on the left, so that the chain can grow to the right, on the carboxylic acid end.
- A countless number of different polypeptides can be formed from the 20 amino acids.

EXERCISE 15: Draw the structure of the peptide: His-Lys-Val.

STRATEGY: Draw the peptide backbone by linking three generic amino acids together. Then, add the R-groups for each amino acid. Remember, amines go on the left and carboxylic acids go on the right.

SOLUTION:

Try It #5: Consider the amino acids used to construct the peptide sequence in Exercise 15. What other sequences of 1 Hys, 1 Lys, and 1 Val exist?

Secondary and Tertiary Protein Structure

Key Terms:
- **Secondary structure** – the manner in which a protein chain folds or coils into a 3-D shape. Two common patterns are:
    - **α-helix** – shape that results from hydrogen bonding between C=O and N-H groups that are three amino acids away from each other on the same protein chain. See Figure 13-37 in the text for an illustration.
    - **β-pleated sheet** – shape that results from hydrogen bonding of C=O and N-H groups in different regions of the protein chain. See Figure 13-38 in the text for an illustration.
- **Tertiary structure** – the overall, unique 3-D shape that a protein molecule forms in order for it to be in its lowest-energy state.
- **Disulfide bridge** – a cross-link involving two cysteine R-groups on a protein, resulting in an S–S single bond.

Key Concepts:
- A protein's tertiary structure is primarily determined by the way in which water interacts with the R-groups on the protein.
- Two classes of proteins are:
    - **Globular proteins** – proteins which have a compact overall 3-D structure. They contain a mixture of β-pleated sheets and α-helices in their secondary structure. Some examples include: enzymes, antibodies, and transport molecules.
    - **Fibrous proteins** – proteins that form cable-like structures. They contain mostly α-helices in their secondary structure. Some examples include hair, silk, and fur.

Helpful Hint
- A phone cord provides a good visual analogy for a protein's structure: If you pull the cord straight, you can see the full length of it and where one particular segment is located in relation to the entire cord (primary structure). If you let the cord relax, it forms into a coil (helix – secondary structure). Overall, the cord sometimes wraps and twists over itself (tertiary structure).

EXERCISE 16: The structure of a person's hair is largely determined by cysteine amino acids and hydrogen-bonding amino acids in the hair's protein structure. a) Which of these – cysteines or hydrogen bonding sites – are responsible for "bad hair" days?

STRATEGY: Bad hair days are temporary problems, so the factor that influences them must be a reversible situation.

SOLUTION: Cysteine forms disulfide bridges (cross-links) between the protein polymers in hair. These are covalent bonds, which shouldn't be influenced by changes in the weather.

Hydrogen bonds, on the other hand, are intermolecular attractions that can break and reform with water vapor in the air. A particularly dry or wet day can have a huge impact on the structure of someone's hair!

Chapter 13 Self-Test

You may use a periodic table for this test. In addition, you may use reference structures of amino acids, monosaccharides, and organic bases used to construct nucleic acids.

1. Identify the functional groups in each molecule:

   Lawsone – the chemical that gives henna its color

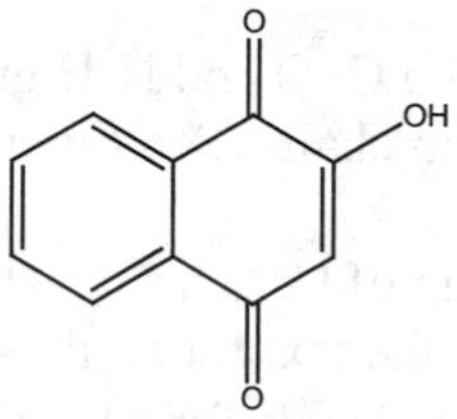

   Benzyl acetate – rose oil

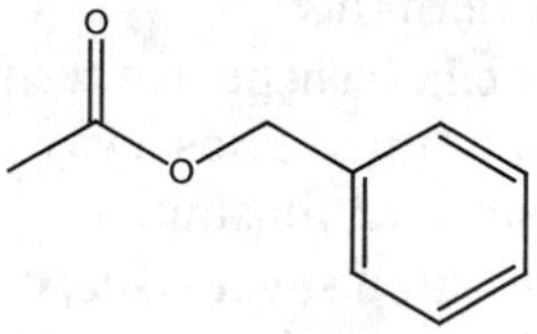

2. Polymethylmethacrylate (Plexiglas) is made by free-radical polymerization of methylmethacrylate. Draw a segment of the polymer. Show three monomer units.

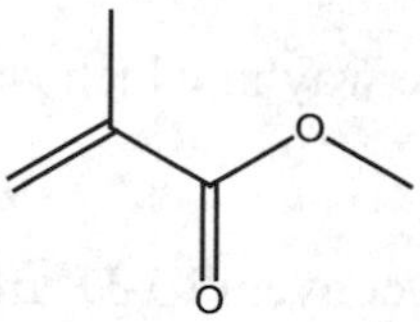

   methyl methacrylate

3. What are the monomers that make up this polymer?

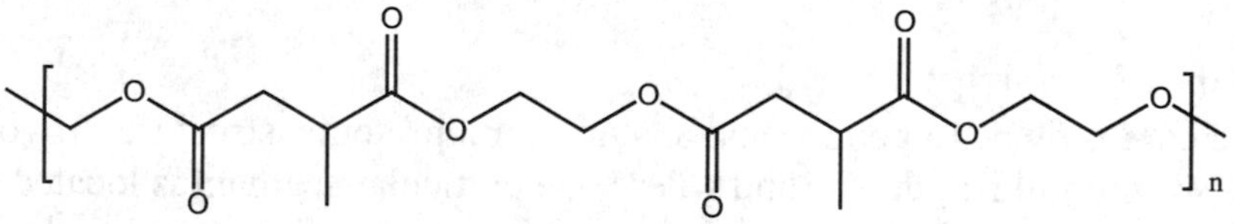

4. a) Propose an explanation for why Teflon (polytetrafluoroethylene) is so slippery.
   b) To what category of polymers does Teflon belong?

5. Compare and contrast the term "primary structure" for nucleic acids and proteins.

6. Answer these questions for this DNA sequence: A-C-T.
   a) Draw the complete structure of the DNA sequence.
   b) Indicate the sites on the bases that will hydrogen-bond to the complementary sequence.
   c) What is its complementary sequence?

Answers to Try Its:

1.

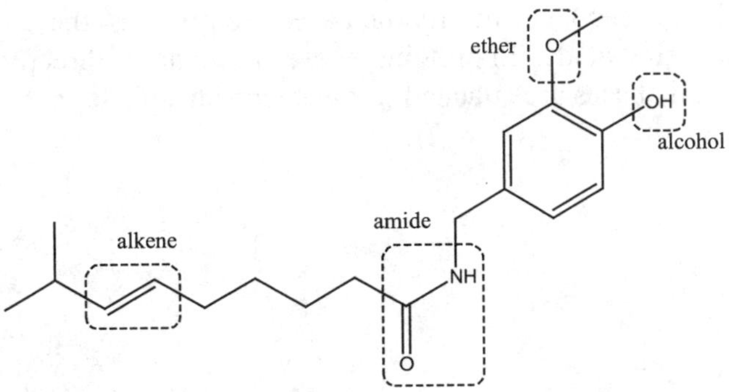

ether

alcohol

OH

amide

alkene

NH

O

2.

Notice: the double bonds in the monomer are not conjugated, so only one of them will be part of the polymer backbone.

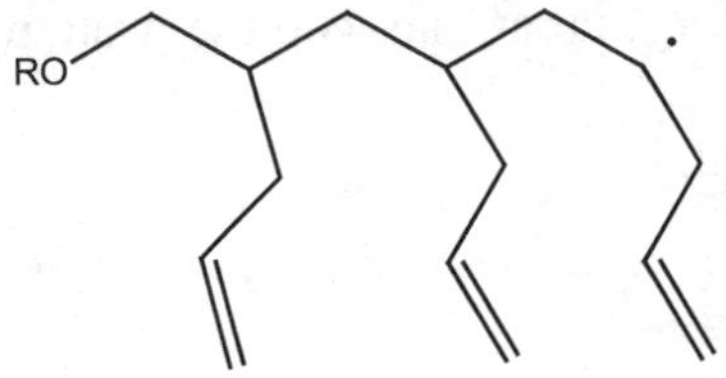

RO

3.

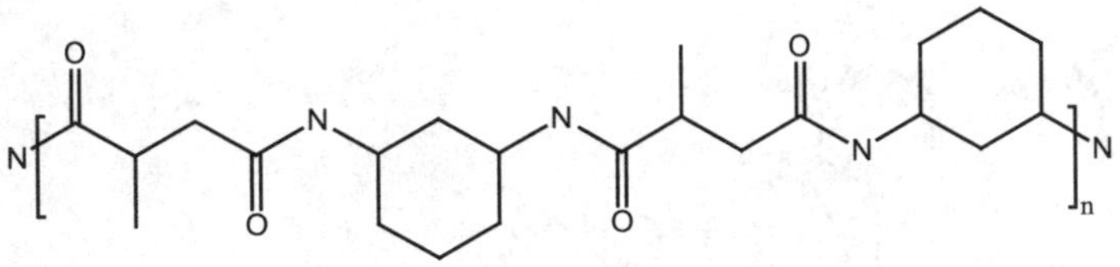

N—O—N—N—O—N—N

4. a) The polymer would be very strong and rigid, not at all flexible. Its density would depend on the lengths of the cross-linkers; the shorter the linkers, the denser the polymer. b) plastic.

5. His-Val-Lys, Val-His-Lys, Val-Lys-His, Lys-Val-His, and Lys-His-Val.

Answers to Self-Test:

1.

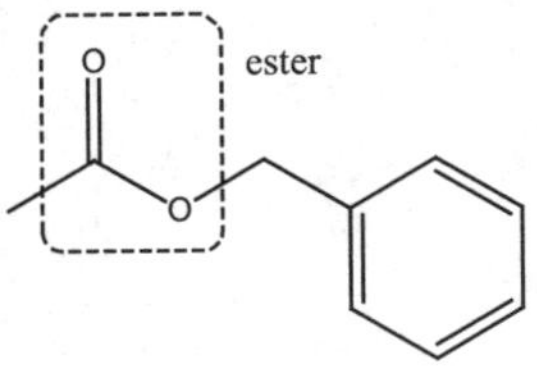

ketone

alcohol

OH

alkene

ketone

ester

O

2.

The –CH_3 groups were drawn in an upward position, so the groups would be less crowded:

$\df"n

3.

It is a copolymer, because two different monomers were used to construct it:

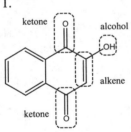

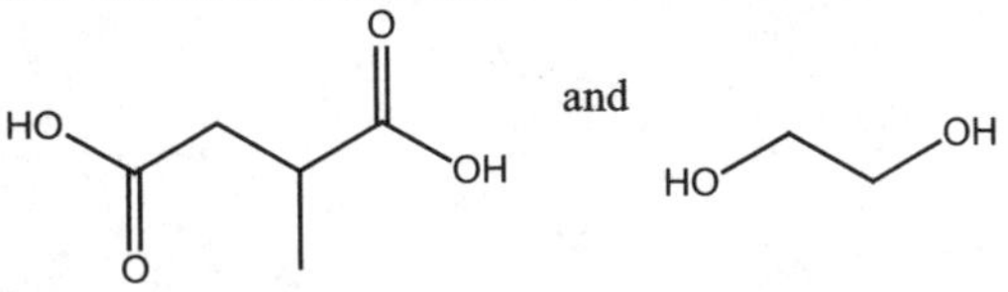

and

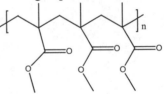

4. a) Polytetrafluorethylene consists of a long chain of carbon atoms singly bonded to each other, in which two fluorine atoms are attached to every carbon on the chain. What is known about the element fluorine? It is an $n=2$ element with 7 valence electrons, it is the most electronegative element on the periodic table, it never forms more than one covalent bond, and its three lone pairs are held tightly to it. The fluorine atoms are already bonded to one atom, so they don't want to bond to anything else. In addition, hundreds of fluorines lined up along the polymer chain cause an incredible amount of repulsion between lone pairs. The fluorines don't want anything else near them, so they repel most foods that come in contact with the polymer surface.

b) Teflon is a thermosetting plastic made from free-radical polymerization.

5. In general, "primary structure" refers to the sequence of monomers within the polymer chain. In nucleic acids, the primary structure is described by the sequence of organic bases. In proteins, the primary structure is described by the sequence of amino acids. In proteins, these amino acids directly make up the polymer chain. In nucleic acids, organic bases are attached to the sugar-phosphate linkages that make up the protein chain.

6. a) and b)

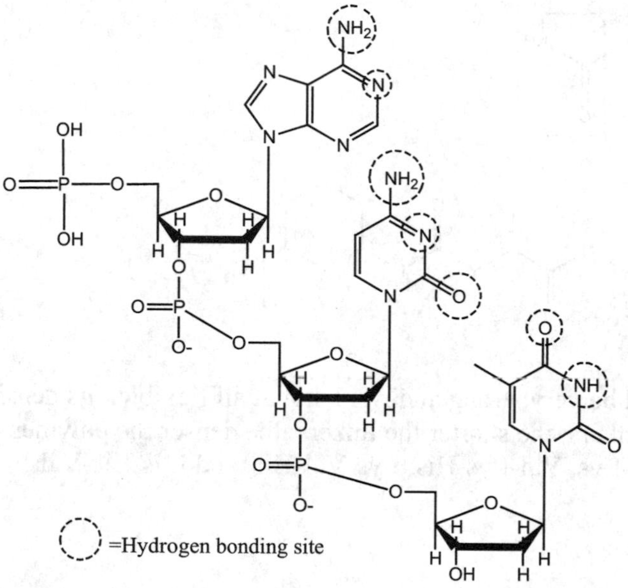

⊂⊃ =Hydrogen bonding site

c) T-G-A

Chapter 14: Spontaneity of Chemical Processes

Learning Objectives

In this chapter, you will learn how to:

- Assess entropy changes within chemical processes.
- Evaluate the two factors that drive a chemical process: energy and entropy.
- Use the Second Law of Thermodynamics to determine if a given reaction is spontaneous.
- Use thermodynamic principles to evaluate whether or not a given process is spontaneous under standard conditions.
- Alter the temperature or concentration of reaction components to influence a reaction's spontaneity.
- Interpret and construct a phase diagram for a substance.
- Drive a non-spontaneous reaction forward by coupling it to a spontaneous process.

Practical Aspects

This chapter is particularly important to chemists, biochemists, biologists, engineers, and physicists. It describes the underlying principles of why a given reaction will or will not run without any external assistance. For example, biological systems use chemical reactions to generate energy stores or to supply energy to other reactions. The fact that you can move or blink or talk is dependent upon spontaneous reactions driving other, non-spontaneous, reactions.

The concepts in this chapter will be used again in Chapters 16, 17, 18, and 19.

14.1 SPONTANEITY

Key Term:
- **Spontaneous** – the direction that a process will take, if left alone and given sufficient time. (Alternate definition – A process that occurs on its own without any external assistance.)

Key Concepts:
- Things naturally tend to become dispersed.
- Two factors influence the spontaneity of a reaction: the dispersal of energy and the dispersal of matter.
- From an energy dispersal standpoint, exothermic processes favor spontaneity, while endothermic processes do not. (Exothermic processes release energy, so nothing is required of the surroundings.)
- From a matter dispersal standpoint, processes that involve the dispersal of matter favor spontaneity, while processes that involve matter becoming more organized do not.
- Spontaneity only indicates *if* the process will occur without external help; it says nothing about the rate. A spontaneous process can be too slow to observe.
- Spontaneity is a state function: a process that is spontaneous in one direction will be non-spontaneous in the reverse direction.

EXERCISE 1: For each pair, determine which situation describes matter more dispersed: a) clean clothes hung in the closet or dirty clothes thrown on the floor; b) 100 parked cars in a parking lot or 100 cars on the freeway; c) Picture A or Picture B, shown below; d) Picture C or Picture D, shown below.

 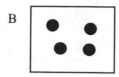  

SOLUTION:
a) dirty clothes thrown on the floor; b) 100 cars on the freeway; c) Picture B; d) Picture C.

EXERCISE 2: Determine, if possible, if each process is spontaneous. Defend your answer with a brief explanation. a) Br (g) + Br (g) → Br$_2$ (l); b) C$_6$H$_{12}$ (g) + 9 O$_2$ (g) → 6 CO$_2$ (g) + 6 H$_2$O (g)

STRATEGY: Both energy and matter dispersal influence spontaneity. If a reaction is exothermic *and* it disperses matter, then it will definitely be spontaneous.

SOLUTION:
a) In this reaction, two atoms are combining to form a molecule. Energy is always released when bonds are formed, so this process releases energy. This process shows 2 moles of gas particles being converted to 1 mole of liquid molecules. The Br is becoming more orderly, because every two disordered gas particles are becoming one liquid particle. The matter is becoming more organized, rather than more dispersed. At this point, we can't determine if the reaction is spontaneous, because the two factors – energy and matter dispersal – oppose each other. The reaction favors spontaneity from an energy standpoint, but not from the standpoint of matter dispersal.

b) This is the combustion reaction of a hydrocarbon fuel. Combustion reactions are highly exothermic, so the reaction favors spontaneity from an energy standpoint. Comparison of the amount of gaseous particles in the reactants (10 mol of gas molecules) to the amount of gaseous particles in the products (12 mol of gas molecules) is a good way to assess relative dispersal of matter. In this case, the process increases in disorder. This process will be spontaneous because both the energy and the matter dispersal factors favor spontaneity. Remember that spontaneity says nothing about the *rate* of the reaction. On its own, this particular spontaneous reaction occurs too slowly at room temperature for us to observe.

Try It #1: A cold pack can be made by mixing solid ammonium nitrate and water. The ammonium nitrate dissolves in the water, and the solution becomes quite cold to the touch, indicating an endothermic, spontaneous process. Assess the energy and matter dispersal factors of the process in the context of the reaction's spontaneity.

14.2 ENTROPY: THE MEASURE OF DISPERSAL

Key Terms:
- **Entropy (S)** – the quantitative measure of disorder.
- **Second Law of Thermodynamics** – the total entropy of the total (system + surroundings) increases in a spontaneous process.

Key Concepts:

- Entropy calculations are greatly simplified by focusing on the *change* in entropy for a process (ΔS) rather than absolute entropies.
- ΔS is a state function. All mathematical manipulations that applied to ΔH will also apply to ΔS.
- $\Delta S_{total} > 0$ for a spontaneous process.
- ΔS can be measured by assessing heat flow for a process at constant temperature.
- It is possible for heat to flow without a change in temperature. Common cases are: 1) during phase changes, and 2) in constant temperature environments (For example: our bodies are maintained at a constant temperature of 37°C.)

Useful Relationship:

- $\Delta S = q/T$. This equation relates ΔS to heat flow *at a constant temperature*.

Units and Conversions:

- ΔS has units of J/K.
- ΔH is measured in kJ, so watch out for unit conversions between kJ and J when relating ΔS to ΔH.

EXERCISE 3: Consider the process occurring inside a container whose sides cannot move:

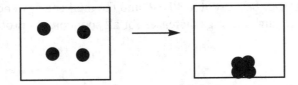

Determine the sign of ΔS *for the surroundings* (+, -, or 0). Briefly explain your choice.

> *STRATEGY AND SOLUTION:* This set of pictures shows a phase transition, which is a constant temperature process. We can find the sign of ΔS_{surr} from the relationship: $\Delta S_{surr} = q_{surr}/T$.
>
> The system lost heat energy as it went from the gas phase to the solid phase (q_{sys} is "-"). There is nothing else inside the container to absorb the heat, and energy cannot be created nor destroyed, so this must be a closed system. The surroundings must have absorbed the heat. The sign for q_{surr} is therefore "+," making the sign of ΔS_{surr} "+".

EXERCISE 4: Diethyl ether's boiling point is 35°C, its ΔH_{vap} is 26.0 kJ/mol, and its molar mass is 74.122 g/mol. Consider 250.0 g of liquid diethyl ether turning into a gas in a room that is 38°C. Calculate the following: a) $\Delta S_{diethyl\ ether}$; b) $\Delta S_{surroundings}$; c) ΔS_{total}.

> *STRATEGY:*
> a) We're asked to find $\Delta S_{diethyl\ ether}$ for the phase change. A phase change is a constant temperature process, so we can use the relationship: $\Delta S = q/T$. During the phase change, $T_{ether} = 35°C = 308$ K.
>
> First, determine the heat required to evaporate the 250.0 g of ether:

$$q_{ether} = \Delta H_{vap\,ether} = 250.0\,g \times \frac{mol}{74.122\,g} \times \frac{26.0\,kJ}{mol} = 87.693\,kJ$$

$$\Delta S_{ether} = \frac{q_{ether}}{T_{ether}} = \frac{87.693\,kJ}{308\,K} \times \frac{1000\,J}{1\,kJ} = 284.72\,J/K = 285\,J/K$$

b) We're asked to find $\Delta S_{surroundings}$. In this case, the room can be considered the surroundings, because from the perspective of the ether, the room represents the rest of the total. A room is a large space relative to 250.0 g of a liquid, so the temperature of the room should not change during this process. The room is at a constant T of 38°C (311 K). We can use the relationship $\Delta S = q/T$ for this process too, if we know q_{room}. The First Law of Thermodynamics tells us that the heat gained by the ether must be lost by the room, so $q_{room} = -87.693$ kJ.

$$\Delta S_{room} = \frac{q_{room}}{T_{room}} = \frac{-87.693\,kJ}{311\,K} \times \frac{1000\,J}{1\,kJ} = -281.97\,J/K = -282\,J/K$$

c) We're asked to find ΔS_{total}.
$$\Delta S_{total} = \Delta S_{system} + \Delta S_{surroundings} = \Delta S_{ether} + \Delta S_{room}$$
$$\Delta S_{total} = 285\,J/K + (-282\,J/K) = 3\,J/K$$

SOLUTION: a) $\Delta S_{diethyl\ ether} = 285$ J/K; b) $\Delta S_{surroundings} = -282$ J/K; c) $\Delta S_{total} = 3$ J/K. These answers seem reasonable. The ether should increase in disorder because it is converting into a gas, the room should decrease in disorder because it lost heat, and the total entropy should increase because the disorder of the system + surroundings increases for all spontaneous processes.

Helpful Hints

- Notice from Exercise 4 that the constant-temperature process must be clearly identified. It helps to include subscripts on each term. For example, "T_{ether}" is more useful than simply "T".
- If you are asked to find ΔS_{total}, simply break the problem down into ΔS_{system} and $\Delta S_{surroundings}$.

14.3 ENTROPIES OF PURE SUBSTANCES

Key Terms:
- **Absolute zero** – temperature at which all molecular motion ceases. (Defined as 0 Kelvin.)
- **Third Law of Thermodynamics** – a pure, perfect crystal at 0 Kelvin has zero entropy.
- **Standard molar entropy** – the entropy of one mole of substance at standard thermodynamic conditions.

Key Concepts:
- Standard thermodynamic conditions are: temperature = 298 K, gas pressure = 1 atm, solution concentration = 1 M for aqueous solutes. Standard conditions are denoted with a "°".
- Entropies of substances are measured relative to absolute zero.
- Standard molar entropy values of pure substances are always positive numbers.
- Two things influence entropy: temperature and position.
  - **Temperature**: For a given substance, the greater the temperature, the greater the molecular motion *and* the greater the distribution of molecular speeds.
  - **Position**: The more locations a substance can be located in, the greater that substance's entropy. This can be thought of in several ways: phase (gas vs. liquid vs. solid); molecular size (electron cloud size, or ability of the molecule to flip around); or concentration (of a gas or aqueous solute).

- Standard entropies can be used to approximate entropies at non-standard temperatures, but a separate calculation must be performed to determine entropies at non-standard concentrations (see Useful Relationships).

Helpful Hint
- Standard molar entropies (S°) of several substances are listed in Appendix D.

Useful Relationships:
- $S = S° - R \ln x$ This equation is used to assess entropy of a gas ($x = P$) or aqueous solute ($x = c$) that is not at standard conditions of 1 atm or 1 M.

- $\Delta S°_{rxn} = \Sigma$ (coeff) $S°$ (products) - Σ (coeff) $S°$ (reactants).

EXERCISE 5: Which has the greater entropy? Why?
     a) 1 mole of gas X at 300 K or 1 mole of gas X at 700 K?
     b) The valence electrons within an atom whose valence shell is $n=6$ or $n=2$?
     c) The reactants or the products in this reaction: $2 NO_2 (g) \rightarrow N_2O_4 (g)$

STRATEGY: Entropy means disorder. Assess which choice has greater disorder.

SOLUTION:
a) Gas X as 700 K will have more entropy than at 300 K. At 700 K, the gas has increased molecular motion *and* has a greater distribution of speeds.
b) The electrons within an atom are highly organized, but the larger atom will house the electrons in a greater variety of locations than the smaller atom. The atom whose valence electrons are in $n=6$ will have greater entropy.
c) The reactant side of the equation shows 2 gas molecules, and the product side shows 1 gas molecule. There are more gaseous reactant than product particles; the reactants are more disordered.

Try It #2: Which has greater entropy? a) 0.050 M NaCl or 2 M NaCl?; b) a hydrocarbon that has 6 carbons singly-bonded in a row, or one with 6 carbons singly-bonded in a ring shape?; c) I_2 (s) or I_2 (g)?

EXERCISE 6: a) What is the entropy of 1.0 M Ag^+ (aq) at 25°C and 1 atm? b) Calculate the entropy of 0.100 M Ag^+ (aq) at 25°C and 1 atm. c) Draw molecular pictures to illustrate the relative entropy values.

STRATEGY:
a) We've been given standard temperature and pressure conditions for a solution that is 1 M (standard concentration). Use Appendix D to find Ag^+ (aq)'s entropy value at standard conditions.
b) The conditions are still standard temperature and pressure, but the concentration is non-standard. Use the relationship for entropy at non-standard concentrations: $S = S° - R \ln c$.
        $S = S° - R \ln c = 73.4$ J/mol K $- (8.314$ J/mol K$)(\ln 0.100) = 91.8$ J/mol K.

SOLUTION:
a) $S°$ for 1.0 M Ag^+ (aq) is 73.4 J/mol K.
b) S for 0.100 M Ag^+ (aq) is 91.8 J/mol K. This answer seems reasonable because the entropy in the diluted solution should be greater than in the concentrated solution. The ions in the diluted solution are located in more possible positions than in the concentrated solution.

c) Molecular pictures of the concentrated solution (left) and the diluted solution (right). The particles in the diluted solution can be located in a greater number of places than the particles in the concentrated solution:

EXERCISE 7: Calculate $\Delta S°_{rxn}$ for:

$$2 \, Na \, (s) \, + \, 2 \, H_3O^+ \, (aq) \; \rightarrow \; H_2 \, (g) \, + 2 \, Na^+ \, (aq) \, + \, 2 \, H_2O \, (l)$$

STRATEGY: No numerical information has been provided, so data from reference tables must be used. We can use standard entropies from Appendix D in the text:

$$\Delta S°_{rxn} = \Sigma \, (coeff) \, S° \, (products) - \Sigma \, (coeff) \, S° \, (reactants)$$

$$\Delta S°_{rxn} = [(1 \, mol \, H_2)(130.680 \, J/mol \, K) + (2 \, mol \, Na^+)(58.5 \, J/mol \, K) + (2 \, mol \, H_2O)(69.95 \, J/mol \, K)]$$
$$- \, [(2 \, mol \, Na)(51.3 \, J/mol \, K) + (2 \, mol \, H_3O^+)(69.95 \, J/mol \, K)]$$
$$387.58 \, J/mol \, K - 242.5 \, J/mol \, K = 145.1 \, J/mol \, K$$

SOLUTION: $\Delta S°_{rxn} = 145.1$ J/mol K. The answer seems reasonable because it is a large, positive number.

14.4   SPONTANEITY AND FREE ENERGY

Key Terms:
- **Free Energy (G)** – the part of a system's energy that is ordered and available to become spontaneously disordered.
- **Standard free energy of formation ($\Delta G°_f$)** – free energy associated with forming one mole of a substance from its elements in their standard states.

Key Concepts:
- ΔG_{sys} is always negative for a spontaneous process under conditions of constant temperature and pressure.
- ΔG is a state function. All mathematical manipulations that applied to ΔH will also apply to ΔG.
- Spontaneity (ΔG) depends upon enthalpy (ΔH) and entropy (ΔS). The quantitative relationship that links these variables is: $\Delta G° = \Delta H° - T\Delta S°$.

Useful Relationships:
- $\Delta G° = \Delta H° - T\Delta S°$ •   The derivation of this equation is in the text.

- $\Delta G°_{rxn} = \Sigma \, (coeff) \, \Delta G°_f \, (products) - \Sigma \, (coeff) \, \Delta G°_f \, (reactants)$.

EXERCISE 8: What is $\Delta G°_{rxn}$ for: $2 Na (s) + 2 H_3O^+ (aq) \rightarrow H_2 (g) + 2 Na^+ (aq) + 2 H_2O (l)$

STRATEGY: Notice that this is the reaction from Exercise 7. No data has been provided, so we will have to use data from Appendix D in the text. $\Delta G°_{rxn}$ can be calculated via one of two methods:

1) $\Delta G° = \Delta H° - T\Delta S°$ or
2) $\Delta G°_{rxn} = \Sigma$ (coeff) $\Delta G°_f$(products) - Σ (coeff) $\Delta G°_f$(reactants)

We'll use method #1. In Exercise 7, we saw that $\Delta S°_{rxn} = 146.3$ J/mol K. Use Appendix D data to calculate $\Delta H°_{rxn}$ and then use $\Delta G° = \Delta H° - T\Delta S°$ to find $\Delta G°_{rxn}$.

$$\Delta H°_{rxn} = \Sigma \text{ (coeff) } \Delta H°_f\text{(products) - } \Sigma \text{ (coeff) } \Delta H°_f\text{(reactants)}$$

$$\Delta H°_{rxn} = [(1 \text{ mol } H_2)(0 \text{ kJ/mol}) + (2 \text{ mol } Na^+)(-240.3 \text{ kJ/mol}) + (2 \text{ mol } H_2O)(-285.83 \text{ kJ/mol})]$$
$$- [(2 \text{ mol } Na)(0 \text{ kJ/mol}) + (2 \text{ mol } H_3O^+)(-285.83 \text{ kJ/mol})]$$
$$-1052.26 \text{ kJ/mol} - (-571.66 \text{ kJ/mol}) = -480.60 \text{ kJ/mol}$$

$$\Delta G° = \Delta H° - T\Delta S° = (-480.60 \text{ kJ/mol}) - (298 \text{ K})(145.1 \text{ J/mol K})(1 \text{ kJ/1000 J}) = -523.8 \text{ kJ/mol}$$

Hint: remember to convert entropy units to kJ before subtracting!

SOLUTION: $\Delta G°_{rxn} = -523.8$ kJ/mol. It seems reasonable to have a large, negative value for $\Delta G°_{rxn}$ because Group 1 metals are highly reactive with aqueous acids. The reaction should be spontaneous under standard conditions.

Try It #3: Use values for $\Delta G°_f$ to determine $\Delta G°_{rxn}$ for the reaction shown in Exercise 8. Compare your result to the result obtained in Exercise 8 to verify that both methods for calculating $\Delta G°_{rxn}$ work equally well.

Summary of Thermodynamic Terms

Term	Symbol	What it measures:	Interpretation:
Enthalpy	ΔH	Heat flow	• "-" = exothermic • "+" = endothermic
Entropy	ΔS	Disorder	• "+" = increase in disorder • "-" = decrease in disorder
Free Energy	ΔG	Spontaneity	• "-" = process will occur without any external assistance • "+" = process will *not* occur without any external assistance

Key Concept:
The sign of ΔG also indicates this information:
- If ΔG is negative, then the process is spontaneous in the forward direction.
- If ΔG is positive, then the process is spontaneous in the *reverse* direction. ($\Delta G_{reverse} = -\Delta G_{forward}$).
- If ΔG is zero, then the process is at equilibrium. There will be no net spontaneous change in either direction.

Free Energy at Non-Standard Conditions

Key Term:
- **Reaction quotient (Q)** – ratio of product to reactant concentrations (c) or gas pressures (p), as shown here:

$$\text{For any reaction: } aA + bB \rightarrow dD + eE, \quad Q = \frac{(c_D)^d (c_E)^e}{(c_A)^a (c_B)^b}$$

Key Concepts:
- Temperature will always influence the degree of spontaneity of reactions.
- Concentration will always influence the spontaneity of a reaction.
- Temperature does not significantly affect enthalpy or entropy, so the "$\Delta G° = \Delta H° - T\Delta S°$" relationship can be used for non-standard temperature conditions.
- Entropy is affected by concentration. Its influence is accounted for in this relationship (which is derived in the text): "$\Delta G_{rxn} = \Delta G°_{rxn} + RT \ln Q$".

Useful Relationships:
- $\Delta G° = \Delta H° - T\Delta S°$ Use this equation for non-standard temperature conditions.
- $\Delta G_{rxn} = \Delta G°_{rxn} + RT \ln Q$ Use this equation for non-standard concentration conditions.

EXERCISE 9: Determine all of the combinations of $\Delta H°$, T, and $\Delta S°$ for which a reaction will be spontaneous.

STRATEGY: $\Delta G°$ must be negative for a reaction to be spontaneous. Use $\Delta G° = \Delta H° - T\Delta S°$ to determine under which conditions $\Delta G°$ is negative. T is reported in units of Kelvin, which are always positive, so the only two terms to assess are $\Delta S°$ and $\Delta H°$.

SOLUTION: The solution has been tabulated here for quick reference.

If: $\Delta H°$	$\Delta S°$	$\Delta G°$ will be negative and the reaction will be spontaneous:	because:
-	-	Only at low temperatures	The $\Delta H°$ term must be greater than the $T\Delta S°$ term for $\Delta G°$ to be negative.
+	+	Only at high temperatures	The $T\Delta S°$ term must be greater than the $\Delta H°$ term for $\Delta G°$ to be negative.
-	+	At all temperatures	Both $\Delta H°$ and $\Delta S°$ favor spontaneity.
+	-	Never	Neither $\Delta H°$ nor $\Delta S°$ favor spontaneity.

Helpful Hint
- To determine the temperature at which a non-spontaneous reaction would become spontaneous, set $\Delta G°$ to zero and solve for T.

EXERCISE 10: Under what temperature conditions will this reaction be spontaneous?

$$Ni\,(s)\ +\ 4\,CO\,(g) \rightarrow Ni(CO)_4\,(g)\quad \Delta H°_{rxn} = -161\ kJ$$

STRATEGY: Use the equation $\Delta G° = \Delta H° - T\Delta S°$ and assess the factors that influence spontaneity. $\Delta H°$ is negative. As we can see from the balanced equation, 4 moles of gas are converted to 1 mole of gas, which indicates an increase in order. $\Delta S°$ is therefore negative. Assessing the signs:

$$\Delta G° = \Delta H° - T\Delta S°\ \ = (-) - (+)(-)$$

we see that $\Delta G°$ will only be negative if the $\Delta H°$ term is larger than the $T\Delta S°$ term. This only occurs at conditions of low temperature.

SOLUTION: At low temperatures this reaction is spontaneous, at high temperatures it is not. *Practical note: this reaction is actually used to purify nickel from ores. At low temperature, any pure nickel present in an ore will react with the CO, creating gaseous $Ni(CO)_4$, which can easily be transferred to a separate container. The new container is then heated and the reaction (which is spontaneous in the reverse direction at elevated temperatures) proceeds in the reverse direction to generate pure nickel.

Try It #4: Calculate the temperature at which the reaction in Exercise 10 ceases to be spontaneous. ($S°$ of $Ni(CO)_4$ (g) is 410.6 J/mol K; it is not in Appendix D.)

EXERCISE 11: Given the reaction: $H_2\,(g)\ +\ I_2\,(g) \rightarrow 2\,HI\,(g)$ $\Delta G°_{rxn} = -15.9\ kJ$

Compare the spontaneity of the reaction under standard conditions to conditions in which the partial pressure of each reactant is 0.10 atm and the partial pressure of the product is 3.80 atm. Is the reaction still spontaneous?

STRATEGY: This reaction is occurring under conditions of non-standard gas pressures, so the relationship below should be used:

$$\Delta G_{rxn} = \Delta G°_{rxn} + RT\,\ln Q\quad \text{where } Q = \frac{p_{HI}^2}{\left(p_{H_2}\right)^1 \left(p_{I_2}\right)^1}$$

$$Q = \frac{(3.80)^2}{(0.10)(0.10)} = 1444;\ \text{so } \Delta G_{rxn} = -15.9\ kJ + \left[\frac{8.314\ J}{mol\ K} \times 298\ K \times \frac{1\ kJ}{1000\ J} \times \ln(1444)\right] = +2.06\ kJ$$

SOLUTION: This reaction will not be spontaneous under these conditions because ΔG_{rxn} is a positive number. The positive value for ΔG_{rxn} indicates that the reaction will be spontaneous *in the reverse direction*. (ΔG_{rxn} in the *reverse* direction is -2.06 kJ).

Try It #5: Given the reaction: $CO\,(g)\ +\ Cl_2\,(g) \rightarrow COCl_2\,(g)$ $\Delta G°_{rxn} = -68.8\ kJ$
Compare the spontaneity of the reaction under standard conditions to conditions in which the partial pressure of each reactant is 0.0100 atm and the partial pressure of the $COCl_2$ product is 1.50 atm. Is the reaction still spontaneous?

14.5 SOME APPLICATIONS OF THERMODYNAMICS

Key Terms:
- **Nitrogen fixation** – process by which molecular nitrogen is converted into a usable form of nitrogen for biological systems.
- **Thermal pollution** – the release of unused, excess heat energy into the environment.

Useful Relationships:
- $\ln vp = -\dfrac{\Delta H_{vap}}{RT} + \dfrac{\Delta S°_{vap}}{R}$ This equation lets you find vapor pressure (*vp*) of a pure substance from thermodynamic data. Its derivation is shown in the text. When you solve for *vp*, the unit for pressure will be the SI unit "bar".

EXERCISE 12: Acetic acid is a fairly volatile chemical, as evidenced by the fact that we can smell it when we open a bottle of vinegar. Determine the vapor pressure of acetic acid at 298 K, given:

	$\Delta H°_f$ (kJ/mol)	$S°$ (J/mol K)
CH_3COOH (l)	− 484.45	159.8
CH_3COOH (g)	− 432.25	282.4

STRATEGY: We're asked to find the vapor pressure of a pure substance. We know that the relationship:

$$\ln vp = -\dfrac{\Delta H_{vap}}{RT} + \dfrac{\Delta S°_{vap}}{R}$$

is used to determine vapor pressure from thermodynamic data. The chemical process that applies to vapor pressure is:

$$CH_3COOH\ (l) \rightarrow CH_3COOH\ (g)$$

Plugging our values into the relationship, we get:

$$\ln vp = -\dfrac{[(-432.25\,kJ/mol) - (-484.45\,kJ/mol)]}{(8.314x10\,J/mol\,K)(298\,K)(1\,kJ/1000J)} + \dfrac{(282.4\,J/mol\,K - 159.8\,J/mol\,K)}{(8.314\,J/mol\,K)}$$

$$\ln vp = -6.242, \text{ so } vp = e^{-6.242} = 1.946x10^{-3}\,bar$$

Converting to torr, we get:

$$1.946x10^{-3}\,bar\ x\ \dfrac{770.\,torr}{1\,bar} = 1.50\,torr$$

SOLUTION: The vapor pressure of acetic acid at 298 K is 1.50 torr. This value may sound low, but we are far below the boiling point temperature of acetic acid (424 K), so the vapor pressure should be much lower than 760 torr.

EXERCISE 13: Explain how any cooling system such as a refrigerator, freezer, or air conditioner, is a source of thermal pollution.

SOLUTION: The First Law of Thermodynamics states that energy cannot be created nor destroyed, just transferred or transformed. Coolant systems remove heat from one area (inside the refrigerator or freezer, for example), and add it to the surroundings. This is evidenced by the fact that the backside of the refrigerator is quite hot. This heat energy is not used for any purpose, so it is simply wasted.

14.6 BIOENERGETICS

Key Term:
- **Coupled reaction** – a spontaneous reaction that results from the addition of a spontaneous reaction to a non-spontaneous reaction. The reactions are linked by a common intermediate that is cancelled out when the two reactions are added together.

Key Concepts:
- A non-spontaneous reaction can occur if it is coupled with a spontaneous reaction and the resulting overall change in free energy is negative.
- Biological systems often use coupled reactions to drive non-spontaneous reactions.
- ATP (adenosine triphosphate) is an energy storage molecule. The conversion of ATP to ADP releases free energy that can be used to drive a non-spontaneous reaction. This ATP to ADP reaction is coupled with many non-spontaneous processes inside our bodies:
$$ATP + H_2O \rightarrow ADP + H_3PO_4 \quad \Delta G°_{rxn} = -30.6 \text{ kJ}$$
- Free energy from spontaneous processes in our bodies (such as food metabolism) creates ATP from ADP in coupled reactions. The synthesized ATP can be stored until it is needed to drive a non-spontaneous reaction.
- The concepts of coupled reactions within biological systems can also be applied to industrial processes.

EXERCISE 14: The decomposition of Fe_2O_3 is a non-spontaneous process:

$$2 Fe_2O_3 \text{ (s)} \rightarrow 4 Fe \text{ (s)} + 3 O_2 \text{ (g)} \quad \Delta G°_{rxn} = +1484.4 \text{ kJ}$$

Nevertheless, pure iron metal can be obtained from the oxide, by coupling the reaction above with a highly spontaneous process, such as:
$$2 CO \text{ (g)} + O_2(g) \rightarrow 2 CO_2 \text{ (g)} \quad \Delta G°_{rxn} = -514.4 \text{ kJ}$$

Write the overall spontaneous reaction that would have to occur in order to be able to make iron metal.

STRATEGY: The spontaneous reaction must occur enough times so that when $\Delta G°_{rxn}$ of the spontaneous reaction is added to $\Delta G°_{rxn}$ of the non-spontaneous iron reaction, the overall $\Delta G°_{rxn}$ will be negative. Recall that if we change the coefficients of a balanced equation, then the state function associated with that reaction will also change by that factor.

We can see that doubling the CO reaction would result in a $\Delta G°_{rxn}$ of -1028.8 kJ, which is not negative enough to overcome the $+1484.4$ kJ of the iron reaction. Tripling the CO reaction yields a $\Delta G°_{rxn}$ of -1543.2 kJ, which will override the $+1484.4$, and result in an overall $\Delta G°_{rxn}$ of -55.8 kJ.

Adding the two reactions together:
$$2 Fe_2O_3 \text{ (s)} \rightarrow 4 Fe \text{ (s)} + 3 O_2 \text{ (g)} \quad \Delta G°_{rxn} = +1484.4 \text{ kJ}$$

$$6 CO (g) + 3 O_2 (g) \rightarrow 6 CO_2 (g) \quad \Delta G°_{rxn} = -1543.2 \text{ kJ}$$

Gives the overall reaction:

$$2 Fe_2O_3 (s) + 6 CO (g) + \cancel{3 O_2 (g)} \rightarrow 4 Fe (s) + \cancel{3 O_2 (g)} + 6 CO_2 (g)$$

O_2 (g) is the common intermediate between the two reactions. Canceling out the O_2 (g) simplifies the reaction to:

$$2 Fe_2O_3 (s) + 6 CO (g) \rightarrow 4 Fe (s) + 6 CO_2 (g) \quad \Delta G°_{rxn} = -58.8 \text{ kJ}$$

or:

$$Fe_2O_3 (s) + 3 CO (g) \rightarrow 2 Fe (s) + 3 CO_2 (g) \quad \Delta G°_{rxn} = -29.4 \text{ kJ}$$

Try It #6: Iron can also be obtained from Fe_2O_3 by coupling the reaction with this reaction:

$$C \text{ (graphite)} + O_2 (g) \rightarrow CO_2 (g) \quad \Delta G°_{rxn} = -394.4 \text{ kJ}$$

Write the overall spontaneous reaction that would occur to produce iron metal.

Chapter 14 Self-Test

You may use a periodic table and a calculator for this test.

This data may also be of use:

$$ATP + H_2O \rightarrow ADP + H_3PO_4 \quad \Delta G°_{rxn} = -30.6 \text{ kJ}$$

Chemical	$\Delta H°_f$ (kJ/mol)	$S°$ (J/mol K)
NO (g)	91.3	210.8
O_2 (g)	0	205.152
NO_2 (g)	33.2	240.1
PCl_3 (l)	-319.7	217.1
PCl_3 (g)	-287.0	311.8

1. Calculate the boiling point of PCl_3.

2. A certain salt has a $\Delta G°_{solution}$ of +15 kJ/mol.

   a) What does this value indicate?
   b) Propose a thermodynamic explanation for why the $\Delta G°$ is positive.
   c) Could this solution process *ever* be spontaneous? If so, under what conditions?

3. Consider one reaction that occurs in the atmosphere to make smog:
$$2 \text{ NO (g)} + O_2 \text{ (g)} \rightarrow 2 \text{ NO}_2 \text{ (g)}$$

   a) Calculate $\Delta S°_{rxn}$ and $\Delta G°_{rxn}$ for this process.

   b) If the pressures of both NO (g) and NO_2 (g) were doubled (relative to standard conditions), how would this affect ΔG? Defend your answer.

4. One process that occurs inside your body has a $\Delta G°_{rxn}$ of -82.4 kJ.

   a) In what way would this reaction couple with the ATP process?
   b) If this particular process occurred once, how many moles of ATP would be made or used? Assume maximum efficiency.

5. The $\Delta G°_f$ for most metal oxides is negative. Which metals do you think would have a positive $\Delta G°_f$ for oxide formation?

Answers to Try Its:

1. When the solid dissolves in the water, the solid becomes much more disordered; this favors spontaneity. Energywise, the process does not favor spontaneity because it is endothermic – it requires heat from the surroundings. The fact that the reaction is spontaneous (i.e., it happened) indicates that the disorder factor (or matter dispersal factor) won out over the energy factor at this temperature.

2. a) 0.050 M NaCl (it's more dilute); b) The molecule with 6 carbons in a row (it has more ways in which the bonds can rotate and the electron cloud can distort); c) the gas (gases are much less ordered than solids).

3. -523.8 kJ/mol The results match.

4. 393 K

5. ΔG_{rxn} = -45.0 kJ. Yes, the reaction is still spontaneous.

6. $2 Fe_2O_3 (s) + 4 C (graph) + O_2 (g) \rightarrow 4 Fe (s) + 4 CO_2 (g)$ $\Delta G°_{rxn}$ = -93.2 kJ

Answers to Self-Test:

1. 72.3°C

2. a) A positive ΔG indicates that the process does not occur spontaneously. In this case, the salt does not dissolve in water under standard conditions.

   b) In a solution process for a salt, a solid forms aqueous ions. This represents an increase in entropy, which is thermodynamically favorable for spontaneity. The fact that the ΔG is positive indicates that the ΔH must be *very* positive. ($\Delta G = \Delta H - T\Delta S$, so if ΔS is positive, then ΔH must be large and positive in order for ΔG to be positive.) This particular salt must have especially strong ion-ion attractive forces within its crystal lattice.

   c) Yes, it could be spontaneous at high temperatures.

3. a) $\Delta S°_{rxn}$ = -146.6 J/K; $\Delta G°_{rxn}$ = -72.5 kJ/mol; b) It will have no effect. $\Delta G_{rxn} = \Delta G°_{rxn} + RT \ln Q$, where Q=1. ln(1) = 0, so that the "RT lnQ" term drops out making $\Delta G_{rxn} = \Delta G°_{rxn}$. Here is how Q=1, given that the pressures of NO_2 and NO were doubled relative to standard conditions (standard pressure is 1 atm, so double would be 2 atm):

$$Q = \frac{\left(p_{NO_2}\right)^2}{\left(p_{NO}\right)^2 \left(p_{O_2}\right)^1} = \frac{(2)^2}{(2)^2 (1)^1} = 1$$

4. a) ATP would be made in this process because this process has a negative ΔG. Its free energy can be stored in ATP.

   b) Two moles of ATP could be formed: 2(30.6 kJ) + (-82.4 kJ) = -21.2 kJ (ΔG is still negative.)

5. Any precious metal (one that resists corrosion) will most likely have a positive $\Delta G°_f$ for the oxide. For example, $\Delta G°_f$ for Au_2O is +163 kJ/mol.

Chapter 15: Kinetics: Mechanisms and Rates of Reactions

Learning Objectives

In this chapter, you will learn how to:
- Calculate the overall rate and relative rates for a given reaction.
- Determine the specific effect of a reactant's concentration on the rate of a reaction.
- Predict the rate law for a given reaction from experimental data.
- Propose a reaction mechanism that is consistent with the rate law.
- Identify reaction intermediates and catalysts.
- Calculate the activation energy for a reaction.
- Draw energy level diagrams to illustrate reaction mechanisms.
- Show how a catalyst speeds up the rate of a reaction.

Practical Aspects

The three big questions a chemist will ask about any given chemical reaction are: 1) will it react?; 2) how fast will it react?; and 3) how much of it will react? The tools to answer questions 1 and 3 are provided in Chapters 14 and 16, respectively. In this chapter, you will learn how to assess question #2.

An understanding of kinetics – the study of reaction rates – is useful on two different levels:

- **Macroscopic** – if you know the factors that influence a reaction's rate, you can control the rate in the most effective manner possible. This is particularly useful for any manufacturing company that wants to get its product on the market in the least amount of time.

- **Molecular** – if you know the factors that influence a reaction's rate, you can predict the step-by-step pathway by which the reactant molecules get converted to products. (This pathway is called a reaction mechanism.) A reaction's mechanism is particularly useful to know, because it can help a chemist make use of that reaction's everyday applications. For example, knowledge of metabolic pathways in your body can help doctors diagnose diseases, and can help chemists to design medications.

15.1 WHAT IS A REACTION MECHANISM?

Key Terms:
- **Reaction mechanism** – the stepwise path that molecules take in converting from reactants to products. Each step represents one molecular event.
- **Elementary reaction** – one step within a reaction mechanism. There are three types of elementary reactions:
    - **Unimolecular** – one molecule fragments into two pieces or undergoes a rearrangement to form a new isomer.
    - **Bimolecular** – two molecules collide and combine or transfer atoms.
    - **Termolecular** – three molecules collide and combine or transfer atoms.

- **Intermediate** – a chemical species that is produced in an early step and consumed in a later step of the reaction mechanism.
- **Reaction rate** – number of molecular events occurring per unit time.
- **Rate-determining step** – the one elementary step in the reaction mechanism that is significantly slower than the others. This step limits or determines the rate of the reaction.

Key Concepts:

- The overall rate of a reaction can be calculated from its rate-determining step.
- The rate-determining step must be determined experimentally.
- The sum of the elementary reactions for a given mechanism must add to give the overall reaction.
- An elementary reaction must show a chemical *change* – bonds must be broken and/or formed.
- Breaking bonds requires energy and making bonds releases energy.
- For a reaction to occur, molecules must collide in the proper orientation and with sufficient energy to break the reactant bonds.

Helpful Hints

- The rate-determining step is also called the rate-limiting step, the slow step, or the bottleneck step.
- Chemical knowledge can sometimes be used to predict the rate-determining step. For example, a highly reactive substance will probably not be a reactant in the rate-determining step, but a very stable molecule might.
- Termolecular reactions are statistically much less likely to occur than unimolecular or bimolecular reactions.

EXERCISE 1: Identify which reactions illustrate elementary processes. Categorize each elementary process as unimolecular, bimolecular or termolecular.

a) $N_2O_5 + O \rightarrow O_2 + N_2O_4$ b) $Br + Br \rightarrow 2Br$

c) $2 C_8H_{18} + 25 O_2 \rightarrow 16 CO_2 + 18 H_2O$ d) $Br_2 \rightarrow 2Br$

STRATEGY: An elementary process shows one step in a reaction mechanism. Try to visualize the change that is taking place in each process.

SOLUTION:

a) This process shows one atom of O colliding with a molecule of N_2O_5. In the process, one bond to O in the N_2O_5 breaks and a new bond with the incoming O atom is formed. This is an elementary process involving the collision of two reactant molecules, so it is classified as bimolecular.

b) This process shows two atoms of Br resulting in two atoms of Br. No chemical change has occurred, so this is not a reaction. It therefore cannot be an elementary process, because an elementary process must show a chemical *change*.

c) This is not an elementary reaction because it would be virtually impossible for 2 molecules of C_8H_{18} to simultaneously collide with 25 molecules of O_2 at the exact proper orientation and with the proper impact. Elementary steps typically involve one or two molecules, but never twenty-seven.

d) This process shows one molecule of Br_2 breaking to form 2 atoms of Br. This is a unimolecular elementary reaction. (Notice how it differs from question "b.")

EXERCISE 2: There are two proposed mechanisms for the reaction:

$$CH_3CH_2Br + OH^- \rightarrow CH_3CH_2OH + Br^-$$

Mechanism I	Mechanism II
Step 1: $CH_3CH_2Br \rightarrow CH_3CH_2^+ + Br^-$ Step 2: $CH_3CH_2^+ + OH^- \rightarrow CH_3CH_2OH$	Step 1: $CH_3CH_2Br + OH^- \rightarrow CH_3CH_2OH + Br^-$

a) For each proposed mechanism, describe in words what is occurring at the molecular level, and then draw molecular pictures to accompany the description. b) Identify any intermediates in Mechanism I. c) Which step do you think is the rate-determining step in Mechanism I?

STRATEGY AND SOLUTION:

a) Each step in a reaction mechanism describes a molecular event in which bonds are broken and/or formed.

Mechanism I: One molecule of CH_3CH_2Br breaks apart into one $CH_3CH_2^+$ and one Br^-. The $CH_3CH_2^+$ then collides with one OH^- to form CH_3CH_2OH.

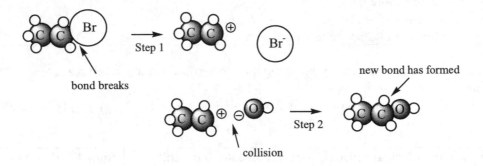

Mechanism II: One molecule of CH_3CH_2Br collides with one OH^-. During the collision, Br^- breaks off and OH^- attaches to the "CH_2" carbon atom to form CH_3CH_2OH.

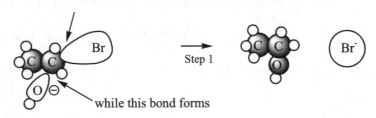

b) A reaction intermediate is a substance that is generated in an early step and consumed in a later step. In Mechanism I, the reaction intermediate is $CH_3CH_2^+$.

c) The rate-determining step is the slow step of the reaction mechanism. Although the rate-determining step must be determined by experiment, we can predict that step 1 will probably be the rate-determining step because it involves breaking a covalent bond to form a positively-charged carbon fragment. Carbon is not stable with a positive charge, so it will be highly reactive; step 2 should be quite fast.

Try It #1: Classify each elementary step in Exercise 2 as unimolecular, bimolecular, or termolecular.

EXERCISE 3: Consider the reaction: $2 NO_2 (g) + F_2 (g) \rightarrow 2 NO_2F (g)$. Write possible mechanisms for the reaction, given these requirements: a) Mechanism I: the first step is a unimolecular process that generates F atoms; b) Mechanism II: the first step is a bimolecular process.

STRATEGY: The mechanism is the series of elementary steps in going from reactants to products. The sum of all steps of the mechanism must equal the overall balanced equation.

SOLUTION:

a) The only way to obtain an F atom from a unimolecular process is to have one F_2 molecule fragment:	**Step 1:** $F_2 \rightarrow F + F$
We want to make NO_2F, so collide an NO_2 with an F that was generated in Step 1:	**Step 2:** $F + NO_2 \rightarrow NO_2F$
To obtain the overall reaction stoichiometry of 2 NO_2 and 1 F_2 producing 2 NO_2F, we still need: • to use the other F from Step 1, • to use one more NO_2, • to make one more NO_2F: We can account for all of these in a third step:	**Step 3:** $F + NO_2 \rightarrow NO_2F$

Quick check: Add the elementary steps together to verify that the overall reaction stoichiometry is obtained:

$$F_2 + \cancel{F} + NO_2 + \cancel{F} + NO_2 \rightarrow \cancel{F} + \cancel{F} + NO_2F + NO_2F = 2 NO_2 (g) + F_2 (g) \rightarrow 2 NO_2F (g)$$

b) This is a bimolecular process, so two molecules must collide:	**Step 1:** $F_2 + NO_2 \rightarrow NO_2F + F$
We generated an F atom in Step 1 that is not in the overall reaction. Use it now by reacting it with the other needed NO_2:	**Step 2:** $F + NO_2 \rightarrow NO_2F$

Quick check: Add the elementary steps together to verify that the overall reaction stoichiometry is obtained:

$$F_2 + NO_2 + \cancel{F} + NO_2 \rightarrow NO_2F + \cancel{F} + NO_2F = 2 NO_2 (g) + F_2 (g) \rightarrow 2 NO_2F (g)$$

Both reaction mechanisms seem reasonable because they both provide the proper overall stoichiometry.

Helpful Hints

Use these steps as we did in Exercise 3 to guide you through writing a reaction mechanism:
• If you generate an intermediate in one step, try to use it as a reactant in the next step.
• After each step, take a tally of what still needs to be used and produced.

Try It #2: Consider the reaction: $AC + B_2 \rightarrow AB + BC$.
a) Write a reaction mechanism for this reaction, given that the first step must involve the unimolecular fragmentation of AC. Identify any reaction intermediates.
b) Write a mechanism for this reaction, given that the first step must involve a bimolecular collision resulting in the formation of AB_2. Identify any reaction intermediates.

15.2 RATES OF CHEMICAL REACTIONS

A Molecular View

Key Term:
- **Kinetics** – the study of reaction rates.

Key Concepts:
- For a unimolecular process, the rate is constant on a *per molecule* basis.
- For a bimolecular process, the rate is constant on a *per collision* basis.

EXERCISE 4: What effect will cutting the initial reactant concentrations in half have on the rate for:
a) a unimolecular process? b) a bimolecular process?

STRATEGY AND SOLUTION:
a) The rate for a unimolecular process is constant on a *per molecule* basis. If the initial number of molecules is cut in half, then the rate will also be cut in half.

b) The rate for a bimolecular process is constant on a *per collision* basis. If the initial number of molecules is cut in half, then each molecule will have ½ the number of chances for *it to collide* with another molecule AND ½ the number of chances for another molecule to *collide with it*. The net result is that the rate will decrease by a factor of four; the rate will be ¼ the rate of the original reference reaction.

A Macroscopic View

Key Concepts:
- Reaction rate is measured as change in concentration over change in time.
- The overall rate of a reaction will decrease as the reaction progresses.
- The rate of a reaction can be discussed from an overall perspective *or* from the perspective of a given reaction component being consumed or generated. For a given reaction, $aA + bB \rightarrow cC + dD$:

$$Overall: \text{ Reaction rate } = -\frac{1}{a}\frac{\Delta[A]}{\Delta t} = -\frac{1}{b}\frac{\Delta[B]}{\Delta t} = \frac{1}{c}\frac{\Delta[C]}{\Delta t} = \frac{1}{d}\frac{\Delta[D]}{\Delta t}$$

or

$$Relative: \text{ Rate of production of } D = \frac{\Delta[D]}{\Delta t} \text{ or Rate of consumption of } A = -\frac{\Delta[A]}{\Delta t}$$

- Values for rates are always positive. Notice in the equations above that a negative sign is placed in front of terms describing reactants being consumed. This is done so that the value for rate will always be a positive number.

Useful Relationships:
- Reaction rate $= -\frac{1}{a}\frac{\Delta[A]}{\Delta t} = -\frac{1}{b}\frac{\Delta[B]}{\Delta t} = \frac{1}{c}\frac{\Delta[C]}{\Delta t} = \frac{1}{d}\frac{\Delta[D]}{\Delta t}$

This shows the relationship for reaction rates in the reaction:
$$aA + bB \rightarrow cC + dD$$

EXERCISE 5: Consider the combustion of gasoline occurring in a 6.0 L engine:
$$2\ C_8H_{18} + 25\ O_2 \rightarrow 16\ CO_2 + 18\ H_2O$$

If the concentration of C_8H_{18} dropped from 0.100 M to 0.086 M during the first 45 seconds of reaction,
   a) what is the rate of O_2 consumption during this time interval?
   b) what is the overall rate of reaction during this time interval?
   c) at what rate is water being produced?
   d) how many grams of carbon dioxide are formed?

STRATEGY: We know that rates for this reaction are related in this manner:

$$\text{Reaction rate} = -\frac{1}{2}\frac{\Delta[C_8H_{18}]}{\Delta t} = -\frac{1}{25}\frac{\Delta[O_2]}{\Delta t} = \frac{1}{16}\frac{\Delta[CO_2]}{\Delta t} = \frac{1}{18}\frac{\Delta[H_2O]}{\Delta t}$$

This basic relationship can be used as a foundation to answer all the questions in this exercise.

SOLUTION:

a) We're asked to find the rate of O_2 consumption during a 45 second time interval. We've been given the balanced chemical equation and concentration data for C_8H_{18}.
$$\text{The rate of } O_2 \text{ consumption} = -\frac{\Delta[O_2]}{\Delta t} \quad \text{and Reaction rate} = -\frac{1}{25}\frac{\Delta[O_2]}{\Delta t} = -\frac{1}{2}\frac{\Delta[C_8H_{18}]}{\Delta t}.$$

If we multiply both sides of the right-hand equation by 25, we can get the rate of O_2 consumption:

$$\textbf{Rate of O}_2 \textbf{ consumption} = -\frac{\Delta[O_2]}{\Delta t} = -\frac{25}{2}\frac{\Delta[C_8H_{18}]}{\Delta t} = -\frac{25}{2}\left(\frac{0.086M - 0.100\ M}{45\ sec}\right) = 3.9 \times 10^{-3}\ M/sec$$

b) We're asked to find the overall reaction rate during this time interval. We can obtain this from the relationship:

$$\textbf{Reaction rate} = -\frac{1}{2}\frac{\Delta[C_8H_{18}]}{\Delta t} = -\frac{1}{2}\left(\frac{0.086\ M - 0.100\ M}{45\ sec}\right) = 1.6 \times 10^{-4}\ M/sec$$

c) We're asked to find the rate at which water is being produced, which can also be found from the rate relationships:

$$\text{Rate of } H_2O \text{ production} = \frac{\Delta[H_2O]}{\Delta t}. \quad \text{Reaction rate} = -\frac{1}{2}\frac{\Delta[C_8H_{18}]}{\Delta t} = \frac{1}{18}\frac{\Delta[H_2O]}{\Delta t}$$

If we multiply both sides of the right-hand equation by 18, we can get the rate of H_2O production:

$$\textbf{Rate of H}_2\textbf{O production} = \frac{\Delta[H_2O]}{\Delta t} = -\frac{18}{2}\frac{\Delta[C_8H_{18}]}{\Delta t} = -\frac{18}{2}\left(\frac{0.086\ M - 0.100\ M}{45\ sec}\right) = 2.8 \times 10^{-3}\ sec$$

d) We're asked to find how many grams of carbon dioxide will form during this time interval, so we should use the relationship that relates octane consumption to carbon dioxide production:

$$\text{Reaction rate} = -\frac{1}{2}\frac{\Delta[C_8H_{18}]}{\Delta t} = \frac{1}{16}\frac{\Delta[CO_2]}{\Delta t}$$

Initially, the concentration of CO_2 was zero. (CO_2 is a product and no products are present at the start of the reaction.) If we find $[CO_2]$ at t = 45 seconds, we can calculate moles of CO_2 from it. Rearranging the above equation, we get:

$$\Delta[CO_2] = -\frac{16}{2}\frac{\Delta[C_8H_{18}]}{\Delta t} \times \Delta t = -8 \times \Delta[C_8H_{18}] = -8(0.086\ M - 0.100\ M) = 0.112\ M$$

$$\Delta[CO_2] = [CO_2]_f - [CO_2]_i \quad \text{so } [CO_2] \text{ at } 45 \text{ sec} = 0.112\ M$$

Working backwards from the goal, we can find mass of CO_2 from moles of CO_2. We can get moles of CO_2 from molarity of CO_2 if we know volume (6.0 L):

$$6.0\ L \times \frac{0.112\ mol}{L} \times \frac{44.01\ g}{mol} = 29.575\ g = 30.\ g\ of\ CO_2$$

The answers seem reasonable because the stoichiometry of 25 O_2 to 18 H_2O indicates that the rate of O_2 consumption should be slightly faster than the rate of H_2O production, and it is. Similarly, the overall rate in this reaction should be slower than the rate of any single component's rate, and it is.

Helpful Hint
- If you encounter a problem that involves calculating relative reaction rates, immediately set up the entire relationship of "Reaction rate $= -\frac{1}{a}\frac{\Delta[A]}{\Delta t} = -\frac{1}{b}\frac{\Delta[B]}{\Delta t} = +\frac{1}{c}\frac{\Delta[C]}{\Delta t} = +\frac{1}{d}\frac{\Delta[D]}{\Delta t}$," for that reaction (just like we did in Exercise 5), so that you'll have a basis from which to work.

Try It #3: Consider the reaction for photosynthesis: $6\ CO_2\ (g) + 6\ H_2O\ (l) \rightarrow C_6H_{12}O_6\ (s) + 6\ O_2\ (g)$
a) How does the rate of O_2 production relate to the rate of CO_2 consumption? b) How does the rate of glucose production relate to the rate of H_2O consumption?

EXERCISE 6: Consider six molecules of CO and four molecules of O_2 involved in the chemical reaction:
$$2\ CO\ (g) + O_2(g) \rightarrow 2\ CO_2\ (g)$$

Within 1.2 min of mixing the chemicals, 2 molecules of CO remain. a) Draw molecular pictures to show 1) initial conditions and 2) the contents of the container after 1.2 minutes; b) Use the pictures to determine which starting material is consumed at a faster rate.

STRATEGY AND SOLUTION:
a) The molecules are in the gas phase, so we should show them taking up the entire space of the container. We can show carbon atoms as black balls and oxygen atoms as white balls. Each CO that reacts uses up ½ of an oxygen molecule.

t = 0 min

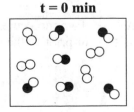

t = 1.2 min

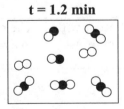

b) Initially, 6 CO molecules and 4 O_2 molecules were present. After 1.2 minutes, 2 CO molecules and 2 O_2 molecules are present. During this time interval, 4 molecules of CO and 2 molecules of O_2 reacted. Therefore, the rate of CO consumption is faster than the rate of O_2 consumption.

Try It #4: Consider the reaction taking place in Exercise 6. Calculate the rates of CO consumption and O_2 consumption during the 1.2 min time interval.

15.3   CONCENTRATION AND REACTION RATES

Key Terms:
- **Rate law** – mathematical expression that links the rate of a reaction to the concentrations of the substances that influence the reaction's rate.
- **Rate constant (k)** – proportionality constant in the rate law for a given reaction.
  - The rate constant changes with temperature.
  - The units for the rate constant for a given reaction are such that the units for rate are "concentration per unit time" (For example: "M/sec").
- **Reaction order** – the exponent on a given concentration term within the rate law.
- **Overall order** – sum of the individual orders in the rate law.

Key Concepts:
- The rate law is written in the form "Rate = $k[A]^y[B]^z$," where A and B are the concentrations of substances A and B, and y and z are their respective orders.
- The exponents in the rate law *depend upon the reaction mechanism*, not reaction stoichiometry, so reaction orders must be determined experimentally.
- Most reaction orders are zero, one or two, but fractional orders are also possible.
- Reaction orders are interpreted as shown in the table below.

Summary Table of Information Derived from Reaction Orders

For the substance "A" in the rate law "Rate = $k[A]^y[B]^z$":

If the exponent of A in the rate law is:	then the reaction is:	which means that A affects the reaction rate in this way:	and means that in the reaction mechanism:
0	Zero-order with respect to A	The rate does not depend upon concentration of A	Zero molecules of A are involved in or prior to the rate-determining step.
1	First-order with respect to A	The rate depends on A on a *per molecule* basis.	One molecule of A is involved in or prior to the rate-determining step.
2	Second-order with respect to A	The rate depends on A on a *per collision* basis.	Two molecules of A are involved in or prior to the rate-determining step.

EXERCISE 7: The rate law for a given reaction is: Rate = $k[R]^2[S]$. The box shown below indicates one set of conditions for the reaction, where R = white dots and S = black dots. Draw a molecular picture that illustrates initial conditions that are three times as fast as the conditions in this box. (Note: the temperature remains constant.)

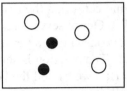

STRATEGY: The rate law relates reaction rate to concentrations. We must alter the concentrations of the reactants to increase the rate by three-fold.

SOLUTION: You may immediately see that one solution to this question is to triple the concentration of S. We'll break the problem down so that you can apply the reasoning to future, more difficult problems.

We're asked to find the conditions that would triple the rate. We've been given this reaction's rate law "Rate = $k[R]^2[S]$" and concentrations of R = 3 dots and S = 2 dots. This means the rate in the original box is:

$$\text{Rate} = k[R]^2[S] = k[3]^2[2] = 18k$$

We need to come up with initial concentrations, that when plugged into the rate law, will yield a value of 54k for the rate. (Three times the original rate would be: 3(18k) = 54k). The original plan to triple [S] works:

$$\text{Rate} = k[R]^2[S] = k[3]^2[6] = 54k$$

For this scenario, the only other simple combination to show in a molecular picture would be to have 1 R and 54 S. All other possible combinations result in non-whole numbers, which would be fine for real situations, but difficult to show in molecular pictures.

Our molecular picture is:

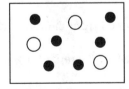

Try It #5: The rate law for the reaction A + 2 B → C + D is: Rate = $k[A]^2[B]^0$. The boxes shown below indicate initial reaction conditions, where A = white dots and B = black dots. a) How fast is the reaction in Box 2, compared to Box 1? b) How fast is the reaction in Box 3, compared to Box 1?

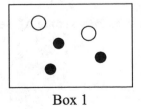

Box 1

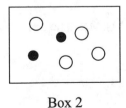

Box 2

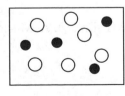

Box 3

EXERCISE 8: Describe the rate law from Exercise 7 in terms of reaction orders.

STRATEGY: The rate law shows the order of each reactant molecule.

SOLUTION: The rate law: "Rate = $k[R]^2[S]$" indicates that the reaction is second-order with respect to R and first-order with respect to S ($[S] = [S]^1$.) Our observations from Exercise 7 are consistent with S being a first-order process, where the rate is constant on a per molecule basis. When we tripled the number of molecules of S, we tripled the rate. The reaction is second-order in R, which means the rate depends on [R] on a per collision basis. Had we tripled the [R], we would have increased the rate by $[3]^2 = 9$ times; the chance of molecules colliding would have increased nine-fold.

EXERCISE 9: In Exercise 3, we had proposed the two mechanisms shown below for this reaction:
$$2 NO_2 (g) + F_2 (g) \rightarrow 2 NO_2F (g).$$

Mechanism I	**Mechanism II**
Step 1: $F_2 \rightarrow F + F$	Step 1: $F_2 + NO_2 \rightarrow NO_2F + F$
Step 2: $F + NO_2 \rightarrow NO_2F$	Step 2: $F + NO_2 \rightarrow NO_2F$
Step 3: $F + NO_2 \rightarrow NO_2F$	

Step 1 in each mechanism shows bonds breaking, while the remaining steps show only bond formation. Step 1 is the rate-determining step in each mechanism. Answer these questions for a) Mechanism I, and b) Mechanism II:
- What rate law would be predicted for this mechanism?
- What is the overall reaction order?
- What are the units for the rate constant if the rate is to be expressed in units of M/sec ($Msec^{-1}$)?

STRATEGY: The rate law and the reaction mechanism must agree.

SOLUTION:
a) Mechanism I shows one molecule of F_2 fragmenting in the slow step. The reaction is therefore first-order with respect to F_2. The reaction rate is independent of $[NO_2]$, so the overall reaction order is one. The predicted rate law for this mechanism is: Rate = $k[F_2]$.

The units for the rate constant can be determined from the rate law relationship:

Rate = $k[F_2]$, so M/sec = (?)(M). The units for k must be 1/sec, or sec^{-1}.

b) Mechanism II shows one molecule each of F_2 and NO_2 colliding with each other in the rate-determining first step. The reaction is first-order with respect to F_2 and first-order with respect to NO_2, which makes the reaction second-order overall. The predicted rate law for this mechanism is:
$$Rate = k[F_2][NO_2]$$

Again, the units for the rate constant can be determined from the rate law relationship:

Rate = $k[F_2][NO_2]$, so M/sec = (?)(M)(M). The units for k must be 1/(Msec) or $M^{-1}sec^{-1}$.

Try It #6: Refer back to the two proposed mechanisms in Try It #2. For each mechanism (assume the first step is rate-limiting): a) write a rate law; b) determine the order with respect to each reactant; c) determine the overall reaction order; d) determine the units for the rate constant.

15.4 EXPERIMENTAL KINETICS

Key Term:
- **Half-life** – time it takes for ½ of a substance to undergo reaction.

Key Concepts:
- One way to determine if a reaction is first or second order is to collect concentration data at several time intervals, and make first- and second-order plots of the data. (See Useful Relationships.) The plot that yields a straight-line relationship provides the correct order.
- The standard convention for collecting kinetics data for plots is to collect data until the concentration of the substance analyzed has fallen to one eighth its original amount (i.e., three half-lives). Only after this amount of time has passed, can linearity or curvature within most plots be detected.
- The half-life for a first-order process is constant.
- The half-life for a second-order process *changes* as concentration changes.

Useful Relationships:

- $\ln\left\{\dfrac{[A]_0}{[A]}\right\} = kt$ — This is the mathematical relationship for a first-order process. If the process is first-order, a plot of $\ln\left\{\dfrac{[A]_0}{[A]}\right\}$ vs t will give a straight line with a slope of k.

- $\dfrac{1}{[A]} - \dfrac{1}{[A]_0} = kt$ — This is the mathematical relationship for a second-order process. If the process is second-order, a plot of $\dfrac{1}{[A]} - \dfrac{1}{[A]_0}$ vs t will give a straight line with a slope of k.

- $t_{1/2} = \dfrac{\ln 2}{k}$ — Use this relationship to determine the half-life for a first-order process. The derivation of this relationship is provided in the text.

EXERCISE 10: The data shown in the table below were collected at some temperature for the reaction: $2\ NOCl\ (g) \rightarrow 2\ NO_2\ (g) + Cl_2\ (g)$. a) What is the overall order of reaction?; b) What is the rate law for this reaction? Include the value of the rate constant at this temperature.; c) If the reaction is first-order overall, what is its half-life?

$$2\ NOCl\ (g) \rightarrow 2\ NO_2\ (g) + Cl_2\ (g)$$

[NOCl]: ($\times 10^{-4}$ M)	5.53	4.45	3.59	2.89	2.33	1.88	1.51	1.22	0.982	0.791	0.637
Time: (sec)	0	15	30	45	60	75	90	105	120	135	150

STRATEGY: We must make first- and second-order plots in order to determine the reaction order. The calculations and plots are shown on the next page.

First-Order Data and Plot

time (sec)	ln(Ao/A)
0	0
15	0.217284
30	0.432036
45	0.648931
60	0.86432
75	1.078916
90	1.298078
105	1.511337
120	1.728352
135	1.944645
150	2.161173

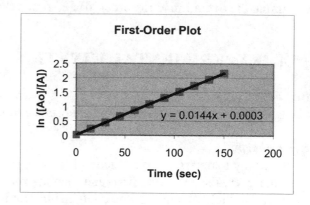

Second-Order Data and Plot

time (sec)	1/A-1/Ao
0	0
15	438.8727
30	977.1971
45	1651.889
60	2483.527
75	3510.831
90	4814.198
105	6388.403
120	8374.981
135	10833.91
150	13890.27

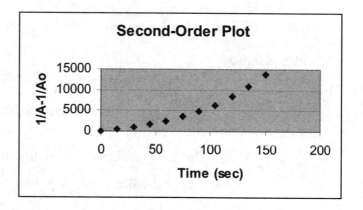

SOLUTION:

a) The first-order plot is linear, so the reaction is first-order in [NOCl].

b) The rate law for the reaction is: Rate = k[NOCl], where k = 0.01444/sec at the temperature the reaction was performed. (Recall that the units of k will result in rate units of M/sec.)

c) The half-life for a first-order process can be found from:

$$t_{1/2} = \frac{\ln 2}{k}, \text{ so t } \frac{1}{2} = (\ln 2)/(0.01444/sec) = 48.0 \text{ seconds}$$

This answer seems reasonable, because the data shows that the concentration of NOCl decreased by nearly ½ every 45 seconds.

Try It #7: The kinetics of the reaction: 2 AB → A$_2$ + 2B was studied at 25°C. The initial [AB] was 0.100 M. Determine the rate law for the reaction from the experimental data:

[AB]: (x10^{-2} M)	6.63	4.96	3.96	3.30	2.82	2.47	2.19	1.97	1.79	1.64	1.52	1.41	1.32	1.23
Time: (min)	20	40	60	80	100	120	140	160	180	200	220	240	260	280

"Isolation" Experiments

Key Term:

- **Isolation experiment** – kinetics experiment that isolates the effect that one particular reactant has on the reaction rate. This is done by making that particular reactant's starting concentration significantly lower than the other reactants' concentrations. This method is only used for reactions that contain more than one reactant. For example:
  - The reaction $aA + bB \rightarrow dD$ has two reactants, so isolation experiments are performed to determine the rate law for the reaction. The rate law for the reaction is: Rate = $k[A]^x[B]^y$, where y and z are the orders of A and B, respectively. If the initial [A] is at least 100 times smaller than [B], then the rate law can be re-written: Rate = $k_{obs}[A]^x$, where $k_{obs} = k[B]^y$. The slope in the straight-line plot for A will be k_{obs}.

Key Concepts:

- If more than one reactant is present, then isolation experiments must be performed in order to determine reaction orders.
- In an isolation experiment, the initial concentration of the reactant being studied is at least 100 times smaller than all other reactants. This way, only the concentration of that particular reactant undergoes a significant change. The other reactant concentrations are essentially constant and can be lumped together with the rate constant, k.

Useful Relationship:

- $k_{obs} = k[B]^y[C]^z$ "k_{obs}" is the *observed* rate constant in an isolation experiment, where [B] & [C]>>>>[A]. B and C represent all other reactants.

EXERCISE 11: Here are experimental results for the reaction: $C_2 + 2B \rightarrow 2\ BC$, studied at 19°C:

Trial	$[C_2]$	[B]	Plot of $\ln\left\{\dfrac{[C_2]_0}{[C_2]}\right\}$ vs t	Plot of $\dfrac{1}{[C_2]} - \dfrac{1}{[C_2]_0}$ vs t
1	0.00100 M	0.125 M	Linear Equation: y = 0.0010775x	Non-linear
2	0.00100 M	0.250 M	Linear Equation: y = 0.002151x	Non-linear

Determine the rate law for the reaction, to include the value for the rate constant at 19°C.

STRATEGY: The rate law for this reaction will be: Rate = $k[C_2]^x[B]^y$. We must determine the values of x, y, and k.

We can see that isolation experiments were performed to determine the effect that $[C_2]$ has on the rate. Notice that the $[C_2]$ is over 100 times smaller than the [B], so that only the $[C_2]$ changes appreciably during the reaction. The rate law can be re-written:

$$\text{Rate} = k_{obs}[C_2]^x, \text{ where } k_{obs} = k[B]^y$$

The results show that the first-order plot of $[C_2]$ is linear, so x = 1, and the reaction is first-order with respect to $[C_2]$.

To determine the order of reaction with respect to [B] from the data provided, we must utilize the fact that k_{obs} may change when [B] is altered, but k won't change as long as temperature remains constant.

For Trial 1, $k_{obs1} = k[B]_1^y$ and for Trial 2, $k_{obs2} = k[B]_2^y$.

$$k = \frac{k_{obs1}}{[B]_1^y} = \frac{k_{obs2}}{[B]_2^y} \text{ ; which rearranges to: } \frac{k_{obs2}}{k_{obs1}} = \frac{[B]_2^y}{[B]_1^y} = \left(\frac{[B]_2}{[B]_1}\right)^y$$

$$\frac{k_{obs2}}{k_{obs1}} = \left(\frac{[B]_2}{[B]_1}\right)^y = \frac{0.002151}{0.0010775} = \left(\frac{0.250 \text{ M}}{0.125 \text{ M}}\right)^y \text{ ; which simplifies to:}$$

$1.996 = (2)^y$, so $y = 1$. The reaction is first-order with respect to [B].

To find the value for k, we can use:
$$k = \frac{k_{obs1}}{[B]_1^y} = \frac{0.0010775 / \sec}{0.125 \text{ M}} = 0.00862 \text{ M}^{-1}\sec^{-1}$$

SOLUTION: The rate law for the reaction is: Rate $= k[C_2][B]$, where $k = 0.000862 \text{ M}^{-1}\sec^{-1}$ at 19°C.

Key Concept:
- The method shown in Exercise 11 for determining the order with respect to [B] is only done if the experimental conditions don't allow for performing an isolation experiment for [B]. If an isolation experiment could be done for [B], then the experiment above would simply be repeated with [B]<<<<[C_2]. Then, first- and second-order plots for [B] would be made.

Initial Rates

Key Concepts:
- Method of Initial Rates –method used to determine the rate law of a reaction in which you determine the rate of the reaction over a very short time interval at the start of the reaction (hence "initial").
- To do method of initial rates, a chemist will systematically vary the concentrations of the reactants to see their concentration influence on initial rate. For example, if the initial rate of reaction doubles when the concentration of a given reactant doubles, then the reaction is first order in that reactant.

EXERCISE 12: Use the experimental data in the table to determine the rate law (including the value for the rate constant, k) for the reaction:

$$3 \text{ R} + \text{ S} \rightarrow \text{R}_3\text{S}$$

Experiment	Temperature (°C)	Initial [R] (M)	Initial [S] (M)	Initial Rate (M/sec)
1	20.0	0.30 M	0.10 M	3.6×10^{-2}
2	20.0	0.10 M	0.10 M	4.0×10^{-3}
3	20.0	0.10 M	0.50 M	2.0×10^{-2}

STRATEGY: The rate law for this reaction will be: Rate = $k[R]^x[S]^y$. We must determine the values of x, y, and k.

To find the value for x, we must compare two experiments that involve a change in the initial concentration of R, while keeping the initial concentration of S constant. (This lets us focus on R's influence on the rate.) Experiments 1 and 2 vary initial concentration of R but not of S, so these experiments are the appropriate ones to compare. Suggestion: put Experiment #1 in the numerator so that when we divide through, we won't have fractional numbers…it's easier to analyze this way.

$$\frac{\text{Initial rate}_1}{\text{Initial rate}_2} = \frac{k[R]^x{}_1[S]^y{}_1}{k[R]^x{}_2[S]^y{}_2}$$

$$\frac{3.6 \times 10^{-2}\,M/sec}{4.0 \times 10^{-3}\,M/sec} = \frac{k[0.30M]^x[0.10M]^y}{k[0.10M]^x\ [0.10M]^y}$$

We can cancel out the k, as well as the $[0.10\,M]^y$ and do the algebra to get:

$$9.0 = \left(\frac{0.30M}{0.10M}\right)^x = (3.0)^x ;\text{ so } x = 2$$

To find the value for y, we must now compare two experiments for which initial concentration of R is now constant, but initial concentration of S is varied. This allows us to hone in on how [S] influences the rate. The two experiments to compare are Experiments 2 and 3. Notice that Experiments 1 and 3 would not be a good pair to choose, because both initial [R] and initial [S] are varied. Suggestion: put Experiment #3 (with the larger numbers) in the numerator to simplify the math:

$$\frac{\text{Initial rate}_3}{\text{Initial rate}_2} = \frac{k[R]^x{}_3[S]^y{}_3}{k[R]^x{}_2[S]^y{}_2} = \frac{2.0 \times 10^{-2}\,M/sec}{4.0 \times 10^{-3}\,M/sec} = \frac{k[0.10M]^x[0.50M]^y}{k[0.10M]^x\ [0.10M]^y}$$

We can cancel out the k, as well as the $[0.10\,M]^x$ and do the algebra to get:

$$5.0 = \left(\frac{0.50M}{0.10M}\right)^y = (5.0)^y ;\text{ so } x = 1$$

Now, to get the rate constant, notice that every experiment is performed at the same temperature, so every experiment should have the same rate constant. Notice, too, that in every experiment we have the initial concentration of each reactant, the initial rate, and the values for x and y. We should be able to plug in all of these values for any one experiment and arrive at the same value for k. Let's calculate k from both Experiment 1 and Experiment 2 data, just to show that we get the same value for k from both:

$$\text{Initial rate}_1 = k[R]^x{}_1[S]^y{}_1, \text{ so } k = \frac{\text{Initial rate}_1}{[R]^2{}_1[S]^1{}_1} = \frac{3.6 \times 10^{-2}\,M/sec}{(0.30\,M)^2(0.10\,M)^1} = 4.0\,M^{-1}\,sec^{-1}$$

$$\text{Initial rate}_2 = k[R]^x{}_2[S]^y{}_2, \text{so } k = \frac{\text{Initial rate}_2}{[R]^2{}_2[S]^1{}_2} = \frac{4.0 \times 10^{-3}\,M/\text{sec}}{(0.10\,M)^2\,(0.10\,M)^1} = 4.0\,M^{-1}\,\text{sec}^{-1}$$

SOLUTION: The rate law for this reaction will be: Rate $= k[R]^2[S]^1$, where $k = 4.0\,M^{-1}\text{sec}^{-1}$ at 20°C. It should seem reasonable that the rate constant is one value at one temperature. It should also seem reasonable that the order with respect to R is 2, because when we tripled the initial concentration of R, the initial rate increased by a factor of 9 ($9 = 3^2$). Finally, it should make sense that the order with respect to S is 1 because when we increased the initial concentration of S by a factor of 5, the initial rate increased by a factor of five.

Try It #8: Consider Exercise #12. Determine the value of the rate constant, k, from Experiment 3.

15.5 LINKING MECHANISMS AND RATE LAWS

Key Concepts:
- A rate expression can be written for any elementary reaction:
  - $A + B \rightarrow$ products; elementary rate $= k[A][B]$
  - $A + A \rightarrow$ products; elementary rate $= k[A]^2$
  - $C \rightarrow$ products; elementary rate $= k[C]$
- The rate law is written from the rate expression for the rate-determining step.
- If the rate law and the proposed mechanism for a reaction do not correlate with one another, then the mechanism is considered invalid and can be eliminated as a possible choice.

Rate-Determining First Step

EXERCISE 13: In Exercise 2, we had proposed two possible mechanisms for the reaction:
$$CH_3CH_2Br + OH^- \rightarrow CH_3CH_2OH + Br^-$$

Mechanism I	Mechanism II
Step 1: $CH_3CH_2Br \rightarrow CH_3CH_2^+ + Br^-$ Step 2: $CH_3CH_2^+ + OH^- \rightarrow CH_3CH_2OH$	Step 1: $CH_3CH_2Br + OH^- \rightarrow CH_3CH_2OH + Br^-$

a) What are the predicted rate laws for the two proposed mechanisms?
b) Given the following experimental results for this reaction, which proposed mechanism is not valid?

- An isolation experiment was performed, in which $[CH_3CH_2Br] <<< [OH^-]$.
  - A plot of $\ln[CH_3CH_2Br]_0/[CH_3CH_2Br]$ vs. time gave a straight-line relationship.
  - A plot of $(1/[CH_3CH_2Br]) - 1/[CH_3CH_2Br]_0$ vs. time was non-linear.

- A second isolation experiment was performed, in which $[CH_3CH_2Br] >>> [OH^-]$.
  - A plot of $\ln[OH^-]_0/[OH^-]$ vs. time gave a straight-line relationship.
  - A plot of $(1/[OH^-]) - 1/[OH^-]_0$ vs. time was non-linear.

STRATEGY: a) Each predicted mechanism shows the first step as rate-determining, so the rate law can be written from the first step. (Notice that no bonds are breaking in Step 2 of Mechanism I, so Step 2 must be fast relative to Step 1.)   b) Interpret the experimental results and compare them to the two mechanisms.

SOLUTION:
a) Mechanism I: Rate = $k[CH_3CH_2Br]$; Mechanism II: Rate = $k[CH_3CH_2Br][OH^-]$.

b) The $\ln[CH_3CH_2Br]_0/[CH_3CH_2Br]$ vs. time plot is linear, which indicates the reaction is first-order with respect to $[CH_3CH_2Br]$. This is consistent with the rate laws for both mechanisms.

The $\ln[OH^-]_0/[OH^-]$ vs. time plot is linear, so the reaction is first-order with respect to $[OH^-]$. The results from this experiment show that Mechanism I is not valid, because Mechanism I's rate law is zero-order with respect to $[OH^-]$, not first-order, as the experimental results indicate. Mechanism II is still a possible mechanism for this reaction, because its rate law correlates to the experimental results.

Try It #9: Consider the isolation experiments performed in Exercise 13. What conclusions would have been drawn if both plots done for the $[CH_3CH_2Br]>>>[OH^-]$ experiment were non-linear?

EXERCISE 14: Propose a mechanism for the overall reaction: $C_2 + 2B \rightarrow 2BC$, given that the rate law for the reaction is: Rate = $k[C_2][B]$.

STRATEGY: The rate law shows that the order with respect to C_2 is 1 and the order with respect to B is 1, which predicts that one C_2 and one B collide in the rate-determining step. An overall reaction order of 2 is reasonable, so it is possible to have these molecules collide in the first step. The other B in the reaction stoichiometry will need to be used in a later step.

SOLUTION:

Step 1: $C_2 + B \rightarrow BC + C$ (slow)
Step 2: $B + C \rightarrow BC$
Quick check: $C_2 + B + B + \cancel{C} \rightarrow BC + \cancel{C} + BC$; the mechanism is reasonable.

Try It #10: If the rate law for the reaction in Try It # 2 is Rate = $k[AC]$, which proposed mechanism is not valid?

Rate-Determining Later Step

Key Concepts:

- Overall reaction orders that are greater than two or contain fractions are indications that the first step in the mechanism is not the rate-determining step.
- Intermediates are so reactive that their concentrations typically cannot be measured. Intermediates, therefore, can never appear in the rate law. *A rate law that shows an intermediate must be re-written to eliminate it from the equation.*
- Any steps prior to the rate-determining step will exist in dynamic equilibrium. (The rate of the forward reaction will equal the rate of the reverse reaction.)

EXERCISE 15: The rate law for the reaction: $A_2 + BC \rightarrow AB + AC$ was determined from experiments to be: Rate = $k[BC][A_2]^{\frac{1}{2}}$. a) Propose a mechanism for the reaction. b) Show that the mechanism and rate law correlate.

STRATEGY AND SOLUTION: (Hint: Notice that the overall reaction order is 1.5, which indicates that the first step is not rate-determining.)

a) The rate law indicates that one BC molecule and ½ of an A_2 molecule collide in the rate-determining step. ½ of an A_2 molecule would be A, so the slow step could be:

$$A + BC \rightarrow AB + C \text{ (slow)}$$

A is not a reactant (A_2 and BC are reactants), so A must be generated *prior to* the slow step: $A_2 \rightarrow A + A$. If this reaction is before the slow step, then its reactants and products will exist in dynamic equilibrium. This is indicated with a double arrow. At this point we have:

Step 1: $A_2 \leftrightarrow A + A$ (fast, equilibrium)
Step 2: $A + BC \rightarrow AB + C$ (slow)

The final step needs to use up the intermediates A and C. This will be a fast reaction, since bonds are forming but no bonds are breaking:

Step 3: $A + C \rightarrow AC$

b) The rate law is written from the slow step of the reaction: Rate = k[A][BC]. However, A is an intermediate, so the rate law must be re-written to eliminate it from the equation. We'll specify the rate constant for elementary step #2 with a numerical subscript: Rate = $k_2[A][BC]$.

Step 1 in the mechanism shows the relationship between the reactant, A_2, and the intermediate, A. Write rate expressions for these elementary steps: $rate_{forward} = k_1[A_2]$; $rate_{reverse} = k_{-1}[A]^2$. Step 1 is in dynamic equilibrium, so the rate of the forward process is equal to the rate of the reverse process:

$$rate_{forward} = rate_{reverse}; \text{ so } k_1[A_2] = k_{-1}[A]^2$$

Solving for [A], we get:

$$[A] = \left(\frac{k_1}{k_{-1}}\right)^{1/2} [A_2]^{1/2}$$

252

We can now eliminate [A] from the originally proposed rate law:

$$\text{Rate} = k_2[A][BC] = k_2 \left(\frac{k_1}{k_{-1}}\right)^{1/2} [A_2]^{1/2}[BC]$$

$$\text{Rate} = k[A_2]^{1/2}[BC]; \text{ where } k = k_2 \left(\frac{k_1}{k_{-1}}\right)^{1/2}$$

This shows that the rate law correlates to the proposed mechanism.

Try It #11: The reaction: $2 \text{ NO (g)} + 2 \text{ H}_2 \text{(g)} \rightarrow \text{N}_2 \text{(g)} + 2 \text{ H}_2\text{O (g)}$ was studied. The rate law for the reaction was found to be: Rate = $k[\text{NO}]^2[\text{H}_2]$. Propose a mechanism for the reaction.

15.6   REACTION RATES AND TEMPERATURE

Key Terms:
- **Activation energy (E_a)** – energy barrier that must be overcome in order for reactants to convert to products.
- **Activation energy diagram** – plot of energy on the y-axis vs. progress of reaction on the x-axis. The x-axis is labeled "reaction coordinate." This is also called a reaction coordinate diagram.
- **Activated complex** – the substance that exists at the highest-energy point on a hump in an activation energy diagram. This substance is in transition between bonds; the bonds forming and bonds breaking are shown with dotted lines. An activated complex is also called a transition state.

Key Concepts:
- Energy is required to break bonds.
- Energy is released when bonds are formed.

Helpful Hint
- Most reactions involving stable molecules have activation energies over 100 kJ/mol.

Useful Relationships:
- $k = Ae^{-E_a/RT}$

  This is the Arrhenius equation, which relates a reaction's rate constant to its temperature. It is used to find E_a for a reaction. A is the value the rate constant would have if all the molecules had enough energy to react. Rearranging the equation shows that a plot of ln k vs. 1/T will give a straight line with a slope of $-E_a/R$. T must be in Kelvin units.

- $E_a = \dfrac{R \ln \dfrac{k_2}{k_1}}{\left(\dfrac{1}{T_1} - \dfrac{1}{T_2}\right)}$

  This is the form of the Arrhenius equation that is typically used. If you have the rate constants at two different temperatures for a reaction, then you can calculate the reaction's activation energy. The derivation of this equation is provided in the text.

EXERCISE 16: Match the picture to the process:

Pictures:

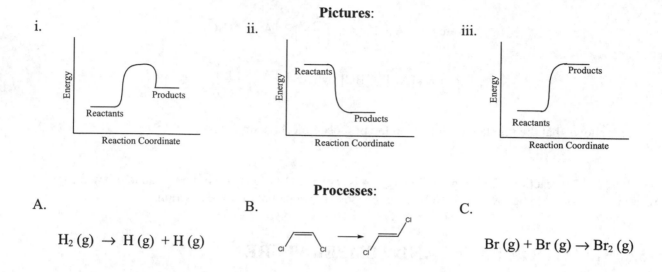

i.   ii.   iii.

Processes:

A.   B.   C.

$$H_2 (g) \rightarrow H (g) + H (g)$$

$$Br (g) + Br (g) \rightarrow Br_2 (g)$$

STRATEGY AND SOLUTION:

Reaction A. One molecule breaks apart into two pieces. A bond is broken and no new bonds are formed. This can be called a unimolecular fragmentation. It requires energy to break bonds and no new bonds were formed, so as we move along the reaction coordinate, we should see an increase in energy, but no drop. The products will be in a higher energy state than the reactants. This is consistent with picture iii.

Reaction B. For this process to occur, the π bond must break, the central σ bond must rotate, and a new π bond must form. This is an example of a unimolecular rearrangement. The activation energy diagram should show a rise in energy for bond breaking followed by a drop in energy for bond formation. The picture that illustrates this is picture i.

Reaction C. This process only shows bond formation. There will be a minimal activation energy. The products will be much more stable than the reactants. This is consistent with picture ii.

EXERCISE 17: Given the rate law, draw an activation energy diagram for the reaction:

$$(CH)_3CCl + H_2O \rightarrow (CH)_3COH + HCl; \quad Rate = k[(CH)_3CCl]$$

Assume the reaction is slightly exothermic. In the diagram, label the activation energy with a single-headed arrow, and draw pictures of any activated complexes.

STRATEGY: We should first propose a reaction mechanism for this reaction. The rate law indicates that the rate-limiting step must be the unimolecular fragmentation of $(CH)_3CCl$. This is a reasonable slow step, so it is probably the first step in the mechanism. HCl is a strong acid, so it will exist as H^+ and Cl^- as products.

Step 1: $(CH)_3CCl \rightarrow (CH)_3C^+ + Cl^-$ (slow)
Step 2. $(CH)_3C^+ + H_2O \rightarrow (CH)_3COH + H^+$

The activation energy diagram should show one hump for each step of the mechanism. The first hump is the rate-determining step, so it will be the largest hump. The reaction is exothermic, so the products should be shown at a lower energy state than the reactants.

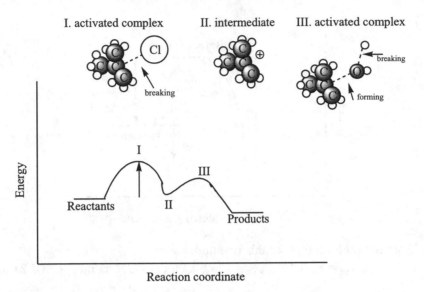

Try It #12: Draw a picture of the activated complex in each step of the proposed mechanism for this reaction: $AC + B_2 \rightarrow AB + BC$. The proposed mechanism is: Step 1: $AC + B_2 \rightarrow AB_2 + C$; Step 2: $C + AB_2 \rightarrow AB + BC$

EXERCISE 18: The rate constant for a given reaction is 0.0265 $M^{-1}sec^{-1}$ at 25.0°C and 0.153 $M^{-1}sec^{-1}$ at 38.0°C. What is the reaction's activation energy?

STRATEGY: Use the relationship for calculating activation energy from two data points. Be sure to convert to Kelvin first.

$$E_a = \frac{R \ln \dfrac{k_2}{k_1}}{\left(\dfrac{1}{T_1} - \dfrac{1}{T_2}\right)} = \frac{\left(\dfrac{8.314 \, J}{mol \, K}\right)\left(\ln \dfrac{0.153 \, M^{-1}sec^{-1}}{0.0265 \, M^{-1}sec^{-1}}\right)}{\left(\dfrac{1}{298 \, K} - \dfrac{1}{311 \, K}\right)} = \frac{\left(\dfrac{8.314 \, J}{mol \, K}\right)(1.75329)}{1.403 \times 10^{-4} \, K} = 103{,}923 \, J/mol$$

SOLUTION: The activation energy for this reaction is 104 kJ/mol. (3 sig figs) The answer seems reasonable; many reactions have activation energies over 100 kJ/mol.

EXERCISE 19: A given aqueous-phase reaction has an activation energy of 126.8 kJ/mol and a ΔH_{rxn} of 43.7 kJ/mol. a) What is ΔH for the reverse reaction? b) What is E_a for the reverse reaction?

STRATEGY: a) ΔH is a state function, so $\Delta H_{reverse} = -\Delta H_{forward} = -43.7$ kJ/mol. b) The activation energy is the energy barrier that must be overcome for the reaction to proceed. It is the difference in energy states between the reactants and the highest-energy activated complex. Use the information provided to draw a general activation energy diagram in order to visualize the relationships between the variables. The picture helps to illustrate that the products for the forward reaction are the *reactants* for the reverse reaction. $\Delta H = \Delta E$ for this reaction. (This is an aqueous-phase reaction, so no volume change occurs. Recall from Chapter 6, that $\Delta H = \Delta E$ if $P\Delta V$ is zero.)

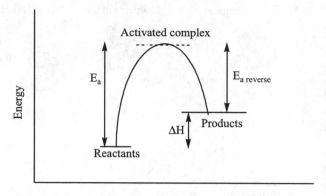

Reaction coordinate

We can see from the picture, that for this reaction:

$$E_{a\ reverse} = E_a - \Delta H_{forward} = 126.8\ kJ/mol - 43.7\ kJ/mol = 83.1\ kJ/mol$$

SOLUTION: a) $\Delta H_{reverse\ rxn} = -43.7\ kJ/mol$. b) $E_{a\ reverse} = 83.1\ kJ/mol$

15.7 CATALYSIS

Key Terms:
- **Catalyst** – substance that speeds up the rate of a reaction by providing a lower-energy mechanism for the reaction. Catalysts can be classified as:
  - **Homogeneous** – the catalyst and the reactants are in the same phase.
  - **Heterogeneous** – the catalyst and the reactants are in different phases. Heterogeneous catalysts are often used in industrial reactions.
- **Adsorption** – the adhering of reactant molecules to the surface of a heterogeneous catalyst.
- **Enzyme** – catalyst for a biochemical process.
  - **Substrate** – the chemical that binds directly to an enzyme. The product is called an enzyme-substrate complex, and is written "E-S."

Key Concepts:
- *A catalyst is consumed in an early step and regenerated in a later step.*
- The general pathways for reactions catalyzed by an enzyme or a heterogeneous catalyst are similar:
  - **Enzyme** – substrate binds to the enzyme, the substrate distorts and reacts with an incoming reactant molecule, then the product leaves the enzyme.
  - **Heterogeneous catalyst** – reactants adsorb onto surface of catalyst, distort, react, and then leave the catalyst surface.
- Many enzymatic processes are first-order in both enzyme and substrate.

Helpful Hint
See Figure 14-21 in the text for an illustration of the general mechanism for heterogeneous catalysis.

EXERCISE 20: Hydrogen peroxide decomposes slowly over time to form water and oxygen:
$$2 \ H_2O_2 \ (aq) \ \rightarrow \ 2 \ H_2O \ (l) \ + \ O_2 \ (g)$$

The activation energy diagram for the reaction can be shown like this:

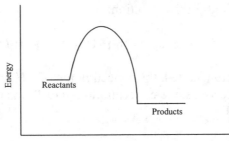

The rate of this reaction speeds up greatly when a small amount of aqueous KI is added to the solution. The proposed mechanism for this process is:

Step 1: H_2O_2 (aq) $+ I^-$ (aq) $\rightarrow$ H_2O (l) $+$ IO^- (aq)   (slow)
Step 2: H_2O_2 (aq) $+$ IO^- (aq) $\rightarrow$ H_2O (l) $+$ O_2 (g) $+ I^-$ (aq)

a) Identify the catalyst and the intermediate in the mechanism. How can they be differentiated from each other?
b) Is the catalyst a homogeneous or heterogeneous catalyst?
c) Draw an activation energy diagram to illustrate the mechanism of the catalyzed reaction. Show the activation energy with an arrow. Show the activated complex(es).

STRATEGY AND SOLUTION:
a) The catalyst is I^- (aq) and the intermediate is IO^- (aq). A catalyst is used in an early step and regenerated in a later step – this describes the behavior of I^-. An intermediate is generated in an early step and consumed in a later step, just as IO^- is doing.

b) I^- (aq) is a homogeneous catalyst because it is in the same phase as the hydrogen peroxide.

c) The activation energy diagram will show 1) a lower activation energy than the original diagram, and 2) two humps to illustrate two steps.

In the activated complex, bonds being formed and broken are shown with dotted lines. We'll distinguish the bonds being formed with stars.

Un-catalyzed reaction (reference)

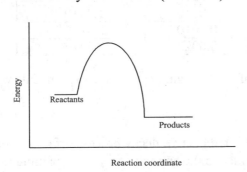

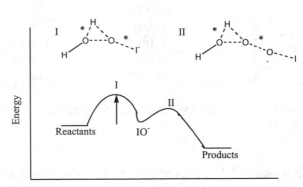

Chapter 15 Self-Test

You may use a periodic table and a calculator for this test.

1. The reaction rate was studied for this reaction:

    $$8 \text{ Al (s) } + 3 \text{ NO}_3^- \text{ (aq) } + 5 \text{ OH}^- \text{ (aq) } + 18 \text{ H}_2\text{O (l) } \rightarrow 8 \text{ Al(OH)}_4^- \text{ (aq) } + 3 \text{ NH}_3 \text{ (aq)}$$

    At 23.8 seconds after the reaction started, the concentration of OH⁻ was 0.247 M. After the reaction had stirred for a total of 69.3 seconds, the concentration of OH⁻ was 0.211 M. The total volume of the reaction mixture was 250.0 mL.

    a) How many grams of Al (s) were consumed during this time interval?
    b) What is the overall reaction rate during this time interval?

2. The reaction: $A_2 + B_2 \rightarrow 2 \text{ AB}$ was studied and the following observations were made:
    *   Data collected in an isolation experiment where $[B_2]$<<<$[A_2]$ showed a straight-line plot for $\ln([B_2]_o/[B_2])$ vs. time
    *   In other experiments performed, when the $[A_2]$ was doubled, the rate increased by $\sqrt{2}$.
    *   The reaction mixture got cool to the touch during each reaction.

    a) What is the rate law for this reaction?

    b) Propose a mechanism for this reaction. Identify any intermediates or catalysts in your mechanism.

    c) Draw a reaction coordinate diagram that is consistent with your mechanism. Include these features in the diagram:
        *   label the activation energy with a single-headed arrow;
        *   label the ΔH with a double-headed arrow;
        *   draw a picture of each activated complex.

3. For initial rate experiments were performed at the same temperature for the reaction:
    $$\text{BrO}_3^- \text{ (aq) } + 5 \text{ Br}^- \text{ (aq) } + 6 \text{ H}^+ \text{ (aq) } \rightarrow 3 \text{ Br}_2\text{(aq) } + 3 \text{ H}_2\text{O (aq)}$$

Experiment #	Initial [BrO₃⁻] (M)	Initial [Br⁻] (M)	Initial [H⁺] (M)	Initial Rate (M/sec)
1	0.0010	0.0010	0.0010	6.0×10^{-7}
2	0.0020	0.0010	0.0010	1.2×10^{-6}
3	0.0020	0.0020	0.0010	2.4×10^{-6}
4	0.0010	0.0010	0.0020	2.4×10^{-6}

    Determine the rate law for the reaction, including the value of k (at the temperature the experiments were performed).

4. Radioactive ^{99m}Tc is used as a medical diagnostic tool. All radioactive decay processes follow first-order kinetics. 8.00 μg of ^{99m}Tc was injected into a patient, and the amount of ^{99m}Tc remaining inside

the patient was monitored at several time intervals (see below table). How much ^{99m}Tc remained in the patient after 24 hours?

Time (hours):	0	3.00	6.00	9.00	12.00
Mass of ^{99m}Tc (μg):	8.00	5.66	4.00	2.83	2.00

5. Two rate experiments were performed for the reaction: $X_2 + 2\,OH^- \rightarrow 2\,XOH^-$. In these experiments, OH^- was the limiting reagent and a huge excess of X_2 was used. The time at which the OH^- was completely consumed could be monitored with an indicator dye. What is the order of reaction with respect to $[OH^-]$? Briefly explain your reasoning.

Initial concentration of OH^- (M):	0.10	0.20
Time it took for indicator to change color (sec):	30	60

6. Of the three elementary processes shown below,

    a) which reaction should have the lowest activation energy?
    b) which reaction should have the highest activation energy? Briefly explain.

        I. CO_2 (g) $\rightarrow CO$ (g) $+ O$ (g)
        II. CH_4 (g) $\rightarrow$ (g) CH_3 (g) $+ H$ (g)
        III. O (g) $+ O$ (g) $\rightarrow O_2$ (g)

Answers to Try Its:

1. Mechanism I: Step 1 = unimolecular; Step 2 = bimolecular. Mechanism II: Step 1 = bimolecular.
2. a) Step 1: $AC \rightarrow A + C$; Step 2: $A + B_2 \rightarrow AB + B$; Step 3: $B + C \rightarrow BC$; Intermediates: A, C & B
    b) Step 1: $AC + B_2 \rightarrow AB_2 + C$; Step 2: $C + AB_2 \rightarrow AB + AC$; Intermediates: AB_2 and C
3. a) Rate of O_2 production = rate of CO_2 consumption.
    b) Rate of glucose production = 1/6 rate of H_2O consumption.
4. Rate of CO consumption = 3.33 molecules/minute; Rate of O_2 consumption = 1.67 molecules/min.
5. a) four times as fast; b) nine times as fast. (Notice: [B] has no effect on the rate!)
6. Mechanism I: Rate = k[AC]; order wrt AC = 1; order wrt B_2 = 0; overall order = 1; units = sec^{-1}.
    Mechanism II: Rate = k[AC][B_2]; order wrt AC = 1, and B_2 = 1; overall order = 2; units = $M^{-1}sec^{-1}$.
7. Rate = $k[AB]^2$, where k = 0.254 $M^{-1}sec^{-1}$ at 25°C.
8. k = 4.0 $M^{-1}sec^{-1}$ at 20°C
9. We would conclude that Mechanism II is not valid, because Mechanism II is first-order wrt [OH⁻].
10. Mechanism "b" would not be valid because that mechanism predicts a rate law of "rate = k[AC][B_2]."
11. Step 1: $NO + NO \leftrightarrow N_2O_2$ (fast, equilibrium);   Step 2: $N_2O_2 + H_2 \rightarrow N_2O + H_2O$ (slow)
    Step 3: $N_2O + H_2 \rightarrow N_2 + H_2O$
12.

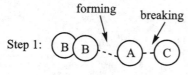   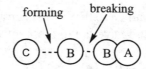

Answers to Self-Test:

1. a) 0.389 g Al; b) 1.58×10^{-4} M/sec
2. a) Rate = $k[A_2]^{1/2}[B_2]$.
    b)
    Step 1: $A_2 \leftrightarrow A + A$ (fast)
    Step 2: $A + B_2 \rightarrow AB + B$ (slow)
    Step 3: $A + B \rightarrow AB$

    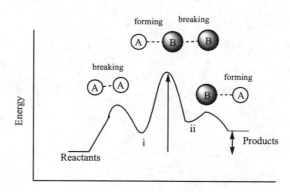

    In diagram,
    "i" = intermediate A has formed
    "ii" = intermediate B has formed

3. Rate = $k[BrO_3^-]^1[Br^-]^1[H^+]^2$, where k = 6.0×105 $M^{-3}sec^{-1}$ at the temperature of the experiments. (hint: for order for [BrO_3^-], compare experiments 1 & 2; for [Br^-], compare 2 & 3; for [H^+], compare 1 & 4.
4. 0.500 μg (hint: did you notice that the half-life was 6.0 hours? This makes k = 0.1155/hr)
5. The reaction is zero order with respect to [OH⁻]. Doubling the concentration had no effect on the *rate*. Rate = $-\frac{1}{2}\Delta[OH^-]/\Delta t$ = 0.00167 M/sec for both experiments.
6. a) O (g) + O (g) $\rightarrow$ O_2 (g): This reaction will have the lowest activation energy of the three reactions shown, because it is the only one that does not show bonds breaking.
    b) We must compare the bond energies of the bonds breaking in Reaction I and II to determine which has the greater activation energy. In Reaction I, a carbon-oxygen double bond within CO_2 is broken. In reaction II, carbon-hydrogen single bond is being broken. The C=O in CO_2 is the stronger bond, so it will be more difficult to break. Therefore, CO_2 (g) $\rightarrow$ CO (g) + O (g) will have the greatest activation energy.

Chapter 16: Principles of Chemical Equilibrium

Learning Objectves

In this chapter, you will learn how to:
- Write equilibrium constant expressions for reactions.
- Assess the value of K_{eq}, the equilibrium constant, for a given reaction.
- Determine the spontaneous direction of a reaction given its initial conditions.
- Use Le Chatelier's Principle to determine the consequences of applying various stresses to a system that is already at equilibrium.
- Categorize equilibrium reactions by type.
- Write appropriate balanced chemical equations to illustrate an equilibrium process (given information on starting materials).
- Calculate equilibrium concentrations of reaction components.

Practical Aspects

This chapter is an introduction to chemical equilibrium. The material in this chapter will build and be used in the next two chapters. It is essential to have a solid understanding of the material in this chapter in order to be able to master Chapters 17 and 18.

Equilibrium is one of the foundational studies in chemistry. As you will see, chemical equilibrium is a factor in every single chemical reaction that takes place. An understanding of the concepts of chemical equilibrium can help an industrial chemist to figure out a way to maximize the yield of a chemical reaction and an environmental chemist to assess the effects of acid rain on plants, lakes, and animals. A medical doctor might use the concepts of chemical equilibrium to understand the delicate balance of the thousands of reactions that take place inside our bodies. A marine biologist could use the knowledge to assess the amount of dissolved oxygen in the ocean, which impacts what life forms can exist in various climates. An engineer would need an understanding of this topic to design a smokestack "scrubber" to minimize air pollution from an industrial plant. These are just a few of the countless examples in which chemical equilibria concepts come into play.

16.1 DESCRIBING CHEMICAL EQUILIBRIA

Key Term:
- **Equilibrium** – condition in which the rate of the forward reaction equals the rate of the reverse reaction.

Key Concept:
- **Equilibrium is dynamic**. At the macroscopic level, a system at equilibrium shows no detectable change, but at the molecular level, reactant molecules convert to product molecules, and vice versa at the same rate.

The Equilibrium Constant & Equilibrium Constant Expression

Key Terms:

- **Equilibrium constant, K_{eq}** – ratio of the rate constants for the forward reaction and the reverse reaction.
- **Equilibrium constant expression** – the mathematical equation which links the equilibrium constants to the concentrations of the chemicals involved in the reaction.

For any reaction: $aA + bB \leftrightarrow cC + dD$, the equilibrium constant expression is:

$$K_{eq} = \frac{[C]_{eq}^c [D]_{eq}^d}{[A]_{eq}^a [B]_{eq}^b}$$

The equilibrium constant and its expression are derived from the rate laws for the forward and reverse reactions. For example:

$$2\ CO\ (g)\ +\ O_2\ (g) \leftrightarrow 2\ CO_2\ (g)$$

Rate of forward reaction = $k_{forward} [CO]_{eq}^2 [O_2]_{eq}$ Rate of reverse reaction = $k_{reverse} [CO_2]_{eq}^2$

At equilibrium, the rate of the forward reaction equals the rate of the reverse reaction:

$$\text{Rate of forward reaction} = \text{Rate of reverse reaction}$$

$$k_{forward} [CO]_{eq}^2 [O_2]_{eq} = k_{reverse} [CO_2]^2$$

$$\frac{k_{forward}}{k_{reverse}} = \frac{[CO_2]_{eq}^2}{[CO]_{eq}^2 [O_2]_{eq}}, \text{ which also} = K_{eq}$$

Helpful Hint

- Note that the equilibrium constant is capital "K" and rate constants are lower case "k." To further emphasize the difference, the equilibrium constant is sometimes given a subscript, "eq."

Key Concepts:

- K_{eq} is constant for a given process at a given temperature. If the temperature changes, the value of K_{eq} changes. This should seem reasonable because rate constants change with temperature. *How* the equilibrium constant changes with temperature will be discussed in Section 16.3.
- For a given process, the value of K_{eq} depends only upon temperature.
- K_{eq} is independent of the direction from which the equilibrium was approached.

EXERCISE 1: Given the reaction: $H_2 (g) + Cl_2 (g) \leftrightarrow 2\ HCl\ (g)$, show the relation of the equilibrium constant to the rate constants for the reaction.

STRATEGY: The equilibrium constant is the ratio of the rate constants for the forward process to the reverse process. At equilibrium, the rate of the forward reaction equals the rate of the reverse reaction, so if we write rate laws for both the forward and reverse reaction and set them equal to each other, we can solve for the ratio of the two rate constants. This will be the equilibrium constant.

SOLUTION:

Rate of forward reaction $= k_{forward}[H_2]_{eq}[Cl_2]_{eq}$ and Rate of reverse reaction $= k_{reverse}[HCl]^2_{eq}$

$$k_{forward}[H_2]_{eq}[Cl_2]_{eq} = k_{reverse}[HCl]^2_{eq}$$

$$\frac{k_{forward}}{k_{reverse}} = \frac{[HCl]^2_{eq}}{[H_2]_{eq}[Cl_2]_{eq}} = K_{eq}$$

EXERCISE 2: Write equilibrium constant expressions for:

　　　a) $2\ HCl\,(g) \leftrightarrow H_2\,(g) + Cl_2\,(g)$ 　　　　& 　　　b) $\frac{1}{2}\ H_2\,(g) + \frac{1}{2}\ Cl_2\,(g) \leftrightarrow HCl\,(g)$

SOLUTION:

　a) $K_{eq} = \dfrac{[H_2]_{eq}[Cl_2]_{eq}}{[HCl]^2_{\ eq}}$ 　　　　　　　　b) $K_{eq} = \dfrac{[HCl]_{eq}}{[H_2]^{1/2}_{eq}[Cl_2]^{1/2}_{eq}}$

Try It #1: Write the equilibrium constant expression for: $2\ H_2O\,(g) \leftrightarrow 2\ H_2\,(g) + O_2\,(g)$

16.2   PROPERTIES OF EQUILIBRIUM CONSTANTS

While working through this section, keep in mind the concepts you learned in Section 16.1:
- K_{eq} applies only at equilibrium.
- K_{eq} is independent of initial conditions.
- K_{eq} is related to the stoichiometry of the balanced reaction.

Units and K_{eq}

Key Concepts:
- **By convention, K_{eq} is unitless.**
- Chemicals in the solid or liquid phase do not appear in the equilibrium constant expression.
- Concentrations in the equilibrium constant expression are *at* equilibrium, so the subscript "eq" is often not written after each concentration.
- Gas concentrations are denoted as partial pressures with the symbol "p" instead of "[ ]."

EXERCISE 3: Write the equilibrium constant expression for each reaction.
　　　a) $CaCl_2\,(s) \leftrightarrow Ca^{2+}\,(aq) + 2\ Cl^-\,(aq)$
　　　b) $H_3O^+\,(aq) + OH^-\,(aq) \leftrightarrow 2\ H_2O\,(l)$
　　　c) $Ag^+\,(aq) + Cl^-\,(aq) \leftrightarrow AgCl\,(s)$

STRATEGY: All of the expressions will show the ratio of products over reactants raised to the power of their respective coefficients. Remember that solids and liquids are not included in the expression.

SOLUTION:

a) $K_{eq} = [Ca^{2+}][Cl^-]^2$ b) $K_{eq} = \dfrac{1}{[H_3O^+][OH^-]}$ c) $K_{eq} = \dfrac{1}{[Ag^+][Cl^-]}$

Try It #2: Write the equilibrium constant expression for: $SO_3\,(g) + H_2O\,(l) \leftrightarrow H_2SO_4\,(l)$

Direction of a Reaction, Effect of Coefficients on Equilibrium Constant

Examples best illustrate the impacts that the direction of reaction and coefficients have on an equilibrium constant.

EXERCISE 4: At some temperature, the equilibrium constant for the reaction below is 25.

$$H_2\,(g) + Cl_2\,(g) \leftrightarrow 2\,HCl\,(g); \quad K_{eq} = 25$$

Use this information to calculate the equilibrium constants for these reactions, assuming the same temperature.

a) $2\,HCl\,(g) \leftrightarrow H_2\,(g) + Cl_2\,(g)$ &   b) $\tfrac{1}{2}\,H_2\,(g) + \tfrac{1}{2}\,Cl_2\,(g) \leftrightarrow HCl\,(g)$

STRATEGY: The three reactions shown here were the reactions used in Exercises 1 & 2. You saw that the equilibrium constant expression for each reaction is:

Original: $K_{eq} = \dfrac{[HCl]^2}{[H_2][Cl_2]}$ a) $K_{eq} = \dfrac{[H_2][Cl_2]}{[HCl]^2}$ b) $K_{eq} = \dfrac{[HCl]}{[H_2]^{1/2}[Cl_2]^{1/2}}$

These are gas phase reactions, so we now know that the equilibrium constant expression should be written in terms of partial pressures, rather than molar concentrations:

Original: $K_{eq} = \dfrac{(p_{HCl})^2}{(p_{H_2})(p_{Cl_2})}$ a) $K_{eq} = \dfrac{(p_{H_2})(p_{Cl_2})}{(p_{HCl})^2}$ b) $K_{eq} = \dfrac{(p_{HCl})}{(p_{H_2})^{1/2}(p_{Cl_2})^{1/2}}$

Notice that the equilibrium constant expression for reaction "a" is the inverse of the original one. Notice, too, that the expression for reaction "b" is the square root of the original. The mathematical outcome of the equilibrium constants must be similarly related.

SOLUTION: a) $K_{eq} = \dfrac{1}{K_{eq\ original}} = \dfrac{1}{25} = 0.040$ b) $K_{eq} = \sqrt{K_{eq\ original}} = \sqrt{25} = 5.0$

Key Concepts:
- Exercise 4 illustrates some important relationships among equilibrium constants:
  - Reversing a reaction inverses the original K_{eq}: $K_{eq\ reverse} = 1/K_{eq\ forward}$
  - Doubling coefficients squares the corresponding K_{eq}; tripling them cubes the K_{eq}; halving them square roots the K_{eq}.
- In general, any changes to the balanced equation will result in a new equilibrium constant whose value can be determined by comparing the two reactions.

Magnitude of Equilibrium Constants

Since the equilibrium constant is a measure of the concentrations of products over reactants, it follows that the more products present relative to reactants, the larger the equilibrium constant. The more reactants relative to products, the smaller the equilibrium constant.

Key Concept:
- The larger the equilibrium constant, the more products (and thus less reactants) are present at equilibrium.

EXERCISE 5: Consider the three processes below. All are at 25°C. Place the reactions in increasing order of which will have the most products at equilibrium.

   a) $PbCl_2$ (s) $\leftrightarrow$ Pb^{2+} (aq) + 2 Cl^- (aq); $K_{eq} = 1.7 \times 10^{-5}$
   b) H_3O^+ (aq) + OH^- (aq) $\leftrightarrow$ 2 H_2O (l); $K_{eq} = 1.0 \times 10^{14}$
   c) Ag^+ (aq) + Cl^-(aq) $\leftrightarrow$ AgCl(s); $K_{eq} = 5.6 \times 10^9$

STRATEGY: The larger the equilibrium constant, the more products are present at equilibrium, so we need to arrange the reactions in order of increasing values of K_{eq}.

SOLUTION: "a" has the smallest K_{eq}, so that reaction will have the fewest products at room temperature. "c" has the next smallest K_{eq}. "b" has the largest K_{eq} and thus the greatest amount of products at room temperature. It is worthwhile to note that reactions "b" and "c" both have very large equilibrium constants, so they will consist of mostly products at equilibrium. Reaction "a," with a K_{eq} of much less than 1, will consist of mostly reactants at equilibrium.

EXERCISE 6: Compare the reactions shown in Exercise 5. Which is less soluble in water, $PbCl_2$ or AgCl?

STRATEGY: Part "a" in Exercise 5 shows $PbCl_2$ dissolving in water with a K_{eq} of 1.7×10^{-5}. The process shown for AgCl in Exercise 5 is the precipitation of AgCl, which is the reverse of dissolving. Thus K_{eq} for dissolving AgCl is $1/K_{precipitation} = 1/5.6 \times 10^9 = 1.8 \times 10^{-10}$.

SOLUTION: Comparison of the two constants – 1.8×10^{-10} for AgCl dissolving and 1.7×10^{-5} for $PbCl_2$ dissolving – shows that AgCl, with a smaller equilibrium constant, is less soluble in water than $PbCl_2$.

Helpful Hint
- One of the most often-used concepts in chemical equilibrium is the comparison of two reactions to determine which will make more (or less) products. The key to comparing reactions is to make sure

you are assessing equilibrium constants *for the same process*. As shown in Exercise 6, assessing the same process sometimes involves reversing a reaction, and adjusting the corresponding equilibrium constant prior to making the comparison.

Try It #3: Which will have a greater concentration of OH^- ions at equilibrium? (Assume reactant concentrations are the same.)

$$C_5H_5NH^+ \text{ (aq)} + OH^- \text{ (aq)} \leftrightarrow C_5H_5N\text{(aq)} + H_2O \text{ (l)}; K_{eq} = 5.9 \times 10^8$$
$$\text{Or}$$
$$NH_3 \text{ (aq)} + H_2O \text{ (l)} \leftrightarrow OH^- \text{ (aq)} + NH_4^+ \text{ (aq)}; K_{eq} = 1.8 \times 10^{-5}$$

16.3   THERMODYNAMICS AND EQUILIBRIUM

In Chapter 14 we learned the effect of concentrations on ΔG. For the general reaction:
$$aA + bB \leftrightarrow cC + dD$$

$$\Delta G_{rxn} = \Delta G^\circ_{rxn} + RT\ln Q \text{ where } Q = \frac{[C]^c[D]^d}{[A]^a[B]^b}$$

Useful Relationship:

- $\Delta G^\circ_{rxn} = -RT\ln K_{eq}$ This relationship is derived from the relationship above, and applies specifically to the situation in which the system is at equilibrium, so $\Delta G_{rxn} = 0$ and $Q = K_{eq}$. The relationship between ΔG° and K_{eq} can also be shown in terms of the equilibrium constant: $K_{eq} = e^{-\Delta G^\circ/RT}$.

  For these relationships, ΔG° is expressed in units of kJ per mole *of reaction as it is written*, so the "e^{-x}" term will be unitless.

Summary of the Relationship Between Q and K_{eq}

If:	Then:	This is the case because:
$Q < K_{eq}$	The system will proceed spontaneously in the forward direction.	$\Delta G_{rxn} < 0$ There are fewer products and more reactants in the Q situation than at equilibrium. To spontaneously proceed toward equilibrium, some reactants must be used up and products must be created.
$Q > K_{eq}$	The system will proceed spontaneously in the reverse direction.	$\Delta G_{rxn} > 0$ There are too many products relative to reactants, so some products must be consumed and reactants created to reach equilibrium.
$Q = K_{eq}$	The system is at equilibrium.	$\Delta G_{rxn} = 0$ The concentrations of reactants and products won't change.

There are many ways an equilibrium constant for a reaction can be determined. From a "big picture" standpoint, it is worth seeing these many ways as interchangeable. K_{eq} can be determined from:

1. The relationship $\Delta G^\circ_{rxn} = -RT \ln K_{eq}$; (calculate ΔG°_{rxn} from thermodynamic data).
2. The ratio of the rate constants for the forward and reverse reactions (recall that is the definition of an equilibrium constant): $K_{eq} = k_f/k_r$.
3. Equilibrium concentrations of reaction components and the equilibrium constant expression.
4. pK values in the back of a textbook or reference text. (We will do this in the next chapter.)

EXERCISE 7: Use thermodynamic data to calculate the K_{eq} for: $CaCl_2$ (s) $\Leftrightarrow$ Ca^{2+} (aq) + 2 Cl^- (aq)

STRATEGY AND SOLUTION: We must first calculate ΔG°_{rxn}, then use the relationship $K_{eq} = e^{-\Delta G^\circ/RT}$.

$$\Delta G^\circ_{rxn} = \Sigma \text{ (coeff) } \Delta G^\circ_f(\text{products}) - \Sigma \text{ (coeff) } \Delta G^\circ_f(\text{reactants})$$

$$= [(1 \text{ mol } Ca^{2+})(-553.6 \text{ kJ/mol}) + (2 \text{ mol } Cl^-)(-131.0 \text{ kJ/mol})] - [(1 \text{ mol } CaCl_2)(-748.8 \text{ kJ/mol})]$$
$$= -66.8 \text{ kJ/mol rxn}$$

The term $\dfrac{\Delta G^\circ_{rxn}}{RT} = \dfrac{(-66.8 \text{ kJ/mol})}{(8.314 \times 10^{-3} \text{ kJ/molK})(298 \text{ K})} = -26.96$

So $K_{eq} = e^{-\Delta G^\circ/RT} = e^{-(-26.96)} = 5.1 \times 10^{11}$

EXERCISE 8: Assume that chemicals are placed in a container so that if no reaction had yet occurred, the concentration of calcium ion is 0.500 M, the concentration of chloride ion is 3.00 M, and 4.05 g of solid calcium chloride are present. Predict the spontaneous direction of reaction.

STRATEGY: The equilibrium reaction from Exercise 7 could be used to solve this problem:

$$CaCl_2 \text{ (s) } \Leftrightarrow Ca^{2+} \text{ (aq) } + 2 Cl^- \text{ (aq); } K_{eq} = 5.1 \times 10^{11}$$

$$Q = [Ca^{2+}][Cl^-]^2 = (0.500)(3.00)^2 = 4.5$$

SOLUTION: $Q << K_{eq}$ so the reaction will proceed in the forward direction to make more aqueous ions.

Equilibrium Constants and Temperature

The relationship of K_{eq} to temperature can be seen in the relationship: $\ln K_{eq} = -\dfrac{\Delta H^\circ}{RT} + \dfrac{\Delta S^\circ}{R}$

Process	Temperature Relationship	Reasoning
Exothermic	K_{eq} decreases as T increases, and vice versa.	ΔH°_{rxn} is "-" for an exothermic process. This makes the first term in the above equation "+," so this term will decrease in magnitude as temperature increases.
Endothermic	K_{eq} increases as T increases, and vice versa.	ΔH°_{rxn} is "+" for an endothermic process. The first term in the above equation decreases in magnitude (becomes less negative) as temperature increases.

EXERCISE 9: Use thermodynamic data to calculate the equilibrium constant at 325 K for

$$NH_4Cl(s) \Leftrightarrow NH_4^+ (aq) + Cl^- (aq)$$

STRATEGY: The thermodynamic relationship between K_{eq} and temperature is $K_{eq} = e^{-\Delta G°/RT}$. We must first find $\Delta G°_{rxn}$. We are at a non-standard temperature, so we must use the relationship:

$$\Delta G° = \Delta H° - T\Delta S°$$

Once we have $\Delta G°$, we can use the relationship: $K_{eq} = e^{-\Delta G°/RT}$ to find the value for the equilibrium constant, K_{eq}.

Note: Appendix D does not contain thermodynamic data for solid ammonium chloride, so it is provided here: $\Delta H°_f$ is –314.4 kJ/mol and $S°$ is –94.6 J/molK.

$$\Delta H°_{rxn} = \Sigma \text{ (coeff) } \Delta H°_f(\text{products}) - \Sigma \text{ (coeff) } \Delta H°_f(\text{reactants})$$

$$= [(1 \text{ mol } NH_4^+)(-133.3 \text{ kJ/mol}) + (1 \text{ mol } Cl^-)(-167.1 \text{ kJ/mol})] - [(1 \text{ mol } NH_4Cl)(-314.4 \text{ kJ/mol})]$$
$$= 14.0 \text{ kJ/mol rxn}$$

$$\Delta S°_{rxn} = \Sigma \text{ (coeff) } S°_f(\text{products}) - \Sigma \text{ (coeff) } S° (\text{reactants})$$

$$= [(1 \text{ mol } NH_4^+)(111.2 \text{ J/mol K}) + (1 \text{ mol } Cl^-)(56.5 \text{ J/mol K})] - [(1 \text{ mol } NH_4Cl)(94.6 \text{ J/mol K})]$$
$$= 73.1 \text{ J/mol K rxn}$$

Now, find $\Delta G°$ from the equation: $\Delta G° = \Delta H° - T\Delta S°$

$$\Delta G° = 14.0 \text{ kJ} - \left[325 \text{ K} \times \frac{73.1 \text{ J}}{K} \times \frac{1 \text{ kJ}}{1000 \text{ J}} \right] = -9.76 \text{ kJ}$$

Now, find K_{eq} from the equation: $K_{eq} = e^{-\Delta G°/RT}$

SOLUTION: $K_{eq} = e^{-\Delta G°/RT} = e^{-[(-9.76 \text{ kJ/mol})/(0.008314 \text{ kJ/molK})(325 \text{ K})]} = 37$. This answer seems reasonable – since $\Delta G°$ is a negative value, K_{eq} should be greater than 2.71 (2.71 is e^0, so if $\Delta G°$ were 0 or negative, then K_{eq} would be smaller than 2.71.) Furthermore, $\Delta G°$ is not large in magnitude, so K_{eq} shouldn't be large in magnitude either.

EXERCISE 10: The reaction in Exercise 9 is endothermic. Qualitatively compare the equilibrium constant at room temperature versus at 325 K.

SOLUTION: The equilibrium constant at room temperature for this process will be smaller than at 325 K. K_{eq} increases with increasing temperature for an endothermic process, as one can see by the relationship of K_{eq} to temperature in the equation $\ln K_{eq} = -\dfrac{\Delta H°}{RT} + \dfrac{\Delta S°}{R}$.

16.4 SHIFTS IN EQUILIBRIUM

**Le Chatelier's Principle: If a stress is applied to a system at equilibrium,
the equilibrium will adjust to compensate for that stress.**

Types of Stresses and Their Impact on the Reaction Mixture:

Type of Stress	What will happen	Reasoning
Change amounts of reagents (Add or remove reactants or products.)	Adding reactants or products will push reaction to opposite side. Removing reactants or products will push reaction to that side.	K_{eq} is a constant at a given temperature. Adding or removing reaction components creates a temporary non-equilibrium state. Comparison of Q to K_{eq} will determine spontaneous direction of reaction to re-establish K_{eq}.
Change temperature	If exothermic: K_{eq} and T are inversely related If endothermic: K_{eq} and T are directly related	K_{eq} changes with temperature according to the equation: $\ln K_{eq} = -\dfrac{\Delta H^\circ}{RT} + \dfrac{\Delta S^\circ}{R}$
Add a catalyst	Nothing	A catalyst speeds up the rate of reaction, but has no effect on the equilibrium state.
Add an inert substance	Nothing	An inert substance is not involved in the K_{eq} expression.

The chemical reasoning behind Le Chatelier's Principle is the fact that an equilibrium constant is *constant* at a given temperature. This is best illustrated through example, as shown in Exercise 11.

EXERCISE 11: Consider the reaction: $2 NO_2 (g) \leftrightarrow N_2O_4 (g)$; where $\Delta H = -55.3$ kJ. Let's say you have a closed container of NO_2 that has been sitting long enough to reach an equilibrium state. The equilibrium constant expression for this reaction is:

$$K_{eq} = \frac{p_{N_2O_4}}{p_{NO_2}^2}$$

- What happens if some extra NO_2 is added to the system?
  STRATEGY: This is an increase in the concentration of NO_2, so the system will temporarily be out of equilibrium. Q must be assessed:

$$Q = \frac{p_{N_2O_4}}{p_{NO_2}^2}$$

An increase in the concentration of NO_2 will make the value of $Q < K_{eq}$, so the reaction will proceed spontaneously in the forward direction to re-establish equilibrium. When the reaction proceeds in the forward direction, more products are formed and some reactants are consumed. *SOLUTION:* More N_2O_4 will be produced and some NO_2 will be consumed.

- Water reacts with NO_2 (g) to make HNO_3 (l). What happens if a little bit of water is added to the container?
  STRATEGY: The net result of the addition of water to the system is that NO_2 is removed, i.e., the concentration of NO_2 decreases. Again, we are at a temporary non-equilibrium state, so a comparison of Q to K_{eq} must be made. Looking at the Q expression, we see that decreasing NO_2 will increase Q. Thus, when NO_2 is removed, $Q > K_{eq}$, so the reaction will proceed spontaneously in the reverse direction to re-establish equilibrium.
  SOLUTION: Some N_2O_4 will decompose to make more NO_2.

- What happens if you add in a chemical that isn't in the equilibrium expression, nor reacts with any of the species involved the equilibrium?
  STRATEGY: Addition of an inert species does not affect temperature or Q, so it will have no effect on the equilibrium.
  SOLUTION: Nothing will happen.

- What happens if a catalyst is added?
  STRATEGY: Addition of a catalyst does not affect temperature or Q, so it will have no effect on the equilibrium. A catalyst only speeds up the rate at which the reaction approaches equilibrium, but has no effect on the equilibrium state.
  SOLUTION: Nothing will happen.

- What happens if the temperature is increased?
  STRATEGY A change in temperature actually changes the value of K_{eq}. This reaction is exothermic because it has a negative ΔH. Therefore, as temperature increases for this process, K_{eq} decreases. Thus, K_{eq} will decrease, which means the ratio of products to reactants will decrease.
  SOLUTION: Some N_2O_4 will break down to create more NO_2.

To Summarize the Effects of Possible Stresses to a Reaction at Equilibrium:
- Any change in conditions that increases the value of Q will cause the reaction to proceed spontaneously in the reverse direction to re-establish K_{eq}. (example: removing reactants).
- Any change in conditions that decreases the value of Q will cause the reaction to proceed spontaneously in the forward direction to re-establish K_{eq}. (example: adding reactants).
- Addition of any substance that is not one of the reaction components and that does not react with a reaction component has no effect.
- A catalyst has no effect.
- A change in temperature will change the value of K_{eq}. The influence will depend upon whether the reaction is exothermic or endothermic.

EXERCISE 12: Magnesium hydroxide is the active ingredient in some antacids. How will the highly acidic environment of a person's stomach affect the solubility of magnesium hydroxide?

STRATEGY: First an appropriate reaction must be written. "Solubility of magnesium hydroxide" indicates the relation between the solid and the aqueous ions: $Mg(OH)_2$ (s) $\leftrightarrow Mg^{2+}$ (aq) $+ 2 OH^-$ (aq)

Ask: Is acid (H_3O^+) one of the components in the equilibrium reaction? NO. Will acid react with any of the components of the equilibrium reaction? Yes, it will react with OH^- (aq) to make water, because acids and bases neutralize each other. The net result is that OH^- (aq) is removed from the system. This means that temporarily $Q < K_{eq}$ and the reaction will proceed in the forward direction to re-establish equilibrium. Reactants will break down to make more products.

SOLUTION: Magnesium hydroxide's solubility will be greater in an acidic environment than in a non-acidic environment.

Overall Summary of Equilibrium Constants:

1. K_{eq} is valid only at equilibrium.
2. K_{eq} is independent of initial conditions.
3. K_{eq} is related to the stoichiometry of the balanced chemical equation.
4. K_{eq} is unitless, by convention.
5. The magnitude of K_{eq} provides an indication of the relative [products] to [reactants].
6. K_{eq} can be determined several ways.
7. Comparison of Q to K_{eq} will indicate the spontaneous direction of reaction.
8. For an exothermic process, K_{eq} decreases as T increases; for an endothermic process, K_{eq} increases as T increases.
9. If a stress is applied to a system at equilibrium, the system will shift to adjust to the stress.

Try It #4: Perfume chemists can synthesize a chemical responsible for the pleasant smell of a rose:

This is an exothermic reaction. At equilibrium, quite a lot of reactants remain. List two things a perfume chemist could do to increase the amount of products.

16.5   WORKING WITH EQUILIBRIA

It is necessary to analyze the chemistry behind what is happening for an equilibrium process prior to doing any calculations. If the chemical changes are not understood, then it is almost impossible to know how to set up a proper calculation. Follow the text's stepwise method for attacking these problems:

Solving Equilibrium Problems
1. Determine what is asked for.
2. Identify the major chemical species.
3. Determine what chemical equilibria exist.
4. Write the K_{eq} expressions.
5. Organize the data and unknowns.
6. Carry out the calculations.
7. Does the result make sense?

Helpful Hint
*   Brush up on your stoichiometry skills from Chapter 4.

EXERCISE 13: The reaction SO_2 (g) + Cl_2 (g) $\leftrightarrow$ SO_2Cl_2 (g) was studied at some temperature. The equilibrium partial pressures of SO_2 and Cl_2 were 0.0525 atm and the partial pressure of SO_2Cl_2 was 0.00780 atm. What is the value of the equilibrium constant at this temperature?

STRATEGY AND SOLUTION: The equilibrium partial pressures have been provided, so the equilibrium constant can be calculated directly from the expression.

$$K_{eq} = \frac{p_{SO_2Cl_2}}{p_{SO_2} p_{Cl_2}} = \frac{0.00780}{(0.0525)^2} = 2.83$$

Initial Conditions and Concentration Tables

Always make a concentration table to compare initial and equilibrium concentrations. The method used to construct a concentration table is illustrated in the next exercise.

EXERCISE 14: HCN can donate a proton to water to generate H_3O^+. What is the equilibrium constant for this process at 25°C, given that a 0.079 M solution of HCN contains 7.00×10^{-6} M H_3O^+ at equilibrium?

STRATEGY: Note: The reasoning for Steps 2 & 3 of this method will be covered in Section 16.6
1. **Determine what is asked for.** The equilibrium constant for the reaction. If the equilibrium concentrations are known, this can be obtained from the equilibrium constant expression.
2. **Identify the major chemical species.** HCN and water. The concentration of HCN is much greater than H_3O^+ because the equilibrium constant for the process is so small.
3. **Determine what chemical equilibria exist.** HCN donates a proton to water, forming H_3O^+. The other product formed (by comparison of reactants to products) is CN^-.
$$HCN \text{ (aq)} + H_2O \text{ (l)} \leftrightarrow H_3O^+ \text{ (aq)} + CN^- \text{ (aq)}$$
4. **Write the K_{eq} expressions.** We have one K_{eq} expression for this problem:

$$K_{eq} = \frac{[H_3O^+][CN^-]}{[HCN]} = 6.2 \times 10^{-10}$$

5. **Organize the data and unknowns.**
Any time that you must relate initial and equilibrium conditions, it is a good idea to make a concentration table to organize the data. The general format for the table is always the same. **First, input the information that was directly provided in the question.**

	HCN (aq) +	H₂O (l) ↔	H₃O⁺ (aq) +	CN⁻ (aq)
Concentrations:				
Initial	0.079			
Change				
Equilibrium			7.00×10^{-6}	

Second, interpret the relationships within the equilibrium to complete the table. H_3O^+ and CN^- were not present initially, so their initial values will be "0" in the table. The stoichiometry of the reaction is 1:1:1, so the $[CN^-]$ at equilibrium must be the same as the $[H_3O^+]$. The $[CN^-]$ increased by 7.00×10^{-6} M as did the $[H_3O^+]$, while the $[HCN]$ decreased by that amount. At equilibrium, then, the $[HCN] = 0.079 - 7.00 \times 10^{-6}$ M. The information fills into the table like this:

	HCN (aq) +	H$_2$O (l) $\leftrightarrow$	H$_3$O$^+$ (aq) +	CN$^-$ (aq)
Concentrations:				
Initial	0.079		0	0
Change	-7.00x10^{-6}		+7.00x10^{-6}	+7.00x10^{-6}
Equilibrium	0.079 – 7.00x10^{-6}		7.00x10^{-6}	7.00x10^{-6}

Key Concept:
- **The "Change" line always indicates the spontaneous direction of the reaction and the reaction stoichiometry.** Notice that the HCN consumed is given a "-7.00x10^{-6}," while the products have "+7.00x10^{-6}." Notice, too, that if the coefficient for HCN had been "2" the change would have been "-2(7.00x10^{-6})" or "-1.40x10^{-5}."

Helpful Hint
- If the spontaneous direction of a reaction is not obvious, calculate Q and compare it to K$_{eq}$.

6. **Carry out the calculations.**

$$K_{eq} = \frac{[H_3O^+][CN^-]}{[HCN]} = \frac{(7.00x10^{-6})(7.00x10^{-6})}{(0.079 - 7.00x10^{-6})} = 6.2x10^{-10}$$

SOLUTION: To answer the question, K$_{eq}$ = 6.2x10^{-10}.

7. **Does the result make sense?** At equilibrium, the [H$_3$O$^+$] and [CN$^-$] were several orders of magnitude smaller than [HCN], so the small value for the equilibrium constant seems reasonable. A small K$_{eq}$ indicates very little products relative to reactants at equilibrium.

Try It #5: Create a concentration table for the process: 2 CO (g) + O$_2$ (g) $\leftrightarrow$ 2 CO$_2$ (g), given that no CO or O$_2$ is initially present, and that the initial partial pressure of CO$_2$ is 3 atm.

EXERCISE 15: What is the equilibrium concentration of C$_2$H$_5$NH$_3^+$ (aq) in a 0.175 M solution of C$_2$H$_5$NH$_2$ (aq)? The reaction is:

C$_2$H$_5$NH$_2$ (aq) + H$_2$O (l) $\leftrightarrow$ C$_2$H$_5$NH$_3^+$ (aq) + OH$^-$ (aq); K$_{eq}$ = 4.5x10^{-4} at 25°C

STRATEGY: Note: The reasoning for Steps 2 & 3 of this method will be covered in Section 16.6.
1. **Determine what is asked for.** We need the equilibrium concentration of C$_2$H$_5$NH$_3^+$. To find this, we must determine how much of the 0.175 M C$_2$H$_5$NH$_2$ dissociated.
2. **Identify the major chemical species.** The major species will be the reactants because the equilibrium constant is so small.
3. **Determine what chemical equilibria exist.** The equilibrium reaction has been provided.
4. **Write the K$_{eq}$ expressions.** We have one K$_{eq}$ expression:

$$K_{eq} = 4.5x10^{-4} = \frac{[C_2H_5NH_3^+][OH^-]}{[C_2H_5NH_2]}$$

5. **Organize the data and unknowns.**
   Initially, there are no products. The stoichiometry of the reaction is 1:1:1, so some amount, "x," of reactants will be lost (negative sign) and that same amount "x" of products will be gained (positive sign). Remember, the "Change" line correlates to the reaction stoichiometry. The table summarizes the data:

$$C_2H_5NH_2 \text{ (aq)} + H_2O \text{ (l)} \quad \leftrightarrow C_2H_5NH_3^+ \text{ (aq)} + OH^- \text{ (aq)}$$

Concentrations:			
Initial	0.175	0	0
Change	-x	+x	+x
Equilibrium	0.175 – x	x	x

6. **Carry out the calculations.**

$$K_{eq} = 4.5x10^{-4} = \frac{[C_2H_5NH_3^+][OH^-]}{[C_2H_5NH_2]} = \frac{(x)(x)}{0.175-x} = \frac{x^2}{0.175-x}$$

$$4.5x10^{-4}(0.175-x) = x^2$$

$$7.88x10^{-5} - 4.5x10^{-4}x = x^2$$

This requires the quadratic equation:
$$0 = x^2 + 4.5x10^{-4}x - 7.88x10^{-5}; \quad a = 1, b = 4.5x10^{-4}, c = -7.88x10^{-5}$$

$$x = \frac{-b \pm \sqrt{b^2 - 4ac}}{2a} = \frac{-4.5x10^{-4} \pm \sqrt{(4.5x10^{-4})^2 - 4(1)(-7.88x10^{-5})}}{2(1)}$$

$$x = 8.65x10^{-3}$$

SOLUTION: To answer the question, $x = [C_2H_5NH_3^+] = 8.65x10^{-3}$ M.

7. **Does the result make sense?** The $[C_2H_5NH_3^+]$ should be smaller than the $[C_2H_5NH_2]$ because the small equilibrium constant indicates that there are few products relative to reactants at equilibrium. The $[C_2H_5NH_2]$ at equilibrium is $0.175 - 8.65x10^{-3} = 0.166$ M, so the answer seems reasonable.

Working with Small Equilibrium Constants

If the equilibrium constant for a given reaction is small, the technique illustrated in the following exercise can be used to simplify the math in the equilibrium calculation.

EXERCISE 16: What is the concentration of hydroxide ion in a 0.125 M solution of methylamine?
    Given: CH_3NH_2 (aq) $+ H_2O$ (l) $\leftrightarrow CH_3NH_3^+$ (aq) $+ OH^-$ (aq); $K_{eq} = 4.6x10^{-4}$ at 25°C

STRATEGY: Note: The reasoning for Steps 2 & 3 of this method will be covered in Section 16.6.
1. **Determine what is asked for.** We need the equilibrium concentration of hydroxide. To find this, we must determine how much of the 0.125 M methylamine dissociated.
2. **Identify the major chemical species.** The major species will be the reactants because the equilibrium constant is so small.
3. **Determine what chemical equilibria exist.** The equilibrium reaction has been provided.
4. **Write the K_{eq} expressions.** There is one K_{eq} expression to write:

$$K_{eq} = 4.6x10^{-4} = \frac{[CH_3NH_3^+][OH^-]}{[CH_3NH_2]}$$

5. **Organize the data and unknowns.**
   Initially, there are no products. The stoichiometry of the reaction is 1:1:1, so some amount, "x," of reactants will be lost (negative sign) and that same amount, "x," of products will be gained (positive sign). Remember, the "Change" line correlates to the reaction stoichiometry. The table summarizes the data:

	CH_3NH_2 (aq) +	H_2O(l)	$\leftrightarrow CH_3NH_3^+$ (aq) +	OH^- (aq)
Concentrations:				
Initial	0.125		0	0
Change	-x		+x	+x
Equilibrium	0.125 – x		x	x

6. **Carry out the calculations.**
   Here's the trick: A small equilibrium constant signifies that very little products are formed and very little reactants are consumed. Thus, the value of "x" must be quite tiny relative to the starting concentration of methylamine. To simplify the math, we can drop the "-x" term from the equilibrium concentration for methylamine.

$$K_{eq} = 4.6 \times 10^{-4} = \frac{[CH_3NH_3^+][OH^-]}{[CH_3NH_2]} = \frac{(x)(x)}{0.125 - x} \cong \frac{x^2}{0.125}$$

Solving for x we get: $\sqrt{(4.6 \times 10^{-4})(0.125)} = x = 7.58 \times 10^{-3}$

To test if our assumption to drop "x" was mathematically acceptable, **plug the calculated value for x back into the location where it was originally removed, and re-solve for x.** If our assumption was acceptable, then the new answer should be close to – if not identical to – the original value obtained for x.

$$4.6 \times 10^{-4} = \frac{(x)(x)}{0.125 - x} = \frac{x^2}{0.125 - 7.58 \times 10^{-3}}; \text{ so } x = 7.35 \times 10^{-3}$$

In this case, the result for x is close, but not equal to, the original x. This indicates that our initial approximation is good, but not perfect. Therefore, the *new* value for x should be tested by plugging it into the equation and re-solving for x. This process must be repeated until the value for x is reproducible.

$$4.6 \times 10^{-4} = \frac{(x)(x)}{0.125 - x} = \frac{x^2}{0.125 - 7.35 \times 10^{-3}}; \text{ so } x = 7.36 \times 10^{-3}$$

Since this value for x differs by only one in the last decimal place, we can stop here.

SOLUTION: To answer the question, x = [OH⁻] = 7.36×10^{-3} M = 7.4×10^{-3} M (2 sig figs).

7. **Does the result make sense?** The concentration of hydroxide should be much lower than that of methylamine, since the equilibrium constant is so small. The concentration of methylamine at equilibrium is 0.125 M – 7.36×10^{-3} M = 0.118 M, which is much higher than the [OH⁻], so the numbers seem reasonable.

This method, called the **Method of Successive Approximations**, is extremely timesaving in solving some chemical equilibria problems. Here is a re-cap:
1. If the change is small, drop it from the equation, and solve for x.
2. To test the validity of the approximation, plug the calculated value for x back into the location where it was originally removed. Re-solve for x.
3. If the new value for x is the same as the original value, then the approximation was good, so stop.
4. If not, then repeat the process of plugging in and resolving for x until a reproducible value for x is obtained.

It often takes only two or three successive approximations to obtain a final value for x, but occasionally it may take more.

Working with Large Equilibrium Constants

Technique for Working with Large K_{eq}s:
A large K_{eq} indicates that mostly products are present at equilibrium. The mathematics of an equilibrium calculation involving a large K_{eq} can therefore be simplified by first assuming the reaction goes to completion, and then back-tracking from that. This method requires two tables. The first table shows 100% reaction. The second table shows products decomposing to the equilibrium state. The method will be illustrated in the following exercise.

EXERCISE 17: Calculate the concentration of silver ions in solution when 100.0 mL of 0.250 M silver nitrate is mixed with 50.0 mL of 0.120 M sodium carbonate. At 25°C, the equilibrium constant for silver carbonate dissolving in water is 8.46×10^{-12}.

STRATEGY: The reasoning for Steps 2 & 3 of this method will be covered in Section 16.6.
1. **Determine what is asked for.** We need the equilibrium concentration of silver ion. To find this, we must determine how much of the silver reacted with carbonate ion to form a precipitate.
2. **Identify the major chemical species.** The major species prior to reaction will be Ag^+, NO_3^-, Na^+, and CO_3^{2-}. Silver carbonate precipitates, so the major species after reaction will be solid Ag_2CO_3 and the spectator ions, NO_3^- and Na^+. Minor species will be Ag^+ and CO_3^{2-}.
3. **Determine what chemical equilibria exist.** The pertinent equilibrium is the two ions forming a precipitate:

$$2\,Ag^+\,(aq) + CO_3^{2-}\,(aq) \leftrightarrow Ag_2CO_3\,(s)$$

This is the reverse of the solid dissolving, so the $K_{eq} = 1/K_{dissolving} = 1/8.46 \times 10^{-12} = 1.18 \times 10^{11}$

4. **Write the K_{eq} expressions.** There is one K_{eq} expression to write:

$$K_{eq} = 1.18 \times 10^{11} = \frac{1}{[Ag^+]^2[CO_3^{2-}]}$$

5. **Organize the data and unknowns.**
Before any reaction takes place, we must account for the fact that each chemical has been diluted by the other upon mixing. The new total volume of solution is 150.0 mL.

[AgNO₃] = (100.0 mL)(0.250 M)/(150.0 mL) = 0.1667 M
[Na₂CO₃] = (50.0 mL)(0.120 M)/(150.0 mL) = 0.0400 M

The initial $[Ag^+] = [AgNO_3] = 0.1667$ M, because there is one silver ion in every silver nitrate. Similarly, the initial $[CO_3^{2-}] = [Na_2CO_3] = 0.0400$ M, because there is one carbonate in every sodium carbonate:

	2 Ag$^+$ (aq)	+	CO$_3^{2-}$ (aq)	↔	Ag$_2$CO$_3$ (s)
Concentrations:					
Initial	0.1667		0.0400		0
Change	-0.0800		-0.0400		+0.0400
At 100% rxn	0.0867		0		+0.0400
"Initial"	0.0867		0		+0.0400
Change	+2x		+x		-x
At equilibrium	0.0867+2x		x		0.0400 – x

Notice that the "Change" lines correlate to the reaction stoichiometry. In the first table, 0.0400 M carbonate reacts with *twice* that amount of silver ion. These quantities are being consumed so the signs are negative, while the product sign is positive. In the second table silver carbonate loses the same quantity that the carbonate gains (1:1 ratio) and the silver ion gains twice as much (1:2 ratio). Again, the loss and gain are shown with +/- signs.

6. **Carry out the calculations.**

$$K_{eq} = 1.18 \times 10^{11} = \frac{1}{[Ag^+]^2[CO_3^{2-}]} = \frac{1}{(0.0867 + 2x)^2(x)} \cong \frac{1}{(0.0867)^2 x}$$

Solving for x we get: $x = \dfrac{1}{(0.0867)^2(1.1 \times 10^{11})} = 1.1274 \times 10^{-9}$

The value of x is so small compared to 0.0867 that plugging x into the term "0.0867+2x" would be insignificant. Our assumption to drop the "+2x" term was therefore valid.

SOLUTION: To answer the question, $[Ag^+] = 2x = 2.25 \times 10^{-9}$ M.

7. **Does the result make sense?** The equilibrium constant for the reaction is an enormous number, indicating mostly products are present at equilibrium. A small concentration of aqueous silver ions makes sense.

Key Concept:
- **When two solutions are mixed, it is necessary to account for each diluting the other to obtain initial concentrations.** See step 5 in Exercise 17 as an illustration.

Recall that the equilibrium state is the same regardless of the direction approached. The answer to Exercise 17 could also have been calculated using the reverse of the reaction shown. Exercise 18 illustrates this alternate process.

EXERCISE 18: Calculate the concentration of silver ions in solution when 100.0 mL of 0.250 M silver nitrate is mixed with 50.0 mL of 0.250 M sodium carbonate. At 25°C, the equilibrium constant for silver carbonate dissolving in water is 8.46×10^{-12}. (Note: this is the same question as Exercise #17, but the answer will be approached from a different angle.)

STRATEGY:

1. **Determine what is asked for.** We need the equilibrium concentration of silver ion.
2. **Identify the major chemical species.** The major species prior to reaction will be Ag^+, NO_3^-, Na^+, and CO_3^{2-}. Silver carbonate precipitates, so the major species after reaction will be solid Ag_2CO_3 and the spectator ions, NO_3^- and Na^+. Minor species will be Ag^+ and CO_3^{2-}.
3. **Determine what chemical equilibria exist.** The equilibrium constant provided is for the reaction:

$$Ag_2CO_3 \text{ (s)} \leftrightarrow 2\,Ag^+ \text{ (aq)} + CO_3^{2-} \text{ (aq)}$$

4. **Write the K_{eq} expressions.** We have one K_{eq} expression to write:

$$K_{eq} = 8.46 \times 10^{-12} = [Ag^+]^2[CO_3^{2-}]$$

5. **Organize the data and unknowns.**
   As noted in the previous exercise, the initial concentrations of the ions involved in the reaction are: $[Ag^+] = 0.1667$ M and $[CO_3^{2-}] = 0.0400$ M.

   Since the equilibrium constant is so small, mostly reactants are present at equilibrium. We can use the same reasoning as for a large equilibrium constant: the change for this process relative to initial conditions will be significant, nearly 100% reaction. We will assume 100% reaction for the formation of reactants, then set up a second table to establish the equilibrium state.

Concentrations:	Ag_2CO_3 (s) $\leftrightarrow$	2 Ag^+ (aq) +	CO_3^{2-} (aq)
Initial	0	0.1667	0.0400
Change	0.0400	-0.0800	-0.0400
At 100% rxn	0.0400	0.0867	0

"Initial"	0.0400	0.0867	0
Change	-x	+2x	+x
At equilibrium	0.0400 − x	0.0867+2x	x

Notice again that the "Change" lines correlate to the reaction stoichiometry.

6. **Carry out the calculations.**

$$K_{eq} = 8.46 \times 10^{-12} = [Ag^+]^2[CO_3^{2-}] = (0.0867 - 2x)^2(x) \cong (0.1667)^2(x)$$

Helpful Hint

- Which x should be dropped? The term that is mathematically less significant must be dropped. The term "0.0867-2x" will approximately equal 0.0867 M in the end (because K_{eq} is so small) so that particular "x" is not very significant. An "x" involved in multiplication is more important in the equation than an "x" that is being subtracted from another number.

Solving for x we get: $x = \dfrac{8.46 \times 10^{-12}}{(0.0867)^2} = 1.1274 \times 10^{-9}$

We can see without even re-plugging this x back into the equation, that it has no impact on the value "(0.0867 – 2x)," so our assumption was valid.

SOLUTION: To answer the question, $[Ag^+] = 2x = 2.25 \times 10^{-9}$ M.

Note that this is the same answer as obtained in Exercise 17. This exercise illustrates the fact that the equilibrium state is the same regardless of the direction approached, thus the calculations can be performed via the most convenient method. In short, there is more than one correct way to do the calculation.

7. **Does the result make sense?** The concentration of silver ion should be miniscule given the small equilibrium constant. It is, so the answer is reasonable.

Try It #6: Acetic acid, $HC_2H_3O_2$, donates a proton to water to form $C_2H_3O_2^-$ and H_3O^+. The equilibrium constant for this reaction is 1.8×10^{-5} at 25°C. Calculate the concentrations of all solute species present at equilibrium in a 0.35 M solution of acetic acid.

16.6   EQUILIBRIA IN AQUEOUS SOLUTIONS

Species in Solution

Key Terms:
- **Major species** – substances that are present in relatively large amounts. Water, the solvent, is always a major species in an aqueous solution, as are the (relatively) more highly concentrated solutes.
- **Minor species** – solutes that are present in minor quantities (usually three orders of magnitude less) than the major species.
- **Dominant equilibrium** – the equilibrium process that plays the most important role in the aqueous solution. This will be the equilibrium with the largest K_{eq}.

Typical substances that generate ions as major species in solution:
- Water-soluble salts (example: KCl, $NaNO_3$)
- Strong acids – react essentially 100% with water to produce H_3O^+ ions
- Strong bases – react essentially 100% with water to produce OH^- ions

Helpful Hint
Table 16.1 in the text lists the common strong acids and bases. LEARN THEM NOW.

EXERCISE 19: Identify the major species in each aqueous solution: a) KCl (potassium chloride); b) HCl (hydrochloric acid); c) $C_6H_5NH_2$ (aniline); d) LiOH (lithium hydroxide); e) C_2H_5OH (ethanol)

STRATEGY AND SOLUTION: Soluble salts, strong acids and strong bases generate ions as the major species. All other substances keep the undissociated substance as the major species. Water would be a major species in every process here, because water is the solvent.

a) KCl is a water-soluble salt, so the major species will be K^+ (aq) and Cl^- (aq).
b) HCl is a strong acid, so the major species will be H_3O^+ (aq) and Cl^- (aq).
c) $C_6H_5NH_2$ is not a salt, a strong acid or a strong base; the major species will be $C_6H_5NH_2$ (aq).
d) LiOH is a water-soluble salt, so its major species will be Li^+ (aq) and OH^-(aq).
e) C_2H_5OH is not a salt, a strong acid or a strong base, so the major species will be C_2H_5OH (aq). Note: this may look like a base because of the "OH" at the end of the compound, but it is not. This is a covalently-bonded compound; the "OH" is an alcohol functional group. It won't break off from the rest of the molecule to produce OH^- ions in water.

Try It #7: Identify the major and minor species present in Try It #6.

Types of Aqueous Equilibria

The various types of aqueous equilibria are shown in a table on the next page. Notice that each is given a specific subscript for K. You will see in the upcoming chapters that these equilibrium constants are tabulated, and can thus be looked up when needed.

EXERCISE 20: What type of equilibrium constant would each of these reactions have?

a) Co^{2+} (aq) $+ 4\ Cl^-$ (aq) $\leftrightarrow CoCl_4^-$(aq)
b) $HC_2H_3O_2$ (aq) $+\ H_2O$ (l) $\leftrightarrow C_2H_3O_2^-$ (aq) $+ H_3O^+$ (aq)
c) OH^- (aq) $+ NH_4^+$(aq) $\leftrightarrow NH_3$ (aq) $+ H_2O$ (l)
d) $MgCl_2$ (s) $\leftrightarrow Mg^{2+}$ (aq) $+ 2\ Cl^-$ (aq)

STRATEGY AND SOLUTION:
a) This reaction shows a metal cation and non-metal anion (that has lone pairs) reacting. The product is still aqueous so it is not a precipitation reaction. This is a complex formation equilibrium, K_f.
b) Water is accepting a proton from $HC_2H_3O_2$ and H_3O^+ is being produced. This is a K_a process.
c) OH^- is present as a reactant. If we were to reverse this reaction, we'd see that NH_3 is accepting a proton from water, a K_b process. Thus, the reaction shown is the reverse of a K_b process. This process is classified as $1/K_b$.
d) This shows a solid dissolving so it is a K_{sp} process.

EXERCISE 21: Write an equilibrium constant expression for each reaction in Exercise 20.

SOLUTION:

a) $K_f = \dfrac{[CoCl_4^-]}{[Co^{2+}][Cl^-]^4}$

b) $K_a = \dfrac{[H_3O^+][C_2H_3O_2^-]}{[HC_2H_3O_2]}$

Chapter 16: Principles of Chemical Equilibrium

Symbol	Type	General Process	K Expression	Notable
K_{sp}	Solubility product: solid dissolving in water	$X(s) \leftrightarrow X(aq)$ Note: it's usually ionic: $YZ_2(s) \leftrightarrow Y^{2+}(aq) + 2\,Z^-(aq)$	$K_{sp} = [X]$ $K_{sp} = [Y^{2+}][Z^-]^2$	A small K_{sp} ($\ll 1$) indicates a water-insoluble compound.
K_a	Acid Dissociation: an acid dissociating into ions (generating H_3O^+)	$H_2O(l) + HA(aq) \leftrightarrow H_3O^+(aq) + A^-(aq)$	$K_a = \dfrac{[H_3O^+][A^-]}{[HA]}$	The larger the K_a, the stronger the acid, the more H_3O^+ is formed.
K_b	Base Dissociation: a base dissociating into ions (generating OH^-)	$A^-(aq) + H_2O(l) \leftrightarrow OH^-(aq) + HA(aq)$	$K_b = \dfrac{[OH^-][HA]}{[A^-]}$	The larger the K_b, the stronger the base, the more OH^- is formed.
K_w	Water Dissociation: water dissociating into ions	$H_2O(l) + H_2O(l) \leftrightarrow H_3O^+(aq) + OH^-(aq)$	$K_w = [H_3O^+][OH^-]$	K_w is 1.0×10^{-14} at 25°C. Thus, $[H_3O^+]$ and $[OH^-]$ are each 1.0×10^{-7} M.
K_f	Complex Formation: a metal cation attaching to a molecule or ion that has a lone pair of electrons	$M^+(aq) + L^-(aq) \leftrightarrow ML(aq)$	$K_f = \dfrac{[ML]}{[M^+][L^-]}$	The charges in the generic K_f expression will depend on the metal and ligand charges.

Key Concept:
The reverse of these reactions can also occur. For example, a precipitation reaction would have a $K_{eq} = 1/K_{sp}$.

281

c) $K_{eq} = \dfrac{1}{K_{b \text{ for } NH_3}} = \dfrac{[NH_3]}{[NH_4^+][OH^-]}$ 　　　　　 d) $K_{sp} = [Mg^{2+}][Cl^-]^2$

Try It #8: What type of equilibrium constant expression would each be?
   a) NO_2^- (aq) $+ H_2O$ (l) $\leftrightarrow OH^-$ (aq) $+ HNO_2$ (aq)
   b) Aqueous silver nitrate is mixed with aqueous sodium sulfide.

Key Terms:
- **Spectator ion** – ion that is present in the solution, but is not involved in the chemical equilibrium.
- **Net ionic reaction** – reaction which illustrates the net change. Any ion not involved in a chemical change is not shown in a net ionic reaction.

All of the reactions you've seen up to this point are net ionic reactions. In reality, the starting materials won't exist as sole ions without counter-ions, but as compounds. For example, if a chemist needs 0.10 M Na^+, he/she will prepare 0.10 M NaCl, which is 0.10 M in Na^+. The counter-ion, Cl^-, will be a spectator ion. In another example, the net ionic reaction:

$$OH^- (aq) + NH_4^+ (aq) \leftrightarrow NH_3 (aq) + H_2O (l)$$

would be the result of mixing two chemicals such as aqueous NaOH and NH_4Cl together. The ions initially present would be Na^+, OH^-, NH_4^+, and Cl^-. Na^+ and Cl^- would be considered spectator ions because they are not involved in the reaction.

EXERCISE 22: Determine the type of process, the net ionic equation, the spectator ions, and the equilibrium constant expression for the reaction: Aqueous hydrochloric acid, HCl, is mixed with aqueous potassium hydroxide, KOH. Draw molecular pictures to illustrate the solutions prior to mixing and after they have been mixed together.

STRATEGY AND SOLUTION:
HCl is a strong acid, so it quantitatively generates H_3O^+ and Cl^- in water. KOH is a strong base, so it completely ionizes in water to K^+ and OH^-. The H_3O^+ (aq) and OH^- (aq) will react, the chloride and potassium ions won't. Cl^- and K^+ are therefore the spectator ions. The net ionic reaction is the reverse of the water dissociation reaction, so the K_{eq} will be $1/K_w$.

$$H_3O^+ (aq) + OH^- (aq) \leftrightarrow 2H_2O (l); \qquad K_{eq} = \dfrac{1}{K_w} = \dfrac{1}{[H_3O^+][OH^-]}$$

The molecular picture shows the spectator ions as well as the reaction. The reactant concentrations were not specified, so we can use any concentration we wish. We'll show 2 HCls and 3 KOHs in slightly different volumes of solution. The final picture should show the total volume of solution.

Initial HCl solution

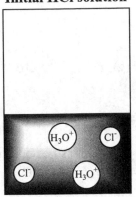

Initial KOH solution

The two solutions after mixing

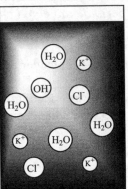

Chapter 16 Self-Test

You may use a periodic table and a calculator for this test.

1. For questions 1 a-e, consider this reaction occurring at 700°C: $I_2 (g) \leftrightarrow 2 I (g)$ $K_{eq} = 3.5 \times 10^{-5}$

   a) If extra I (g) was added into the container, what would occur?
   b) This process is endothermic. What is the relationship between K_{eq} and temperature for this process?
   c) Some gaseous I_2 (g) and I (g) were placed in a container at 700°C. The partial pressures before any reaction occurred were: Pressure of I_2 (g) = 0.050 atm and Pressure of I (g) = 0.025 atm. Predict the spontaneous direction of the reaction.
   d) What is K_{eq} for the reaction: $4 I (g) \leftrightarrow 2 I_2 (g)$?
   e) If the value for K_{eq} was not provided, what is one way the value could be determined?

2. The following reaction illustrates how wine gets oxidized to vinegar (CH_3CO_2H):
   $$C_2H_5OH (aq) + O_2 (aq) \leftrightarrow CH_3CO_2H (aq) + H_2O (l) \quad K_{eq} = 1.2 \times 10^{82} \text{ at } 25°C$$

   Identify the major and minor species in this reaction mixture at equilibrium.

3. Consider the process: $HOAc(aq) + OH^- (aq) \leftrightarrow OAc^- (aq) + H_2O (l)$

   a) Write an equilibrium constant expression for this process.
   b) What type of process is this?
   c) If HCl is added to the chemical system above, what will happen to:

The rate constant, k_f :	increase	decrease	nothing
[HOAc]:	increase	decrease	nothing

4. Determine the equilibrium concentration of all solute species present in 0.050 M aniline (aq) at 25°C. The K_b for aniline is 7.4×10^{-10}.

5. Consider the reaction that occurs when aqueous lead (II) nitrate is mixed with aqueous potassium chloride. Determine the following for the process:
   a) the type of process,
   b) the net ionic equation,
   c) the spectator ions,
   d) the equilibrium constant expression for the reaction.

   e) Draw molecular pictures to illustrate the two solutions prior to mixing and after they have been mixed together. In your pictures, show 2 Pb^{2+} ions and 2 K^+ ions.

Answers to Try Its:

1. $K_{eq} = \dfrac{[H_2]^2_{eq}[O_2]_{eq}}{[H_2O]^2_{eq}}$

2. $K_{eq} = \dfrac{1}{p_{SO_3}}$

3. The ammonia reaction. To compare the equilibrium constants, make sure the two processes are the same. Reverse the $C_5H_5NH^+$ (aq) reaction to compare the two reactions. Its equilibrium constant is $<<$ than that of the ammonia reaction. *Note: You will see in later sections that the relative reactant concentrations impact the answer to this question.*

4. Decrease temperature to increase K_{eq}; remove water from the container by adding a chemical that reacts only with water.

5.

	2 CO (g)	+	O_2 (g)	$\leftrightarrow$	2 CO_2 (g)
Concentrations:					
Initial	0		0		3
Change	+2x		+x		-2x
Equilibrium	2x		x		3-2x

6. $[HC_2H_3O_2] = 0.347\ M \approx 0.35\ M$; $[C_2H_3O_2^-]$ and $[H_3O^+]$ are each 0.0025 M

7. Major: H_2O and $HC_2H_3O_2$; minor: $C_2H_3O_2^-$ and H_3O^+ (and very minor, OH^-)

8. a) K_b b) $1/K_{sp}$ for Ag_2S; Ag_2S is not water-soluble

Answers to Self-Test:

1. a) The reaction would proceed to the left to use up some products and make more reactants (to re-establish K_{eq}); b) Increasing temperature will increase K_{eq}, decreasing temperature will decrease K_{eq} for this endothermic process. c) The reaction will proceed in the reverse direction. Comparison of Q to K_{eq} shows that $Q > K_{eq}$. $Q = 0.0125$; d) 8.2×10^8 (reaction was reversed and coefficients doubled); e) one way (of several) is to calculate ΔG_{rxn} from thermodynamic data, and then calculate K_{eq} from the relationship $\Delta G = -RT\ln K_{eq}$.

2. major species = water and CH_3CO_2H; minor species = C_2H_5OH and O_2

3. a) $K_{eq} = \dfrac{[OAc^-]}{[HOAc][OH^-]}$ b) $1/K_b$ for OAc^-

c) k_f – nothing; [HOAc] would increase because HCl would react with OH^-

4. $[\text{aniline}] \cong 0.050\ M$, $[\text{aniline}H^+] = [OH^-] = 6.1 \times 10^{-6}\ M$

5. a) This is a precipitation reaction, the reverse of a solubility equilibrium, so the K_{eq} will be $1/K_{sp}$.

b) Pb^{2+} (aq) + 2 Cl^- (aq) $\leftrightarrow$ $PbCl_2$ (s)

c) The spectator ions are NO_3^- (aq) and K^+ (aq) because they are not involved in the reaction.

d) $K_{eq} = \dfrac{1}{K_{sp\ \text{for PbCl}_2}} = \dfrac{1}{[Pb^{2+}][Cl^-]^2}$

e) The molecular picture shows the spectator ions as well as the reaction. $PbCl_2$ precipitates.

Initial lead nitrate solution   **Initial KCl solution**   **The two solutions after mixing**

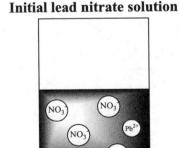
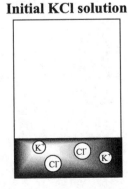
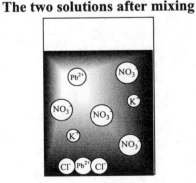

Chapter 17: Aqueous Acid-Base Equilibria

Learning Objectives

In this chapter, you will learn how to:
- Identify strong and weak acids and bases.
- Determine the pH of an aqueous solution (strong acid, strong base, weak acid, weak base).
- Assess factors which influence the strength of acids and bases.
- Calculate equilibrium concentrations of solutes in acidic and basic solutions.
- Determine which equilibrium process within a solution will predominate.

Practical Aspects

Acids and bases are found everywhere in life: biological systems depend upon delicate balances of them; toxicologists are concerned with their corroding capabilities; environmental scientists study their impact on pollution. Acids and bases are so commonly encountered in everyday life, that the pH scale – a logarithmic scale used to describe the concentration of H_3O^+ and OH^- in a system – is well known, even among non-scientists.

This chapter on acids and bases builds on the concepts covered in Chapter 16, where you learned how to calculate equilibrium concentrations of solutes in an aqueous reaction. It also lays the foundation for Chapter 18. Several concepts learned in this chapter will be used extensively in other chemistry courses, such as organic and biochemistry. Of particular interest to both biochemists and organic chemists is the ability to recognize acids and bases and to assess relative acid or base strength. Biochemists use the term "pH" on a daily basis. Learn how to calculate pH now. Save your class notes from this chapter for later chemistry courses…they will prove to be a tremendous asset.

17.1 PROTON TRANSFERS IN WATER

Dissociation of Water

Water dissociates according to the reaction: $H_2O\ (l) + H_2O\ (l) \leftrightarrow H_3O^+\ (aq) + OH^-\ (aq)$

Useful Relationship:

- $K_w = [H_3O^+][OH^-] = 1 \times 10^{-14}$ at 25°C

  This is the equilibrium constant expression for the dissociation of water. If either the $[H_3O^+]$ or $[OH^-]$ is known, then the concentration of the other can be determined using this relationship.

Key Concept:
- $[H_3O^+]$ and $[OH^-]$ are inversely related.

Strong Acids and Bases

Key Terms:
- **Acid** (Bronsted-Lowry definition) – a proton donor.
- **Base** (Bronsted-Lowry definition) – a proton acceptor.
- **Proton** – H^+ ion – a hydrogen ion with a 1+ charge.
- **Strong** – dissociation occurs essentially 100% (in other words, the reaction goes to nearly 100% completion; K_{eq} is near infinity).
- **Weak** – dissociation does not occur 100%, it is usually significantly less than 100%. (In other words, K_{eq} is small. At equilibrium, only a small amount of the chemical has dissociated into ions.) This topic will be covered in detail in section 17.3.
- **Amphiprotic** – substance that is capable of acting as an acid or a base, depending upon its environment. Water is amphiprotic. Here, water acts as a base; it accepts a proton from $HC_2H_3O_2$:

$$HC_2H_3O_2\ (aq)\ +\ H_2O\ (l)\ \leftrightarrow C_2H_3O_2^-(aq)\ +H_3O^+\ (aq)$$

Here, water acts as an acid; it donates a proton to NO_2^-:
$$NO_2^-\ (aq)\ +H_2O\ (l)\ \leftrightarrow OH^-\ (aq)\ +HNO_2\ (aq)$$

Helpful Hint
- Be able to readily identify strong acids and bases.
  - There are six common strong acids: HCl, HBr, HI, H_2SO_4, HNO_3, $HClO_4$. Memorize them.
  - Strong bases are usually the group I and II metal hydroxides (examples: KOH, $Mg(OH)_2$).

EXERCISE 1: Calculate the concentrations of all ions in a 0.055 M aqueous solution of HNO_3.

STRATEGY AND SOLUTION: HNO_3 is one of the six strong acids, so it ionizes completely in water to make H_3O^+ and NO_3^- ions. The $[NO_3^-]$ and $[H_3O^+]$ are each therefore 0.055 M. The $[OH^-]$ can be determined from the relationship: $K_w = [H_3O^+][OH^-]$

$$[OH^-] = \frac{K_w}{[H_3O^+]} = \frac{1.0\times10^{-14}}{0.055} = 1.8\times10^{-13}\,M$$

These answers seem reasonable because the $[H_3O^+]$ should be greater than the $[OH^-]$, and it is.

EXERCISE 2: a) How many grams of KOH would be needed to make 250.0 mL of an aqueous solution that is 0.0100 M in hydroxide ion? b) What would be the $[H_3O^+]$ in this solution?

STRATEGY: a) The $[OH^-]$ will equal the $[KOH]$ because this is a strong base, so each KOH will dissociate to form K^+ and OH^-.

$$\text{grams of KOH needed} = 0.2500\ L \times \frac{0.0100\ \text{mol OH}^-}{L} \times \frac{1\ \text{mol KOH}}{1\ \text{mol OH}^-} \times \frac{56.1049\ g}{\text{mol}} = 0.140\ \text{g KOH}$$

b) The $[H_3O^+]$ in this solution will $= K_w/[OH^-] = 1.00\times10^{-14}/0.0100 = 1.00\times10^{-12}$ M

SOLUTION: a) 0.140 g of KOH is needed. b) $[H_3O^+] = 1.00\times10^{-12}$ M.

Try It # 1: Calculate the $[H_3O^+]$ and $[OH^-]$ in a 2.50×10^{-5} M aqueous solution of $Mg(OH)_2$ at 25°C.

17.2 THE pH SCALE

Key Concepts:
- "pH" is the negative log of the hydronium ion concentration, or "$-log[H_3O^+]$."
- At 25°C, a pH of 7 is neutral, which indicates that the concentrations of hydroxide ion and hydronium ion are equal.

pH<7	pH=7	pH>7
Solution is acidic	Solution is neutral	Solution is basic
$[H_3O^+] > [OH^-]$	$[H_3O^+] = [OH^-]$	$[H_3O^+] < [OH^-]$

Helpful Hints
- "p" means "-log."
- To convert from pH to the $[H_3O^+]$, take 10^{-pH}.

EXERCISE 3: Calculate the pH of these HCl (aq) solutions: 0.000010 M, 0.00010 M, 0.0010 M, 0.010 M, 0.10 M.

STRATEGY: HCl is a strong acid, so it will dissociate 100% in water:

$$H_2O \text{ (l)} + HCl \text{ (aq)} \rightarrow H_3O^+ \text{ (aq)} + Cl^- \text{(aq)}$$

$$[H_3O^+] = [HCl], \text{ so pH} = -log[HCl]$$

The [HCl] does not need to be converted to scientific notation before calculating pH, but it will be done here to emphasize the trend between [HCl] and pH.

SOLUTION:

$[H_3O^+] = [HCl]$ (M)	$[H_3O^+]$ rewritten in scientific notation (M)	pH
0.000010	1.0×10^{-5}	5.00
0.00010	1.0×10^{-4}	4.00
0.0010	1.0×10^{-3}	3.00
0.010	1.0×10^{-2}	2.00
0.10	1.0×10^{-1}	1.00

The results from the table above illustrate several important relationships between pH and $[H_3O^+]$:
- **The greater the $[H_3O^+]$, the lower the pH (or the lower the $[H_3O^+]$, the higher the pH).**
- **A pH change of one unit indicates a 10-fold change in hydronium ion concentration.**
- **Significant figures on log scales include only the numbers *after* the decimal.** The number(s) before the decimal in a pH reading signify the order of magnitude. A pH of 5.00 has two sig figs: "00."

Practical Tip: The greater the $[H_3O^+]$, the more corrosive the substance is.

EXERCISE 4: What is the [OH⁻] in a solution that has a pH of 3.60?

STRATEGY AND SOLUTION: We want the [OH⁻], which can be calculated from $[H_3O^+]$.

$$pH = -\log [H_3O^+] = 3.60, \text{ so the } [H_3O^+] = 10^{-pH} = 10^{-3.60} = 2.51188 \times 10^{-4}$$

$$K_w = 1 \times 10^{-14} = [H_3O^+][OH^-]; \text{ so } [OH^-] = 3.98107 \times 10^{-11} = 4.0 \times 10^{-11} \text{ (2 sig figs in final answer)}$$

The answer seems reasonable because a pH < 7 indicates an acidic solution, where the $[H_3O^+] > [OH^-]$.

Other p's

The term "-log" is not reserved for $[H_3O^+]$. "-log" is a convenient way to describe very large and small values, because the numbers are quick and easy to record. Here are some other commonly used values that utilize "-log":

$$\textbf{pOH = -log[OH⁻]} \qquad \textbf{pK}_\textbf{a} \textbf{= -log of K}_\textbf{a} \qquad \textbf{pK}_\textbf{b}\textbf{= -log of K}_\textbf{b} \qquad \textbf{pK}_\textbf{sp} \textbf{= -log of K}_\textbf{sp}$$

Helpful Hints
- Appendix E in the text contains tables of pK values for acid dissociation, base dissociation, and solubility products.
- Just like with pH, for any term "px," take 10^{-px} to obtain "x."

Key Concepts:
- For whatever system you're studying, **as p_ decreases, the value it is assessing increases**. For example, a small pK_a for a given acid indicates a large K_a, which means a large percentage of that acid will be dissociated into ions at equilibrium.

  - The smaller the pK_a, the stronger the acid.
  - The smaller the pK_b, the stronger the base.
  - The smaller the pK_{sp}, the more solid will dissolve in water (covered in Chapter 18).

- The water dissociation constant, K_w is related to K_a and K_b:
  - Consider the K_a process for some acid: $H_2O \text{ (l)} + HA \text{ (aq)} \leftrightarrow H_3O^+ \text{ (aq)} + A^- \text{(aq)}$
  - Now consider the K_b process for the anion of that acid. $A^- \text{ (aq)} + H_2O \text{ (l)} \leftrightarrow OH^- \text{ (aq)} + HA \text{ (aq)}$
  - Adding the two reactions results in the K_w process: $H_2O \text{ (l)} + H_2O \text{ (l)} \leftrightarrow H_3O^+ \text{ (aq)} + OH^- \text{ (aq)}$

Useful Relationships:
- $\textbf{pK}_\textbf{w} \textbf{= 14 = pH + pOH}$ This relationship is derived from taking the "-log" of the equation: $K_w = [H_3O^+][OH^-]$. *If you know the pH for a solution, you can find its pOH and vice versa.*

- $\textbf{K}_\textbf{w} \textbf{= K}_\textbf{a} \textbf{K}_\textbf{b} \textbf{= 1x10}^\textbf{-14}$ Recall that when two reactions are added together, their equilibrium constants are multiplied. *If you know the K_a for a substance, you can find its K_b and vice versa.*

- $\textbf{pK}_\textbf{w} \textbf{= 14 = pK}_\textbf{a} \textbf{+ pK}_\textbf{b}$ This relationship is obtained from taking the "-log" of the equation: $K_w = K_aK_b = 1 \times 10^{-14}$. *If you know the pK_a for a substance, you can find its pK_b and vice versa.*

EXERCISE 5: Rank the acids from weakest to strongest: HNO_2 with a pK_a of 3.25, $HClO_2$ with a K_a of 1.1×10^{-2}, arsenic acid with a pK_a of 2.26, phenol with a K_a of 1.0×10^{-10}. All values are at 25°C.

STRATEGY: To compare acid strength, we'll need to convert all the values to either pK_a values (the smaller the pK_a the stronger the acid) or to K_a values (the larger the K_a the stronger the acid). Let's convert each value to a pK_a.

HNO_2 with a pK_a of 3.25

$HClO_2$ with a K_a of 1.1×10^{-2}: The $pK_a = -\log$ of $K_a = 1.96$

Arsenic acid with a $pK_a = 2.26$

Phenol with a K_a of 1.0×10^{-10}: $pK_a = -\log$ of $K_a = 10.00$

SOLUTION: The smaller the pK_a the stronger the acid, so the ranking from weakest to strongest is: phenol, HNO_2, arsenic Acid, $HClO_2$.

Try It #2: Repeat Exercise 5, but compare K_a values instead of pK_as. (Start with the data provided within the question.) Verify that you obtain the same trend.

EXERCISE 6: Consider Solution A which has a pH of 8.27 and Solution B, which has a pOH of 3.65. Which solution has a greater $[H_3O^+]$? Both solutions are at the same temperature.

STRATEGY: Covert both to $[H_3O^+]$ and compare.

Solution A: $[H_3O^+] = 10^{-pH} = 10^{-8.27} = 5.4 \times 10^{-9}$ M

Solution B: $[OH^-] = 10^{-pOH} = 10^{-3.65} = 2.24 \times 10^{-4}$; $[H_3O^+] = 1 \times 10^{-14}/[OH^-] = 4.5 \times 10^{-11}$ M

SOLUTION: Solution A has a greater $[H_3O^+]$ than Solution B.

Try It # 3: Repeat Exercise 6, but compare pH values rather than $[H_3O^+]$. (Start with the data provided within the question.) Verify that Solution A has a greater $[H_3O^+]$ based on its pH.

EXERCISE 7: Use Appendix E to calculate the value of K_{eq} at 25°C for the reaction:
$$OH^- (aq) + NH_4^+ (aq) \leftrightarrow NH_3 (aq) + H_2O (l)$$

STRATEGY AND SOLUTION: This is the reverse of a K_b process for NH_3. Appendix E contains tables of pK values, so find pK_b for ammonia: 4.75.

$$K_b = 10^{-4.75} = 1.8 \times 10^{-5}$$

$$K_{reaction} = 1/K_b = 1/1.8 \times 10^{-5} = 5.6 \times 10^4$$

17.3   WEAK ACIDS AND BASES

As mentioned in Section 17.1, the difference between a strong acid/base and a weak acid/base is the extent to which the acid/base dissociates in water. Molecular pictures help illustrate the concept. Let's say you have 1000 molecules of HCl (a strong acid) versus 1000 molecules of HA (some generic weak acid). If we were to place each of them into separate containers, each of which contain the same quantity of water, what would occur?

HCl (aq) **Scenario**	**HA (aq)** **Scenario**
1000 H_3O^+ 1000 Cl^-	1 H_3O^+ 1 A^- 999 HA

These pictures illustrate several key concepts:
- At equal concentrations, a strong acid generates much more H_3O^+ than a weak acid.
- At equal concentrations, a strong acid will have a lower pH than a weak acid.
- At equal concentrations, a strong acid will be significantly more corrosive than a weak acid (recall it is the H_3O^+ that acts corrosively). This shows why we can eat chemicals that contain weak acids such as citrus fruits and tomatoes, but not strong acids.

Key Concepts:
- **pH does NOT measure acid strength.**   This is a common misconception for students. Remember, pH is a measure of $[H_3O^+]$. K_a is a measure of acid strength. It is entirely possible to have a concentrated, weak acid solution that has a lower pH than a dilute, strong acid solution.
- One may compare pH's to assess acid strength *only if* the two solutions are the same concentration.
- The trends illustrated above also occur for strong and weak bases, with respect to $[OH^-]$.

Try It #4: A 1.0 M solution of the weak acid CH_3CO_2H has a pH of 2.4. What concentration of HCl would yield this pH? Compare the two acid concentrations.

Weak Acids: Proton Transfer to Water

Key Concept:
- Equilibrium calculations must be performed to determine the concentrations of solute species and pH in weak acid solutions. This is the case because weak acids do not dissociate 100% in water.

Useful Relationship:
- $$\text{Percent HA ionized} = \frac{\left[H_3O^+\right]_{eq}}{\left[HA\right]_{initial}} \times 100$$

This relationship is used to determine the percent ionization of a weak acid (which is another way of showing the extent to which a given weak acid dissociates.)

EXERCISE 8: A chemist noticed that part of a label on a reagent bottle had corroded off. She could still read "0.100 M" on it, but the identity of the chemical was no longer legible. A pH reading of the solution indicated that its pH was 2.379. She knew from her inventory that the solution must be one of three acids: acetic acid ($pK_a = 4.75$), formic acid ($pK_a = 3.75$), or benzoic acid ($pK_a = 4.20$). Which one is it?

STRATEGY: The seven-step method would work well here.

1. **Determine what is asked for.** If we can calculate the pK_a of the acid, we can determine the identity of the acid from the choices given. To calculate the K_a of the acid, we would need equilibrium concentrations of solute species in solution.
2. **Identify the major chemical species.** This is a K_a process for an unknown acid, so the general major chemical species will be the weak unknown acid – call it HA – and the ions it produces when it reacts with water: A^- and H_3O^+.
3. **Determine what chemical equilibria exist.** The equilibrium reaction will be:
$$H_2O \text{ (l)} + HA \text{ (aq)} \leftrightarrow H_3O^+ \text{ (aq)} + A^- \text{(aq)}$$
4. **Write the K_{eq} expressions.**

$$K_a = \frac{[H_3O^+][A^-]}{[HA]}$$

5. **Organize the data and unknowns.**
The pH of the solution is 2.379, so the $[H_3O^+]$ is $10^{-2.379} = 4.18 \times 10^{-3}$. According to the 1:1 reaction stoichiometry, this must also be the $[A^-]$. The [HA] at equilibrium must be the initial 0.100 M *minus* the amount that was converted to A^- and H_3O^+:

	HA (aq)	+	H$_2$O (l)	↔	H$_3$O$^+$ (aq)	+	A$^-$(aq)
Concentrations:							
Initial	0.100				0		0
Change	-4.18×10^{-3}				$+4.18 \times 10^{-3}$		$+4.18 \times 10^{-3}$
Equilibrium	$0.100 - 4.18 \times 10^{-3}$				4.18×10^{-3}		4.18×10^{-3}

6. **Substitute and calculate.**

$$K_a = \frac{[H_3O^+][A^-]}{[HA]} = \frac{(4.18 \times 10^{-3})^2}{0.100 - 4.18 \times 10^{-3}} = 1.82 \times 10^{-4}$$

$$-\log(1.82 \times 10^{-4}) = 3.739 = 3.74$$

SOLUTION: To answer the question, the unknown acid is formic acid, with a pK_a of 3.75.

7. **Does the result make sense?** The pK_a calculated nearly matched one of the unknowns' pK_as, so the answer seems reasonable.

EXERCISE 9: Calculate the percent ionization of a 0.100 M aqueous solution of formic acid. (Note: formic acid was analyzed in the previous exercise.)

STRATEGY: Given the information from the previous exercise and the mathematical relationship for percent ionization, we can easily solve this question.

$$\text{Percent formic acid ionized} = \frac{[H_3O^+]_{eq}}{[\text{formic acid}]_{initial}} \times 100 = \frac{4.18 \times 10^{-3} \text{ M}}{0.100 \text{ M}} \times 100 = 4.18\%$$

SOLUTION: The percent ionization of 0.100 M formic acid is 4.18%. The answer seems reasonable because formic acid is a weak acid, so it will not dissociate to a great extent.

EXERCISE 10: What is the $[H_3O^+]$ in a 0.20 M weak acid solution that has a pH of 4.25?

STRATEGY AND SOLUTION: The extraneous information in this problem might make it appear more difficult than it actually is. The pH = $-\log[H_3O^+]$, so $[H_3O^+] = 10^{-pH} = 10^{-4.25} = 5.6 \times 10^{-5}$ M.

Take-home message: pH = $-\log[H_3O^+]$, no matter if it is a strong acid or a weak acid.

Weak Bases: Proton Transfer From Water

Key Concept:
- Equilibrium calculations must be performed to determine the concentrations of solute species and pH in weak base solutions. This is the case because weak bases do not dissociate 100% in water.

EXERCISE 11: What is the pH of a 0.0200 M aqueous solution of aniline, $C_6H_5NH_2$?

STRATEGY: The seven-step method works well here.

1. **Determine what is asked for.** We need to find the pH of the solution. This is a weak base, so equilibrium calculations are necessary. We can get pH from $[OH^-]$ at equilibrium.
2. **Identify the major chemical species.** This is a K_b process for aniline, $C_6H_5NH_2$, so mostly aniline will be present at equilibrium.
3. **Determine what chemical equilibria exist.** The equilibrium reaction will be:
$$H_2O \text{ (l)} + C_6H_5NH_2 \text{ (aq)} \leftrightarrow C_6H_5NH_3^+ \text{ (aq)} + OH^- \text{(aq)}$$
   Looking in Appendix E, we find pK_b of aniline is 9.13 so $K_b = 7.4 \times 10^{-10}$
4. **Write the K_{eq} expressions.**
$$K_b = \frac{[C_6H_5NH_3^+][OH^-]}{[C_6H_5NH_2]} = 7.4 \times 10^{-10}$$
5. **Organize the data and unknowns.**
   The starting concentration of aniline is 0.0200 M and some quantity of it will convert to $C_6H_5NH_3^+$.

Concentrations:	$C_6H_5NH_2$ (aq) +	H_2O (l) ↔	$C_6H_5NH_3^+$ (aq)	+ OH^-(aq)
Initial	0.0200		0	0
Change	-x		+x	+x
Equilibrium	0.0200 − x		x	x

6. **Substitute and calculate.**

$$K_b = \frac{[C_6H_5NH_3^+][OH^-]}{[C_6H_5NH_2]} = \frac{x^2}{0.0200 - x} = 7.4 \times 10^{-10}$$

Assume x is small relative to "0.0200" and drop it. $x = 3.85 \times 10^{-6}$, which is indeed $<<< 0.0200$

$[OH^-] = x$; so pOH = -log(3.85×10^{-6}) = 5.4149
pH = 14 − pOH = 8.585

SOLUTION: To answer the question, pH of the solution is 8.59. (2 sig figs)

7. **Does the result make sense?** This is a basic solution and pH>7, so the answer seems reasonable.

17.4 RECOGNIZING ACIDS AND BASES

Both acids and bases can be categorized by type. The table shown here summarizes the general types of acids that you'll encounter in Chapter 17.

Types of Acids in Chapter 17

Type of Acid	How to Identify It	Format for Writing the Formula	Example
Oxyacid	Contains an inner atom which is bonded to a various number of oxygen atoms. One or more of these oxygens will have H attached to them (acidic proton).	Acidic proton(s) is/are written first, then polyatomic anion	$HClO_4$ 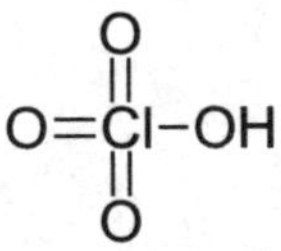
Carboxylic Acid	Contains a carboxylic acid functional group: 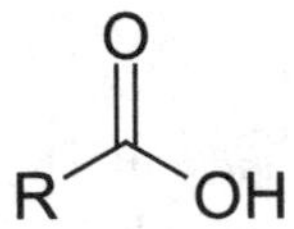 Found in carbon-based compounds	Carboxylic acid group is often written last: $C_6H_5CO_2\mathbf{H}$ But is sometimes written like this to indicate acidic H: $\mathbf{H}C_7H_5O_2$	$C_6H_5CO_2H$ 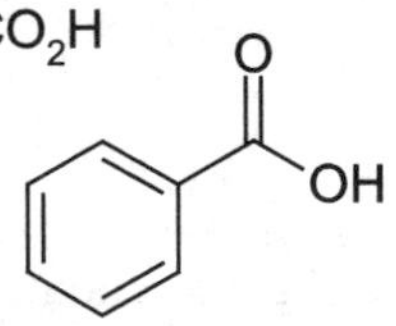
Polyprotic Acid	Contains more than one acidic H.	Depends on if it is an oxyacid or carboxylic acid: H_3PO_4 or $C_2H_4(CO_2H)_2$ or $H_2C_4H_4O_4$	H_2SO_4 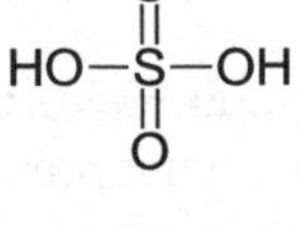
Other Acids	The acidic proton is written first. Usually these are binary acids.	Acidic hydrogen, followed by other atom(s)	HCl, HBr, HI, HCN

Type of Acid (continued)	How to Identify It	Format for Writing the Formula	Example
*Salts containing the conjugate acid of a weak base	Ask yourself: "does it look like a weak base with a proton added to it?" (This will be the cation portion of the salt.)	A salt	$C_6H_5NH_3Cl$ (structure of benzene ring with $-NH_3+$ and Cl^-)

* Will be covered in Section 17.5.

Notice that in all the acid structures in the table, only the H atoms attached to the most polar bonds are acidic. The hydrogen atoms bonded to carbon won't be acidic.

Structures and Names of Oxyacids

The name of the polyatomic ion in the acid is changed in this manner: "-ate" becomes "-ic" and "-ite" becomes "-ous." The word "acid" is added to the end of the name. For example, $HClO_4$ contains the perchlor*ate* ion. The name of the acid is then "perchlor*ic* acid."

Weak Bases

As mentioned in Section 17.1, strong bases are typically Group I or II metal hydroxides. Weak bases, on the other hand, typically contain a nitrogen with single bonds to hydrogen or carbon.

Ammonia   Aniline   Methylamine

Organic chemicals that contain nitrogen with all single bonds attached to hydrogens and/or singly-bonded carbons are called "amines." The nitrogen atom is the basis for the amine functional group. It is the lone pair on this nitrogen atom that is capable of accepting a proton from another molecule, just like in ammonia.

Conjugate Acid-Base Pairs

Conjugate acid/base pairs differ by one proton. For example: HCl is an acid. After HCl donates a proton, what remains is its conjugate base, Cl^-. Here is another example: NH_3 is a base. After it accepts a proton, it becomes NH_4^+. NH_4^+ is the conjugate acid of NH_3.

EXERCISE 12:   a) What is the conjugate acid of HPO_4^{2-}? b) What is the conjugate acid of HCO_3^-?

c) What is the conjugate base of HCO_3^-?

STRATEGY AND SOLUTION: Conjugates differ by one proton.
a) The question "what is the conjugate acid of HPO_4^{2-}?" indicates that the HPO_4^{2-} must be acting as a base. After it accepts a proton, it will be $H_2PO_4^-$. Thus, the conjugate acid of HPO_4^{2-} is $H_2PO_4^-$.
b) Similar reasoning as in "a" applies. The conjugate acid of HCO_3^- is H_2CO_3.
c) The conjugate base of HCO_3^- is what remains after the substance acted as an acid: CO_3^{2-}.

Try It # 5: Label each substance as an acid or a base. Determine its conjugate.
  a) $CH_3CH_2CH_2CO_2H$ b) CH_3NH_2 c) $HClO_2$ d) H_2NNH_2

Naming Organic Conjugates

Key Concepts:
- When an amine compound picks up a proton, the resulting conjugate acid is called an "ammonium" ion. For example: methylamine = CH_3NH_2, methylammonium ion = $CH_3NH_3^+$.
- When a carboxylic acid compound loses a proton, the resulting conjugate base is given the suffix "-ate ion" in place of the "-ic acid" suffix. For example, benzoic acid = $C_6H_5CO_2H$, benzoate ion = $C_6H_5CO_2^-$.

17.5 ACIDIC AND BASIC SALTS

Basic and Acidic Salts

Key Concepts:
- The conjugate acid of a weak base is weak.
- The conjugate base of a weak acid is weak.
- The conjugate of a strong species is neutral. For example, the conjugate base of the strong acid HCl is Cl^-, which is a neutral species.

Helpful Hints
- **The pK_b for the conjugate base of a weak acid can be found from the acid's pK_a using the relationship "$14 = pK_a + pK_b$."** This is also the case for finding the pK_a of the conjugate acid of a weak base.
- **If you cannot recognize weak acids and bases just from sight, then look at Appendix E in the text.** If you don't see the substance you're looking for, then determine its conjugate and look for that. If the substance *or* its conjugate is not in Appendix E, then it is probably strong. (Appendix E will list most of the weak species we'll work with in General Chemistry.)
- **For an acid or base that contains many atoms: identify the acid and call it "HA." Identify its conjugate base and call it "A⁻."** This can save time in calculations. The HA and A⁻ designations are relative to each other, so it doesn't matter if your A⁻ is neutral and HA is positively-charged. Relative to *each other* they'll always be HA vs. A⁻.

EXERCISE 13: Label each aqueous solution as acidic, basic, or neutral. For the solutions that create weak acid/base equilibria, write the appropriate equilibrium reaction, the corresponding equilibrium constant expression, and use Appendix E to calculate the value of K_{eq}.

    a) KCN   b) Triethylammonium chloride   c) KBr

STRATEGY AND SOLUTION:

a) KCN is an ionic compound, so we must assess the ions it creates: K^+ and CN^-. K^+ is a neutral ion – it would not want to snatch up an OH^- in water because KOH is a strong base and would immediately dissociate again. CN^- is not a strong base BUT its conjugate acid is HCN, a weak acid. Therefore, CN^- is a weak base. A salt solution containing CN^- will be slightly basic.

$$CN^- \text{ (aq)} + H_2O \text{ (l)} \leftrightarrow OH^- \text{ (aq)} + HCN \text{ (aq)}$$

The equilibrium expression will be: $K_b = \dfrac{[OH^-][HCN]}{[CN^-]}$

The pK_a of HCN is 9.21 (from Appendix E)

We can use $pK_w = 14 = pK_a + pK_b$ to find pK_b of CN^-, which will be 4.79.

$$K_b \text{ is } 10^{-4.79} = 1.6 \times 10^{-5}$$

b) Triethylammonium chloride contains triethylammonium ion, the conjugate acid of the weak base, triethylamine. Chloride ion is not attracted to water because the conjugate acid would be the strong acid HCl. The pH of the solution will be determined by the triethylammonium ion. If we call the triethylammonium ion "HA" for short, then the conjugate base, triethylamine, will be "A⁻":

$$H_2O \text{ (l)} + HA \text{ (aq)} \leftrightarrow H_3O^+ \text{ (aq)} + A^- \text{(aq)}$$
$$K_a = \dfrac{[H_3O^+][A^-]}{[HA]}$$

The K_a for this process is not (directly) listed in Appendix E, but can be obtained from the pK_b of triethylamine, which is 3.25.
The pK_a of HA is $14 - pK_b = 10.75$.
The K_a for the reaction is: $10^{-10.75} = 1.8 \times 10^{-11}$

c) KBr is an ionic compound so it will dissociate if dissolved in water into: K^+ and Br^-. K^+ is not attracted to water, so it will not alter the pH of the water. Br^- is the conjugate base of the strong acid HBr, so it will not pick up a proton from water (if it tried to, it would immediately dissociate 100%, because HBr is a strong acid). Br^- won't alter the pH of the water either. Therefore KBr is a neutral salt.

Try It # 6: Label each aqueous solution as acidic, basic, or neutral. For the solutions that create weak acid/base equilibria, write the appropriate equilibrium reaction, corresponding equilibrium constant expression, and use Appendix E to calculate the value of K_{eq}.

    a) NaClO   b) $MgCl_2$ c) ethylammonium chloride – $(C_2H_5NH_3Cl)$

EXERCISE 14: Calculate the pH of a 0.500 M aqueous solution of hydroxylammonium sulfate: $(HONH_3)_2SO_4$ (aq)

STRATEGY: Use the seven-step method.

1. **Determine what is asked for.** We need to find the pH of the solution. To determine the pH of a solution, we must first determine if the solute species present are acidic, basic or neutral.
2. **Identify the major chemical species.** Hydroxylammonium ion ($HONH_3^+$) and SO_4^{2-}. $HONH_3^+$ is the conjugate acid of the weak base hydroxylamine, $HONH_2$. Thus, $HONH_3^+$ will be a weak acid. Sulfate ion is a spectator ion here ($K_b = 9 \times 10^{-13}$).
3. **Determine what chemical equilibria exist.** The equilibrium reaction will be the acid dissociation of $HONH_3^+$:
$$H_2O \text{ (l)} + HONH_3^+ \text{ (aq)} \leftrightarrow H_3O^+ \text{ (aq)} + HONH_2 \text{ (aq)}$$

4. **Write the K_{eq} expressions.** First, we need K_a for the process. This isn't directly available in Appendix E, but the pK_b of hydroxylamine is: 8.06.
$$pK_a \text{ for } HONH_3^+ = 14 - 8.06 = 5.94$$
$$K_a = 1.15 \times 10^{-6}$$

$$K_a = \frac{[HONH_2][H_3O^+]}{[HONH_3^+]} = 1.15 \times 10^{-6}$$

5. **Organize the data and unknowns.**
   Note this: the initial $[HONH_3^+]$ is $2 \times [(HONH_3)_2SO_4]$ because there are 2 $HONH_3^+$ for every one $(HONH_3)_2SO_4$.

	$HONH_3^+$ (aq) +	H_2O (l) $\leftrightarrow$	H_3O^+ (aq)	+ $HONH_2$(aq)
Concentrations:				
Initial	1.00		0	0
Change	-x		+x	+x
Equilibrium	1.00 – x		x	x

6. **Substitute and calculate.**
$$K_a = \frac{[HONH_2][H_3O^+]}{[HONH_3^+]} = \frac{(x)^2}{1.00 - x} = 1.15 \times 10^{-6}$$

Use Method of Successive Approximations to solve for x.
(Drop "x" from the term "1.00-x")
$$x = [H_3O^+] = 1.07 \times 10^{-3}$$
We can see that $1.07 \times 10^{-3} << 1.00$, so no testing is needed. Our approximation was valid.
$$-\log(1.07 \times 10^{-3}) = 2.97$$

SOLUTION: To answer the question, pH of the resulting solution is 2.97.

7. **Does the result make sense?** Hydroxylammonium ion acted as an acid. The final pH seems reasonable because it is less than 7 (acidic).

Here is a summary of the different scenarios that exist for pH calculations in Chapter 17:

Possible Scenarios	Process	Appropriate Mathematical Relationship(s) to Use
A strong acid	None – complete dissociation	$[H_3O^+]$ = [acid solution] if monoprotic; can get pH directly from [acid solution].
A strong base	None – complete dissociation	Have $[OH^-]$. Can get pH two ways: 1. $14 = pH + pOH$. 2. $K_w = 10^{-14} = [H_3O^+][OH^-]$.
A weak acid (this includes the conjugate acid of a weak base in a salt)	K_a process	Write a K_a expression. Make an equilibrium table and calculate $[H_3O^+]$, then pH.
A weak base (this includes the conjugate base of a weak acid in a salt)	K_b process	Write a K_b expression. Make an equilibrium table and calculate $[OH^-]$. Calculate pH from $[OH^-]$.

17.6 FACTORS AFFECTING ACID STRENGTH

Effect of Charge & Structural Factors

Consider these factors when assessing acid strength:

- **The charge of the acid** (Opposite charges attract, like charges repel, so a proton will leave a cation more easily than a neutral compound or anion.)
- **The polarity of the H-X bond** (The electron pair within the bond will be pulled towards the X to the extent that the proton could break off, resulting in two ions. Recall that a polar bond is part-way between a perfect covalent bond and an ionic bond.)
- **The size of X in the H-X bond** (Acidity increases down a column because of the decrease in overlap between orbitals that make up the sigma bond.)
- **The stability of the acid's conjugate base** (The more stable the conjugate base, the stronger the acid. A chemical is not going to act as an acid if its only option is to generate an unstable base.)

Factors that stabilize a conjugate base – and increase acid strength – include:
- **Electron withdrawing groups nearby**. A highly electronegative atom near the negative charge will pull electron density towards it. (This is sometimes called an "inductively electron withdrawing" group.)
- **Resonance structures**. The more resonance structures that can be drawn for an anion, the stronger its conjugate acid will be. This is due to the fact that π electrons in the resulting anion won't be stuck in one place, but will be delocalized.

EXERCISE 15: Draw structures of HNO_2 and HNO_3. Why is HNO_2 a weaker acid than HNO_3?

SOLUTION: The structures show that HNO_3 has one more O attached to the central N atom than HNO_2 does. The extra oxygen is able to pull even more electron density away from the H than in HNO_2.

HNO_2 $\qquad\qquad\qquad\qquad\qquad\qquad\qquad$ HNO_3

EXERCISE 16: Propose a rationale based on the structures of the acids to explain the trend in pK_as.

Trichloroacetic acid $\qquad$ Chloroacetic acid $\qquad$ Acetic acid
$(pK_a = 0.66)$ $\qquad\qquad\quad$ $(pK_a = 2.87)$ $\qquad\qquad$ $(pK_a = 4.75)$

SOLUTION: All three structures contain a carboxylic acid group. When the acidic proton is removed in each case, the resulting conjugate base can be stabilized by resonance because the "-" charge can reside on either oxygen. The differences in the structures lie in the number of chlorine atoms attached to the left-most carbon in each structure. Cl, which is more electronegative than C, will withdraw electron density from the left carbon, which in turn, will withdraw electron density from the right carbon. The more chlorine atoms present, the more intensely this will occur, and the more stable the resulting anion will be.

Base Strength

This topic is not covered in the text, but the concepts are similar to acid strength.

Base strength can be assessed by determining how willing a molecule or ion is to donate a lone pair to making a bond with an incoming proton. How easily it accepts the proton is based on how willing that lone pair is to bond with it. **The better the lone pair donor, the stronger the base.** One must consider the environment of the lone pair to assess the base's strength.

For example, consider NH_3 versus H_2O. The lone pairs on both central atoms are in the same valence shell ($n=2$). The key difference between the two substances is that nitrogen's lone pair is held less tightly than oxygen's because nitrogen's nucleus contains seven protons instead of oxygen's eight. All other factors being equal, **nitrogen within a neutral compound is a better base than oxygen**.

In another example, consider O^{2-} versus OH^-. Opposite charges attract, so the O^{2-} would be a better base even than OH^-. (The conjugate acid of O^{2-} is OH^-, so it follows that O^{2-} will be the stronger base.) All other factors being equal, **the greater the difference in charge between the proton and the base, the stronger the base will be.** An anion makes a better base than a neutral compound or cation (opposite charges attract).

Another factor to consider is that **a lone pair that's stuck on one atom will be a better base than a similar lone pair that is delocalized in resonance structures.** A lone pair used in resonance structures won't spend all of its time on the one atom, so it won't be as good of a base.

EXERCISE 17: NH_3 contains protons. Why does it act as a base when dissolved in water? Why does it not act as an acid? Provide a structural explanation.

SOLUTION: Both N and O are $n=2$ elements, so the important factor in this assessment is bond polarity. The bond between O-H is more polar than the bond between N-H, so the oxygen atom pulls electron density within that bond more strongly than nitrogen does. The proton can thus more easily break off the O than the N. Furthermore, the O (with an extra proton in the nucleus relative to N) can better stabilize the resulting anion. It is worth mentioning that nitrogen is a better proton acceptor than water is because its lone pair is not held as tightly to it as oxygen's is (nitrogen is a better base than water).

EXERCISE 18: Could NH_3 *ever* act as an acid?

SOLUTION: Acid-base reactions are competitions for protons. NH_3 could act as an acid only if it was in the presence of a chemical that is a better base than it AND no better acids were in the container. Since water is a stronger acid than NH_3, this could only occur in the absence of water, i.e., in a non-aqueous environment.

17.7 MULTIPLE EQUILIBRIA

We learned in Section 17.5 that the dominant equilibrium is the one with the largest equilibrium constant. In reality, we must keep in mind that several equilibria can be present simultaneously within a given system. Occasionally it is necessary to determine the concentration of a component involved in a minor equilibrium.

Key Concept:
• Acid-base reactions are a competition for protons.

EXERCISE 19: Write equations for the equilibria that are present in an aqueous solution of H_3PO_4. Look up their pK_a values in Appendix E of the text. Identify the dominant equilibrium.

SOLUTION:

$$H_3PO_4 \text{ (aq) } + \text{ } H_2O \text{ (l) } \leftrightarrow H_2PO_4^- \text{(aq) } + H_3O^+ \text{ (aq); } pK_a = 2.16$$
$$H_2PO_4^- \text{ (aq) } + \text{ } H_2O \text{ (l) } \leftrightarrow HPO_4^{2-} \text{(aq) } + H_3O^+ \text{ (aq); } pK_a = 7.21$$
$$HPO_4^{2-} \text{ (aq) } + \text{ } H_2O \text{ (l) } \leftrightarrow PO_4^{3-} \text{(aq) } + H_3O^+ \text{ (aq); } pK_a = 12.32$$

A solution of a polyprotic acid contains multiple equilibria, but only one equilibrium will dominate at a time. In the solution of H_3PO_4, the equilibrium reaction with the smallest pK_a will have the largest equilibrium constant. The first reaction is the dominant equilibrium. This means that all of the H_3PO_4 molecules will react with base before any of the $H_2PO_4^-$ molecules will start reacting. Similarly, all of the $H_2PO_4^-$ ions will react with base before any HPO_4^{2-} ions begin reacting. This should make sense because the H_3PO_4 is a stronger acid than $H_2PO_4^-$, which in turn is a stronger acid than HPO_4^{2-}. The stronger acid will react preferentially with added base until it is completely consumed.

Key Concept:
- In a polyprotic acid, every molecule of the neutral compound will react with added base before any of the anion will start to react.

EXERCISE 20: Is an aqueous solution of NH_4NO_2 acidic, basic, or neutral? Briefly explain why. Include any applicable equilibrium reaction(s).

STRATEGY: The major solute components are NH_4^+ and NO_2^-. NH_4^+ is the conjugate acid of the weak base NH_3. NO_2^- is the conjugate base of the weak acid HNO_2. Thus, the equilibria taking place in this solution are:

$$H_2O \text{ (l)} + NH_4^+ \text{ (aq)} \leftrightarrow NH_3 \text{ (aq)} + H_3O^+ \text{ (aq)}$$

$$NO_2^- \text{ (aq)} + H_2O \text{ (l)} \leftrightarrow OH^- \text{ (aq)} + HNO_2 \text{ (aq)}$$

Notice that the salt contains NH_4^+ (not NH_3) and NO_2^- (not HNO_2) so the NH_4^+ and NO_2^- must be written as reactants. Equilibrium constants must be assigned for each process to determine the dominant equilibrium.

The pK_b of NH_3 is 4.75, so pK_a of NH_4^+ is $14 - 4.75 = 9.25$
The pK_a of HNO_2 is 3.25, so pK_b of NO_2^- is $14 - 3.25 = 11.75$

SOLUTION: The relative sizes of the equilibrium constants show that NH_4^+ is a stronger acid than NO_2^- is a base. Thus, the solution will be acidic.

EXERCISE 21: Determine the concentration of all solutes in a 0.125 M aqueous solution of H_2S.

STRATEGY: Use the seven-step method.
1. **Determine what is asked for.** We need to find the concentrations of all solute species in a polyprotic acid solution.
2. **Identify the major chemical species.** H_2S, a weak acid, is the major species. It will be a reactant in the dominant equilibrium. HS^-, its conjugate base, will be a minor species. It, too, will dissociate, but to a much smaller degree.
3. **Determine what chemical equilibria exist.** The equilibria will be the two sequential acid dissociations of H_2S:
$$H_2O \text{ (l)} + H_2S \text{ (aq)} \leftrightarrow H_3O^+ \text{ (aq)} + HS^- \text{ (aq)}; \quad pK_{a1} = 7.05$$
$$H_2O \text{ (l)} + HS^- \text{ (aq)} \leftrightarrow H_3O^+ \text{ (aq)} + S^{2-} \text{ (aq)}; \quad pK_{a2} = 19$$

4. **Write the K_{eq} expressions.** In this problem, we have two weak acid equilibria expressions:
$$K_{a1} = 10^{-pK_{a1}} = 10^{-7.05} = 8.91 \times 10^{-8} = \frac{[HS^-][H_3O^+]}{[H_2S]}$$

and

$$K_{a2} = 10^{-pK_{a2}} = 10^{-19} = 1.00 \times 10^{-19} = \frac{[S^{2-}][H_3O^+]}{[HS^-]}$$

5. **Organize the data and unknowns.**
    Assess the dominant equilibrium first.

	H_2S (aq) +	H_2O (l) $\leftrightarrow$	H_3O^+ (aq)	+ HS^- (aq)
Concentrations:				
Initial	0.125		0	0
Change	-x		+x	+x
Equilibrium	0.125 − x		x	x

6. **Substitute and calculate.**

$$K_{a1} = 10^{-pK_{a1}} = 10^{-7.05} = 8.91 \times 10^{-8} = \frac{[HS^-][H_3O^+]}{[H_2S]} = \frac{(x)^2}{0.125 - x}$$

Use the Method of Successive Approximations to solve for x.
(Drop "x" from the term "0.125-x")
x = 1.055×10^{-4}; in a second approximation, the same number is obtained.

$0.125 - x = [H_2S] = 0.1249 \approx 0.125$ M; x = 1.055×10^{-4} M = $[H_3O^+] \approx [HS^-]$,
but the HS^- will dissociate to a small degree.

We need to make an equilibrium table for HS^- to determine its actual concentration and the concentration of S^{2-} that it generates. Note that we must account for the H_3O^+ present in the container.

	HS^- (aq) +	H_2O (l) $\leftrightarrow$	H_3O^+ (aq)	+ S^{2-} (aq)
Concentrations:				
Initial	1.055×10^{-4}		1.055×10^{-4}	0
Change	-x		+x	+x
Equilibrium	$1.055 \times 10^{-4} - x$		$1.055 \times 10^{-4} + x$	x

$$K_{a2} = 10^{-pK_{a2}} = 10^{-19} = 1.00 \times 10^{-19} = \frac{[S^{2-}][H_3O^+]}{[HS^-]} = \frac{x(1.055 \times 10^{-4} + x)}{1.055 \times 10^{-4} - x}$$

Since K_{a2} for this process is so incredibly small, $1.055 \times 10^{-4} + x \approx 1.055 \times 10^{-4} - x$, so the above equation simplifies to:

$$K_{a2} = x = 1.0 \times 10^{-19} = [S^{2-}]$$

SOLUTION: The equilibrium concentrations of all solute species are: $[H_2S]$ = 0.125 M, $[HS^-] = [H_3O^+] = 1.1 \times 10^{-4}$ M, $[S^{2-}] = 1 \times 10^{-19}$ M, (and $[OH^-] = 9.1 \times 10^{-11}$ M). (2 sig figs for equilibrium answers involving pK_{a1}; 1 sig fig for $[S^{2-}]$).

7. **Does the answer make sense?** This is a weak acid solution, so the $[H_3O^+]$ should be greater than the $[OH^-]$ and it is. The $[H_2S] > [HS^-] \gg [S^{2-}]$, which makes sense, because of the relative magnitudes of the equilibrium constants for the acid dissociations. The H_2S should be the major component of the various weak acid species, and it is. The $[S^{2-}]$ should be miniscule, and it is.

Helpful Hints
- The K_{a2} for a weak polyprotic acid is usually several orders of magnitude smaller than the K_{a1}.
- The math for determining the concentrations of minor species in polyprotic acids usually simplifies to K_{a2} = [products of second dissociation process]. See the exercise above, which illustrates this fact.

Salts of Polyprotic Acids

Key Concept:
- The salt of a polyprotic acid contains the conjugate base of a weak polyprotic acid and one or more spectator counter ions. A salt composed of a fully deprotonated acid will be weakly basic.
- Multiple equilibria must be analyzed in order to determine the concentrations of all species present in the salt of a polyprotic acid. Note:
  - Calculations involving monoanions of polyprotic acids are beyond the scope of the course, so you will only see calculations involving fully deprotonated salts.
  - The notation "K_{b1}" is used for the K_b process of the conjugate base of the K_{a1} process. Using H_2A as an example, the K_{b1} process would be:

$$HA^- + H_2O \leftrightarrow H_2A + OH^-$$

The "K_{b2}" process would then be the K_b process for the base: A^{2-}.

EXERCISE 22: Determine the concentration of all solutes in a 0.0850 M aqueous solution of $Na_2C_4H_4O_6$, given that the acid $H_2C_4H_4O_6$ has a K_{a1} of 1.0×10^{-3} and a K_{a2} of 4.6×10^{-5}.

STRATEGY: Use the seven-step method.
1. **Determine what is asked for.** We need to find the concentrations of all solute species in a solution composed of the dianion ($C_4H_4O_6^{2-}$) salt of a polyprotic acid.
2. **Identify the major chemical species.** $C_4H_4O_6^{2-}$ (a weak base) and Na^+ (a spectator ion) are the major species. $C_4H_4O_6^{2-}$ will be a reactant in the dominant equilibrium. The conjugate acid of this weak base, $HC_4H_4O_6^-$, will be a minor species. It, too, will dissociate, but to a much smaller degree.
3. **Determine what chemical equilibria exist.** The equilibria will be the two sequential base dissociations of $C_4H_4O_6^{2-}$:
$$H_2O \text{ (l)} + C_4H_4O_6^{2-} \text{ (aq)} \leftrightarrow OH^- \text{ (aq)} + HC_4H_4O_6^- \text{ (aq)}$$
$$H_2O \text{ (l)} + HC_4H_4O_6^- \text{ (aq)} \leftrightarrow OH^- \text{ (aq)} + H_2C_4H_4O_6 \text{ (aq)}$$

We are going to need the equilibrium constants for these reactions. The first reaction is a "K_{b2}" process. We know that this is a K_b process because OH^- is generated when the $C_4H_4O_6^{2-}$ ion reacts with water. We know that it is classified as "K_{b2}" because the acid generated during this K_b process is the acid for the K_{a2} process. The value of K_{b2} will therefore be based on the K_{a2} of the conjugate acid:

$$K_{b2} = \frac{K_w}{K_{a2}} = \frac{1.0 \times 10^{-14}}{4.6 \times 10^{-5}} = 2.2 \times 10^{-10}$$

The value of the equilibrium constant for the second reaction is a K_{b1} process for similar reasoning:

$$K_{b1} = \frac{K_w}{K_{a1}} = \frac{1.0 \times 10^{-14}}{1.0 \times 10^{-3}} = 1.0 \times 10^{-11}$$

4. **Write the K_{eq} expressions.** In this problem, we have two weak base equilibria expressions:

$$K_{b2} = 2.2 \times 10^{-10} = \frac{[HC_4H_4O_6^-][OH^-]}{[C_4H_4O_6^{2-}]}$$

and

$$K_{b1} = 1.0 \times 10^{-11} = \frac{[H_2C_4H_4O_6][OH^-]}{[HC_4H_4O_6^-]}$$

5. **Organize the data and unknowns.**
   Assess the dominant equilibrium first.

	$C_4H_4O_6^{2-}$ (aq) + H_2O (l) + $\leftrightarrow$	OH^- (aq) +	$HC_4H_4O_6^-$ (aq)
Concentrations:			
Initial	0.0850	0	0
Change	-x	+x	+x
Equilibrium	0.0850 – x	x	x

6. **Substitute and calculate.**

$$K_{b2} = 2.2 \times 10^{-10} = \frac{[HC_4H_4O_6^-][OH^-]}{[C_4H_4O_6^{2-}]} = \frac{(x)(x)}{0.0850 - x}$$

The equilibrium constant is quite small, so use the Method of Successive Approximations to solve for x. (Drop "x" from the term "0.0.0850-x".)

$$[C_4H_4O_6^{2-}] = 0.0850 \text{ M} - x = 0.085 \text{ M (2 sig figs from } K_{eq})$$
$$x = 4.32 \times 10^{-6} = [OH^-] \text{ and } [HC_4H_4O_6^-].$$

We know that the $HC_4H_4O_6^-$ will dissociate to a small degree, so we will need to analyze its actual concentration based on its equilibrium reaction:

	$HC_4H_4O_6^-$ (aq) + H_2O (l) $\leftrightarrow$	OH^- (aq) +	$H_2C_4H_4O_6$ (aq)
Concentrations:			
Initial	4.32×10^{-6}	4.32×10^{-6}	0
Change	-x	+x	+x
Equilibrium	4.32×10^{-6} – x	4.32×10^{-6} + x	x

Note that we accounted for the 4.32×10^{-6} M OH^- already present in the container on our "initial" line.

$$K_{b1} = 1.0 \times 10^{-11} = \frac{[H_2C_4H_4O_6][OH^-]}{[HC_4H_4O_6^-]} = \frac{(x)(4.32 \times 10^{-6} + x)}{4.32 \times 10^{-6} - x}$$

Since K_{b1} for this process is so incredibly small, x will be small relative to 4.32×10^{-6}, so the above equation simplifies to:

$$K_{b2} = x = 1.0 \times 10^{-11} = [H_2C_4H_4O_6]$$

We see here that x is indeed quite small relative to 4.32×10^{-6}, so this second base dissociation process does not alter the concentration of OH^- or $HC_4H_4O_6^-$.

The other solute species present in solution that have not yet been determined are H_3O^+ and the spectator ion, Na^+.

$$[H_3O^+] = \frac{K_w}{[OH^-]} = \frac{1.0 \times 10^{-14}}{4.32 \times 10^{-6}} = 2.3 \times 10^{-9}$$

$$[Na+] = 2 \times [Na_2C_4H_4O_6]_{initial} = 2 \times 0.0850\ M = 0.170\ M$$

SOLUTION: The equilibrium concentrations of all solute species are: $[Na^+] = 0.170\ M$ (3 sig figs because it is based only on the initial concentration of the original solution), $[OH^-]$ and $[HC_4H_4O_6^-] = 4.3 \times 10^{-6}\ M$, $[H_2C_4H_4O_6] = 1.0 \times 10^{-11}\ M$, $[H_3O^+] = 2.3 \times 10^{-9}\ M$, and $[C_4H_4O_6^{2-}] = 0.085\ M$ (2 sig figs for answers involving equilibrium constants).

7. **Does the answer make sense?** This is a weak base solution, so the $[H_3O^+]$ should be smaller than the $[OH^-]$ and it is. The $[C_4H_4O_6^{2-}] > [HC_4H_4O_6^-] > [H_2C_4H_4O_6]$, which makes sense, because of the relative magnitudes of the equilibrium constants for the base dissociations. The $C_4H_4O_6^{2-}$ should be the major component of the various weak acid species, and it is. The $[H_2C_4H_4O_6]$ should be quite small, and it is.

Chapter 17 Self-Test

You may use a periodic table, a calculator, and Appendix E in the text for this test. Assume all solutions in this test are at 25 °C.

1. Write a chemical equation to show the equilibrium process that is established when CH_3NH_3Cl is dissolved in water. Use Appendix E from the text to determine the value of the equilibrium constant for this reaction.

2. What is the pH of 0.20 M sodium hydroxide (aq)? What is the hydronium ion concentration in this solution?

3. What is the pH of 0.25 M aqueous formic acid?

4. Describe how to prepare 250.0 mL of a pH 8.55 solution from solid $NaNO_2$.

5. Will a solution of ammonium cyanide be acidic, basic, or neutral? Defend your answer quantitatively.

6. Which will be a stronger acid? Briefly explain.

   a) H_3PO_3 vs. $H_2PO_3^-$ b) H_2Se vs. H_2O c) CH_3CH_2OH vs. CH_3CO_2H

7. Explain the relative pK_a values:

   $CH_3CH_2CH_2CO_2H$ $CH_3CHClCH_2CO_2H$ $CH_3CH_2CHClCO_2H$

   $pK_a = 4.81$ $pK_a = 4.05$ $pK_a = 2.86$

Answers to Try Its

1. 5.00×10^{-5} M OH$^-$; 2.00×10^{-10} M H$_3$O$^+$

2. The ranking from weakest to strongest would be (with K$_a$s in parentheses): Phenol (1.0×10^{-10}), HNO$_2$ (5.6×10^{-4}), Arsenic (5.5×10^{-3}), HClO$_2$ (1.1×10^{-2})

3. pH of Solution A = 8.27, pH of Solution B = 10.35. The lower the pH, the greater the [H$_3$O$^+$].

4. [HCl] = 0.0040 M, [CH$_3$CO$_2$H] = 1.0 M. A strong acid completely dissociates and a weak acid barely dissociates. It therefore takes much less strong acid (HCl) – than weak acid (CH$_3$CO$_2$H) – to generate the [H$_3$O$^+$] corresponding to this pH.

5. a) CH$_3$CH$_2$CH$_2$CO$_2$H is an acid and its conjugate base is CH$_3$CH$_2$CH$_2$CO$_2^-$
   b) CH$_3$NH$_2$ is a base and its conjugate acid is CH$_3$NH$_3^+$
   c) HClO$_2$ is an acid and its conjugate base is ClO$_2^-$
   d) H$_2$NNH$_2$ is a base and its conjugate acid is H$_2$NNH$_3^+$

6. a) basic; ClO$^-$ (aq) + H$_2$O (l) $\leftrightarrow$ OH$^-$ (aq) + HClO (aq)

$$K_b = \frac{K_w}{K_{a \text{ for HClO}}} = \frac{[OH^-][HClO]}{[ClO^-]} = 2.5 \times 10^{-7}$$

   b) neutral
   c) acidic; H$_2$O (l) + C$_2$H$_5$NH$_3^+$ (aq) $\leftrightarrow$ H$_3$O$^+$ (aq) + C$_2$H$_5$NH$_2$ (aq)

$$K_a = \frac{K_w}{K_{b \text{ for } C_2H_5NH_2}} = \frac{[H_3O^+][C_2H_5NH_2]}{[C_2H_5NH_3^+]} = 2.2 \times 10^{-11} \text{ or } K_a = \frac{[H_3O^+][A^-]}{[HA]} = 2.2 \times 10^{-11}$$

Answers to Self-Tests

1. H$_2$O (l) + CH$_3$NH$_3^+$ (aq) $\leftrightarrow$ H$_3$O$^+$ (aq) + CH$_3$NH$_2$ (aq), K$_a$ = 2.2×10^{-11}

2. pH = 13.30; 5.0×10^{-14} M

3. pH = 2.18 (hint: successive approximations needed)

4. Dissolve 12.2 g of NaNO$_2$ in water in a 250 mL volumetric flask.

5. The K$_a$ of NH$_4^+$ is 5.6×10^{-10}. The K$_b$ of CN$^-$ is 1.6×10^{-5}. CN$^-$ is a stronger base than NH$_4^+$ is an acid, so the solution will be basic.

6. a) H$_3$PO$_3$ will be a stronger acid than its conjugate base, H$_2$PO$_3^-$. Opposite charges attract, so it will be more difficult for the anion to lose a proton than it will be for a neutral compound.
   b) The overlap of orbitals that make up the H-Se sigma bond is much poorer than the overlap of orbtials that make up the H-O sigma bond. It is therefore easier to remove the proton completely from the H$_2$Se.
   c) CH$_3$CO$_2$H has a carboxylic acid functional group, CH$_3$CH$_2$OH does not. The two oxygens in the carboxylic acid group will withdraw electron density from the proton much better than the O in CH$_3$CH$_2$OH.

7. The strongest acid contains an electronegative Cl atom attached to the carbon adjacent to the carboxylic acid carbon. This Cl atom will inductively withdraw electron density from the carbon and stabilize the anion that will result when the proton leaves. The next strongest acid has a Cl atom on the second carbon from the carboxylic acid carbon. It will also be able to inductively withdraw electron density from the carboxylic acid group, but not as effectively as a Cl that's right next to the carboxylic acid. The weakest acid has no electron withdrawing Cl atoms on it to stabilize the anion.

Chapter 18: Applications of Aqueous Equilibria

Learning Objectives

In this chapter, you will learn how to:
*   Identify a buffer.
*   Calculate the pH of a buffered system.
*   Calculate the pH of a reaction mixture at strategic points in an acid/base titration.
*   Choose a proper indicator for an acid/base titration.
*   Determine how much solid will precipitate and the concentration of ions remaining in solution when a precipitation reaction occurs.
*   Write the formula for a complex ion.
*   Calculate concentrations of ions and complex ions in solution.

Practical Aspects

This chapter builds on the concepts in Chapters 16 and 17. The four main *new* topics in this chapter are buffers, acid/base titrations, the common ion effect, and complex ion equilibria. These last two topics have everyday applications ranging from treating hazardous wastes to treating a person for heavy-metal poisoning. Acid-base titrations are performed routinely in laboratory-based professions. Buffers are found everywhere in biological systems. Anyone studying to be a chemist, biochemist, biologist, environmental scientist, and anyone intending to work in a medical profession should have a solid understanding of buffers and how they work. You will see buffers again in later courses, guaranteed.

By the time you're done studying this chapter, you should be able to calculate the pH of an aqueous buffered or non-buffered system, calculate concentrations of components present at various stages of a titration, assess solubility equilibria, and use the common ion effect and complex ion equilibria concepts to control solute concentrations.

18.1 BUFFER SOLUTIONS

Key Terms:
*   **Buffer** – a mixture of a weak conjugate acid/base pair in roughly equal proportions.
*   **Buffer components** – the weak conjugate acid/base pair.
*   **Buffer ratio** – ratio of weak acid to conjugate base.
*   **Buffer capacity** – the amount of weak acid and conjugate base present in the buffer.

Key Concepts:
*   A buffer resists changes in pH because the conjugate acid/base pair can react with either added base or acid.
*   A buffer has both a weak acid and its conjugate base as *major* components in the solution. (Note: The weak acid and conjugate base are typically within a 10-fold concentration range of each other.)
*   Buffer components are often abbreviated "HA" and "A⁻" to describe the conjugate acid/base pair.
*   The concentration of the total buffer components is the sum of the concentration of HA and A⁻.

A buffer works because whatever strong acid or base is added to it will react with one of the buffer components essentially 100%, until that entire buffer component has been consumed. For example, if a strong acid is added, it will react with the weak base component:

$$H_3O^+ (aq) + A^- (aq) \rightarrow H_2O (l) + HA (aq)$$

Notice that this reaction is the reverse of a weak acid dissociation: $K_{eq} = 1/K_a$. The value of this reaction's equilibrium constant will be *huge*. The reaction will occur essentially 100%.

Key Concepts:
- A strong acid (H_3O^+) mixing with a weak base (A^-) will react essentially 100%.
- A strong base (OH^-) mixing with a weak acid (HA) will react essentially 100%.

A buffer is like a sponge for acids and bases. It can be tailored to your needs:

- A big sponge: 2 mol HA & 2 mol A^-

- A small sponge: 0.00010 mol HA & 0.00010 mol A^-

- A sponge that specializes in absorbing added *acid:* 0.1 mol HA & 0.5 mol A^-

- A sponge that specializes in absorbing added *base:* 0.5 mol HA & 0.1 mol A^-

To put the above "sponge" analogies into chemistry terms,
- the big sponge has a much greater *buffer capacity* than the small sponge – it can absorb 20,000 times more added acid or base than the small sponge.
- The big sponge has a *buffer ratio* of 1:1 A^-/HA. So does the little sponge. The sponge specializing in absorbing acid has a *buffer ratio* of 5:1 A^-/HA. What's the buffer ratio of the sponge that specializes in absorbing added base?

EXERCISE 1: Use the pictures shown to answer the questions. Color key: black = A^-, white = H^+, grey = oxygen. Only major species are shown. Water molecules have been omitted for clarity. a) Identify whether or not each is a buffer. b) For the boxes that contain buffers, what is the buffer ratio for each? c) Which buffer is capable of absorbing the most H_3O^+?

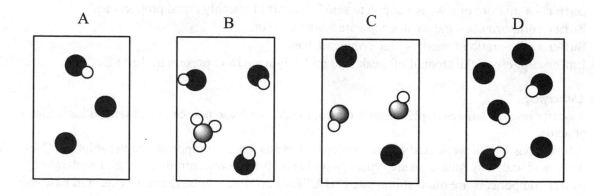

STRATEGY AND SOLUTION:

a) A buffer contains both a weak acid and its conjugate base within a roughly 10-fold concentration range of each other. In these systems, we see that "HA" is being used to describe a weak acid and "A⁻" is that acid's conjugate base. Solution A is a buffer because both weak acid and its conjugate base are present in measurable quantities: Solution A contains 1 HA and 2 A⁻. Solution B is not a buffer. It contains the weak acid, HA, but no amount of conjugate base is shown. Solution C is not a buffer because it contains the conjugate base of HA, but it does not contain HA. Furthermore, Solution C contains excess OH⁻, which would not be present if a buffer were present because the buffer would consume all of the OH⁻. Solution D is a buffer because it contains 3 HA and 3 A⁻ (the conjugate acid/base pair is present.)

b) Solutions A and D contain buffers, so only these two solutions must be addressed in this question. The buffer ratio of Solution A is expressed as a ratio of A⁻/HA is 2:1 and the buffer ratio of Solution D is 1:1.

c) A⁻ will react with any H_3O^+ that is added to the container. If we add H_3O^+ to Solutions A and D, we would see that Solution A can absorb 2 A⁻s while Solution D can absorb 3 A⁻s. Solution D therefore has a larger capacity to absorb H_3O^+ than Solution A. Notice that to assess the buffer's capacity to absorb added H_3O^+, we must assess the initial number of moles of weak base present, not the buffer ratio.

EXERCISE 2: Let's say that you have a buffer composed of 0.50 mol of some acid HA and 0.45 mol of its conjugate base, A⁻. How many moles of strong acid (H_3O^+) could be added to the buffer before the pH would change appreciably?

STRATEGY AND SOLUTION: The strong acid would react in a 1:1 ratio with A⁻. As long as there is A⁻ present, there won't be an excess of H_3O^+ in the solution. You could therefore add 0.45 mol of H_3O^+ before a large change in pH would result.

EXERCISE 3: Consider the buffer in Exercise 2. How many moles of each solute would be present after 0.10 mol of NaOH was added to it? Is the solution still a buffer after the addition of this NaOH?

STRATEGY: Take an inventory of the chemicals after they react. The OH⁻ will react with the HA in a 1:1 ratio to completion:

$$OH^- (aq) + HA (aq) \rightarrow A^- (aq) + H_2O (l)$$

OH⁻ will be the limiting reagent, since we're starting with 0.10 moles of OH⁻ and 0.50 moles of HA. All of the OH⁻ will be consumed, 0.10 moles of HA will be consumed (leaving 0.40 moles HA), and 0.10 moles of A⁻ will be formed (making a total of 0.10 mol + 0.45 mol = 0.55 mol A⁻). *SOLUTION:* The amount of each solute remaining in solution is 0.55 mol A⁻ and 0.40 mol HA. There is still an appreciable amount of the conjugate acid/base pair, so the mixture is still a buffer.

Try It #1: Consider the buffer in Exercise 2 (0.50 mol HA and 0.45 mol A⁻). How many moles of each solute would be present after the addition of 0.55 moles of HCl? Is the mixture still a buffer?

EXERCISE 4: Which of these mixtures will make a buffer solution?
Mixture A: 45.0 mL of 0.10 M HOCl + 20.0 mL of 0.10 M NaOH

Mixture B: 45.0 mL of 0.10 M NaOH + 20.0 mL of 0.10 M HOCl

STRATEGY: Take an inventory of the chemicals in each mixture.

Mixture A:
HOCl: 0.0450 L x 0.10 mol/L = 0.0045 mol HOCl to start
NaOH: 0.0200 L x 0.10 mol/L = 0.0020 mol OH^- to start (strong base)

These chemicals will react completely according to the reaction:

$$OH^- (aq) + HOCl (aq) \leftrightarrow OCl^- (aq) + H_2O (l)$$

OH^- is the limiting reagent with 0.0020 mol initially present. It will be completely consumed in the reaction.

Inventory after reaction:
HOCl: 0.0045 mol to start – 0.0020 mol reacted with OH^- = 0.0025 mol remaining.
OCl^-:   0 mol to start + 0.0020 mol generated.
OH^-:   0 moles remaining. It was completely consumed by the HOCl.

SOLUTION: Mixture A is a buffer because the mixture will contain a weak acid and its conjugate base as major components: 0.0025 mol HOCl and 0.0020 mol OCl^-.

Mixture B:
HOCl: 0.0200 L x 0.10 mol/L = 0.0020 mol HOCl to start
NaOH: 0.0450 L x 0.10 mol/L = 0.0045 mol OH^- to start

These chemicals will react completely: $OH^- (aq) + HOCl (aq) \rightarrow OCl^- (aq) + H_2O (l)$

HOCl is the limiting reagent with 0.0020 mol initially present. It will be completely consumed.

SOLUTION: Since the HOCl is completely consumed, Mixture B cannot be a buffer. We need *both* HOCl and ClO^- as major components for the solution to be a buffer.

Try It #2: Which solution is a buffer? Mixture A: 0.050 mol acetic acid and 0.038 mol of acetate ion, Mixture B: 1.00 mol of HCl and 2.00 moles of KCN, Mixture C: 25.0 mL of 0.250 M NaOH and 30.0 mL of 0.500 M H_2S, Mixture D: 10 mL of 1.0 M NaOH and 10 mL of 0.70 M NH_4Cl.

The Buffer Equation

Useful Relationship:

- $pH = pK_a + \log\left(\dfrac{[A^-]}{[HA]}\right)$ This is the Henderson-Hasselbalch equation, which is used to determine the pH of a buffer solution. The equation is derived from the K_a expression. (The derivation is provided in the text.)

Key Concepts:
- HA and A^- *always* refer to the components of the conjugate weak acid/base pair.

- The equation *always* utilizes pK_a of the acid component of the buffer system. (This is apparent when you look at the derivation of the Henderson-Hasselbalch equation.) Don't use pK_b in this equation.
- **pH = pK_a when the buffer ratio is 1:1.** When the mole ratio of the weak acid to its conjugate base is 1:1, the log term drops out because log(1) = 0.

EXERCISE 5: In Exercise 4, we saw that Mixture A was a buffer solution: 45 mL of 0.10 M HOCl mixed with 20 mL of 0.10 M NaOH produced 0.0025 mol of HOCl and 0.0020 mol of OCl⁻. What is the pH of this buffer solution at 25°C?

STRATEGY: This is a buffer solution, so use the Henderson-Hasselbalch equation to find pH. The pK_a of HOCl from Appendix E of the text is 7.40. The total volume of solution is 65 mL – divide moles by this volume to find concentration units.

$$pH = pK_a + \log\left(\frac{[A^-]}{[HA]}\right) = 7.40 + \log\left(\frac{0.0020\,mol / 65mL}{0.0025\,mol / 65mL}\right) = 7.30$$

It makes sense that the buffer pH is slightly less than 7.40 because there is more of the HA component than the A⁻ component.

Helpful Hint
- Here's a shortcut: Notice from Exercise 5 that since the buffer components are in the same container, the total volume for both components is the same. The [HA] and [A⁻] terms can really be based on the number of moles of each. You don't have to convert them to concentrations for this calculation.

Buffer Action

EXERCISE 6: Consider the HOCl/ OCl⁻ buffer we've been working with: 0.0025 mol of HOCl and 0.0020 mol of OCl⁻ in 65.0 mL of solution. Determine the pH of the solution if 5.1×10^{-4} mol of KOH is added to it.

STRATEGY: Take an inventory of the mixture. KOH (a strong base) will react with the HOCl in a 1:1 ratio. Since the KOH is the limiting reagent, it will be completely consumed. In the process, 5.1×10^{-4} moles of HOCl will be consumed and 5.1×10^{-4} moles of OCl⁻ will be generated:

Moles of HOCl: $0.0025 - 5.1\times10^{-4} = 0.00199$
Moles of OCl⁻: $0.0020 + 5.1\times10^{-4} = 0.00251$
Both the weak acid and its conjugate base are still present as major components, so we still have a buffer. Use the Henderson-Hasselbalch equation to calculate pH:

$$pH = pK_a + \log\left(\frac{[A^-]}{[HA]}\right) = 7.40 + \log\left(\frac{0.00251}{0.00199}\right) = 7.50$$

SOLUTION: The pH of the solution is 7.50. This number seems reasonable for two reasons: 1) the pH did not change significantly (we have a buffer), and 2) the pH is slightly higher than the pK_a of the acid component (the buffer ratio is weighted slightly with extra OCl⁻).

The true power of a buffer solution cannot fully be appreciated until its working ability is compared to a non-buffered environment. Exercise 7 illustrates this.

EXERCISE 7: Calculate the pH of a solution resulting from the addition of 5.1×10^{-4} mol of KOH to 65 mL of water. (Notice that these numbers are identical to those in the previous exercise, but this is a non-buffered solution.)

> *STRATEGY:* The pH can be found from pOH. We need $[OH^-]$ first.
> $$[OH-] = 5.1 \times 10^{-4} \text{ mol} / 0.065 \text{ L} = 7.84615 \times 10^{-3} \text{ M}$$
> $$pOH = 2.1053 \text{ so } pH = 11.89 \text{ (2 sig figs)}$$

> *SOLUTION:* The pH of the non-buffered water changed from 7.00 to 11.89. Remember that a 1-unit pH change signifies a 10-fold change in $[H_3O^+]$, so this represents an enormous change in $[H_3O^+]$.

EXERCISE 8: Consider the original HOCl/OCl⁻ buffer again (0.0025 mol of HOCl and 0.0020 mol of OCl⁻ in 65 mL). What volume of 0.0050 M HCl would we need to add to this solution to lower the pH by 0.30 pH units?

> *STRATEGY:* The pH of the original buffer solution was 7.30 so we want the pH to decrease to 7.00. The question simplifies to "how many moles of HCl should be added?" The HCl and the ClO⁻ will react in a 1:1 ratio to decrease the amount of ClO⁻ and increase the amount of HClO:

$$pH = pK_a + \log\left(\frac{[A^-]}{[HA]}\right) \quad \text{so} \quad 7.00 = 7.40 + \log\left(\frac{0.0020 - x}{0.0025 + x}\right) \quad \text{so} \quad 10^{-0.400} = 0.398 = \left(\frac{0.0020 - x}{0.0025 + x}\right)$$

Solving for x, we get $x = 7.19 \times 10^{-4} =$ moles of HCl to add

7.19×10^{-4} mol x 1 L/0.0050 mol = 0.144 L = 140 mL (2 sig figs)

> *SOLUTION:* 140 mL (or 0.14 L) of 0.0050 M HCl would need to be added to make the pH change to 7.00.

18.2   CAPACITY AND PREPARATION OF BUFFER SOLUTIONS

Buffer Capacity

Key Term:
- **Buffer capacity** – (first defined in section 18.1, but here is an alternate definition) - the amount of added strong acid or base that the buffer is capable of absorbing without being destroyed.

Key Concept:
- It is customary to say that a buffer is destroyed if its ratio exceeds 1:10 or 10:1.

EXERCISE 9: Consider the HOCl buffer one last time (0.0025 mol of HOCl and 0.0020 mol of OCl⁻ in 65 mL). Is the buffer capacity destroyed by adding 30.0 mL of 0.100 M NaOH to it? What is the pH of the resulting solution?

STRATEGY: Take an inventory of the components. NaOH (a strong base) will react with the HOCl in a 1:1 ratio. We originally had 0.0025 moles of HOCl. The amount of NaOH added is:

$$0.0300 \text{ L} \times 0.100 \text{mol/L} = 0.00300 \text{ mol OH}^-$$

In this scenario, HOCl is the limiting reagent, so it will be completely consumed. We no longer have a buffer. 0.0025 mol of OH⁻ will react with the 0.0025 mole of HOCl and we'll be left with 0.0005 mol of OH⁻ in a total volume of 95.0 mL (65.0 mL + 30.0 mL). The pH of the solution will be dependent upon the [OH⁻]:

$$[\text{OH}^-] = 0.0005 \text{ mol}/0.0950 \text{ L} = 5.26 \times 10^{-3} \text{ M}; \text{ so pOH} = 2.28 \text{ and pH} = 11.72$$

SOLUTION: The buffer is extinguished. The pH of the solution is 11.7. (1 sig fig)

Try It #3: Repeat Exercise 9, but use a buffer with a larger concentration of buffer components: 65.0 mL of solution containing 0.025 mol of HOCl and 0.020 mol of OCl⁻. Would the buffer be extinguished by adding the 30.00 mL of 0.100 M NaOH to it?

Buffer Preparation

In order to prepare a buffer solution, two questions must be answered:

1. What is the desired pH of the buffer?

The desired pH of the buffer will determine which buffer components to use. Typically, a chemist wants the buffer to react equally well to added acid or base, so a buffer ratio of 1:1 is ideal.

Using Henderson-Hasselbalch: $\text{pH} = \text{pK}_a + \log\left(\dfrac{[\text{A}^-]}{[\text{HA}]}\right)$, we see that the pH of the buffer will equal the

pK$_a$ of the acid component in the buffer (log of 1 = 0). This allows us to choose the conjugate weak acid/base pair of the buffer system. Choose a weak acid that has a pK$_a$ approximately equal to the desired pH of the buffer. A combination of this weak acid and its conjugate base will make up the buffer.

2. What is the overall desired buffer concentration?

The buffer capacity will be dictated by the concentration of the buffer components. We need to know this to determine *how much* weak acid and conjugate base to use. There are three ways a buffer can be made to achieve this mixture of HA and A⁻. These are general recipes that must be fine-tuned for the details of the actual desired buffer.

1. Mix approximately equimolar amounts of the acid/base conjugate pair. The substance that is charged will be a salt of some type. Non-reactive counter ions typically used are Na⁺, K⁺, Cl⁻.

2. Mix a given amount of HA with approximately ½ the number of moles of strong base.

3. Mix a given amount of A⁻ with approximately ½ the number of moles of strong acid.

EXERCISE 10: Select a weak conjugate acid/base pair to use in making: a) a pH 8.60 buffer, b) a pH 6.00 buffer, c) a pH 4.30 buffer. Assume all buffers are at 25°C.

STRATEGY: For each, use Appendix E from the text to find an acid with a pK_a near the desired pH value. That acid and its conjugate base will constitute the buffer.

SOLUTION: a) pH 8.60: HBrO ($pK_a = 8.55$) and BrO^-
b) pH 6.00: $HONH_3^+$ ($pK_a=5.94$) and $HONH_2$
c) pH 4.30: Benzoic acid ($pK_a = 4.20$) and benzoate ion

EXERCISE 11: Make 500.0 mL of a pH 8.75 buffer with a total buffer concentration of 0.250 M. The chemicals available to use in making the buffer are NaBrO and 1.00 M HCl (aq).

STRATEGY: Given that we have NaBrO and HCl to work with, all of the buffer components will have to originate from the conjugate base portion of the buffer, BrO^-. Thus, we need:

0.5000 L x 0.250 mol/L = 0.125 mol of BrO^- or NaBrO to start.

NaBrO is an ionic compound, so it will be a solid at room temperature:
0.125 mol x 118.893 g/mol = 14.86 g

The amount of HCl to add will be dictated by the buffer ratio. Whatever HCl is added will react with BrO^- (decreasing its concentration) and form HBrO (increasing its concentration). This amount we can call "x." Determine the number of moles of each buffer component needed to establish the desired pH using the Henderson-Hasselbalch equation.

$$pH = pK_a + \log\left(\frac{[A^-]}{[HA]}\right)$$

$$8.60 = 8.55 + \log\left(\frac{\text{moles of } BrO^-}{\text{moles of HBrO}}\right) = 8.55 + \log\left(\frac{0.125 - x}{x}\right)$$

$$0.05 = \log\left(\frac{0.125 - x}{x}\right); \quad 10^{0.05} = 1.122 = \left(\frac{0.125 - x}{x}\right)$$

$$1.122x = 0.125 - x; \quad 2.122x = 0.125; \quad x = 0.0589$$

If we add 0.0589 moles of HCl to the NaBrO, we'll generate the proper A^-/HA ratio. The HCl solution is 1.00 M.
0.0589 mol x 1.00 L/1.00 mol = 0.0589 L = 58.9 mL of 1.00 M HCl.

SOLUTION: Mix 14.86 g of NaBrO with 58.9 mL of 1.00 M HCl in a 500.00 mL volumetric flask and dilute to the mark with water.

EXERCISE 12: Prepare the same buffer as in Exercise 11, starting with NaBrO(s) and 2.00 M HBrO.

STRATEGY: In Exercise 11, we saw that the total moles of buffer components had to be 0.125 moles, or we needed 0.0589 moles of HBrO and 0.0661 moles of BrO^- (0.125 – 0.0589 = 0.0661 moles).

We need:

HBrO: 0.0589 mol x 1 L/2.00 mol = 0.02945 L = 29.5 mL of 2.00 M HBrO
NaBrO: 0.0661 mol x 118.893 g/mol = 7.86 g of NaBrO

SOLUTION: Mix 7.86 g of NaBrO with 29.5 mL of 2.00 M HBrO in a 500.00 mL volumetric flask and dilute to the mark with water.

Try It #4: Prepare 250.0 mL of a pH 6.00 buffer with a total buffer concentration of 0.100 M, given these chemicals to work with: solid $HONH_3Cl$ and solid NaOH.

18.3   ACID-BASE TITRATIONS

Key Terms:
- **Titration (acid/base) -** the gradual addition of an acid to a base (or vice versa) to determine the exact amount of material required to react in stoichiometric proportions.
- **Stoichiometric point –** point in titration at which the exact stoichiometric amount of titrant has been added to react with the chemical being titrated. This is also called the equivalence point.
- **Indicator (acid/base) –** chemical which changes colors to signal the endpoint of a titration.
- **Titration curve –** plot of pH vs. volume of titrant added for a titration.

Acid/base titrations are often used to determine the concentration of some acid (or base) solution by reacting it with a stoichiometric amount of base (or acid). The endpoint can be observed by color change. Scenarios 1 and 2 shown below illustrate that pH calculations for acid-base titrations depend upon the types of chemicals involved.

Scenario 1: A Strong Acid with A Strong Base (For example, HCl and NaOH)
- No weak species are present, so pH of solution at any point during titration can be determined from stoichiometry. Be sure to account for V_{total} when assessing molarities.
- The equivalence point will be at pH = 7 because neutral salt and water will be present.
- The pH at the endpoint will be basic because one extra drop of OH^- will be present.

Scenario 2: A Weak Acid with A Strong Base (For example, Acetic Acid and NaOH)
1. **Before any base is added:**
   - solution contains a weak acid. Use K_a process to get pH.
2. **During titration but before equivalence point:**
   - every OH^- that is added reacts with the acid. The solution is a buffer.
   - Use Henderson-Hasselbalch equation to get pH.
3. **At "Half-Titration":**
   - when half the acid has reacted with OH^-, the buffer ratio is 1:1 so pH = pK_a of the acid.
4. **At the equivalence point:**
   - all of the acid has reacted and has been converted to the salt, sodium acetate. At this point, we have the anion of a weak base, so use the K_b process to get pOH then pH.
   - Note: the pH at the equivalence point for this process is *greater than* 7.

5. After the equivalence point:

- excess OH^- is present and the pH can be calculated directly from $[OH^-]$. (Important note: it is possible that the weak base can contribute to the pH of the final solution, but this method of calculating pH directly from $[OH^-]$ typically provides a good approximation.)

As you can see from Scenarios 1 and 2 above, the pH calculations in acid/base titrations are exactly like those you have already performed.

Summary of How to Determine pH of a Solution When Two Solutions Have Been Mixed Together

1. Determine moles of all starting materials.
2. Determine which chemicals will react and account for them using the reaction's stoichiometry.
3. Take an inventory of the chemicals in the container *after* the reaction to see what remains.
4. Use the table here to determine how to assess pH.

Inventory after reaction shows this present as major component(s):	Process	Appropriate Mathematical Relationships to Use
A mixture of a weak acid & its conjugate base	Buffer solution	• Henderson-Hasselbalch equation • Be sure to use pK_a for weak acid
Strong acid	Strong acid dissociation. This will be the dominant equilibrium (100% dissociation)	• $[H_3O^+]$ = [acid solution]; can get pH directly from [acid solution]
Strong base	Strong base dissociation. This will be the dominant equilibrium (100% dissociation)	• Calculate pH from $[OH^-]$
Weak acid	K_a process for a weak acid	• Write a K_a expression • Make an equilibrium table and calculate $[H_3O^+]$, then pH
A weak base	K_b process for a weak base	• Write a K_b expression • Make an equilibrium table and calculate $[OH^-]$ • Calculate pH from $[OH^-]$

Key Concepts:

- The pH at the equivalence point for a strong acid/strong base titration is 7.00.
- The pH at the equivalence point for a strong/weak titration is NOT 7.00.
- The pH at half-titration equals pK_a of the weak acid (if weak species are involved).

Helpful Hint

- Remember to divide by the total volume present when calculating molar concentrations.

EXERCISE 13: Consider the titration of a weak acid, HA, with strong base. Use the pictures shown to answer the questions. Color key: black = A^-, white = H^+, grey = oxygen. Only major species are shown. Water molecules have been omitted for clarity. a) Which box illustrates the titration at the equivalence point? b) Which box illustrates the titration at the start of the titration? c) What is the pK_a of the weak acid, HA? d) Which box illustrates the titration *after* the equivalence point has been passed?

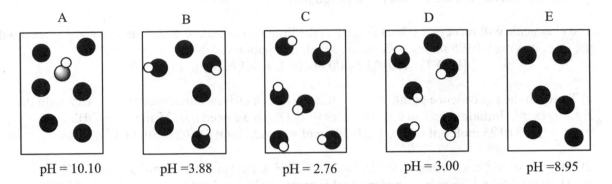

<table>
<tr><td>A</td><td>B</td><td>C</td><td>D</td><td>E</td></tr>
<tr><td>pH = 10.10</td><td>pH =3.88</td><td>pH = 2.76</td><td>pH = 3.00</td><td>pH =8.95</td></tr>
</table>

STRATEGY AND SOLUTION: The overall reaction that is taking place in this titration process is:
$$HA\ (aq) + OH^-\ (aq) \leftrightarrow H_2O\ (l)\ +\ A^-\ (aq)$$

This is a weak acid/strong base titration.

a) At the equivalence point, all of the HA has reacted in a perfect 1:1 ratio with added OH^-, so no excess OH^- exists, no HA remains, and only A^- will be present as a major species. Box E illustrates this scenario.

b) At the start of the titration, no OH^- has been added, so the only major species present is the weak acid HA. Box C is the only box that shows HA as the only major species, so Box C represents the start of the titration.

c) We know that $pH = pK_a$ of the weak acid when we're at the half-titration point. (Recall that at this point, half of the original HA has been converted to A^- and the other half remains unreacted. This creates a buffer with a 1:1 ratio, making the "log" term in the Henderson-Hasselbalch equation = 0, and $pH = pK_a$.) Box B illustrates the situation where half of the HA has been converted to A^-, so the pK_a of HA = 3.88.

d) After the equivalence point has been passed, we will have an excess of OH^- present. Furthermore, there will be no HA present, because all of it had reacted by the time we reached the equivalence point. Box A illustrates this scenario.

EXERCISE 14: 50.00 mL of 0.250 M HCl (aq) is titrated with 0.200 M NaOH.
a) What is the original pH of the solution?
b) What is the pH after 10.00 mL of NaOH has been added?
c) What volume of NaOH is needed to reach the equivalence point?
d) What is the pH at the equivalence point?
e) What is the pH of the solution after a total of 64.00 mL of NaOH has been added?

STRATEGY: The overall reaction in this process is: $H_3O^+\ (aq) + OH^-\ (aq) \leftrightarrow 2\ H_2O\ (l)$

a) The original solution is 0.250 M HCl, a strong acid, so $pH = -\log(0.250) = 0.602$

b) Two chemicals have been mixed:

H_3O^+:   0.05000 L x 0.250 mol/L = 0.0125 moles of H_3O^+ to start
OH^-:   0.0100 L x 0.200 mol/L = 0.00200 moles of OH^- added

The chemicals will react in a 1:1 ratio. OH^- is the limiting reagent. 0.00200 moles of H_3O^+ react with the OH^-, leaving 0.0105 moles H_3O^+ in 60.00 mL of solution (50.00 mL + 10.00 mL).
$[H_3O^+]$ = 0.0105 mol/0.06000 L = 0.175 M, so pH = 0.757

c) To reach the equivalence point, we need to add enough OH^- to stoichiometrically react with the H_3O^+ present. Initially, we had 0.0125 moles of H_3O^+, so we need 0.0125 moles of OH^-.
0.0125 moles of OH^- x 1 L/0.200 mol = 0.0625 L OH^- = 62.5 mL OH^- solution

d) This is a strong acid/strong base titration, so the pH at the equivalence point equals 7.00.
e) pH after 64.00 mL have been added. At this point, we have added a 1.5 mL excess of OH^-. Determine moles of OH^- in the 1.5 mL excess, and divide by the total volume of solution:
(1.5 mL x 0.200 mol/L) / (64.00 mL + 50.00 mL) = 2.63×10^{-3} M; pOH = 2.58; pH = 11.42

SOLUTION: a) initial pH = 0.602; b) pH after 10.00 mL OH^- = 0.757; c) volume needed to reach equivalence point = 62.5 mL; d) pH at equivalence point = 7.00; e) pH = 11.42

EXERCISE 15: 50.00 mL of 0.250 M formic acid is titrated with 0.200 M NaOH.
a) What is the original pH of the solution?
b) What is the pH after 10.00 mL of NaOH has been added?
c) What volume of NaOH is needed to reach the equivalence point?
d) What is the pH at the equivalence point?
e) What is the pH of the solution after a total of 64.00 mL of NaOH has been added?

STRATEGY: Formic acid has the formula HCO_2H, which can be abbreviated "HA." The formate ion, HCO_2^-, can be abbreviated "A^-." The overall reaction in this process is:
HA (aq) + OH^- (aq) $\leftrightarrow$ A^-(aq) + H_2O (l). Any HA and OH^- mixed together will react essentially 100%.

a) The original solution is 0.250 M weak acid, so the pH is determined by the K_a process for formic acid. The pK_a of formic acid is 3.75, so $K_a = 1.8 \times 10^{-4}$.

	HA (aq) +	H_2O (l) $\leftrightarrow$	A^- (aq) +	H_3O^+ (aq)
Concentrations:				
Initial	0.25		0	0
Change	-x		+x	+x
Equilibrium	0.250-x		x	x

$$K_a = \frac{[H_3O^+][A^-]}{[HA]} = \frac{x^2}{0.250 - x} = 1.8 \times 10^{-4}$$

$x = 6.62 \times 10^{-3} = [H_3O^+]$ (successive approximations required); pH = 2.18

b) Two chemicals have been mixed:
HA: 0.05000 L x 0.25 mol/L = 0.0125 moles of HA to start
OH^-:   0.0100 L x 0.200 mol/L = 0.00200 moles of OH^- added

The chemicals will react in a 1:1 ratio. OH^- is the limiting reagent. 0.00200 moles of HA react, leaving 0.0105 moles HA and forming 0.0020 moles of A^-. This is a buffer solution, so the Henderson-Hasselbalch equation is needed.

$$pH = pK_a + \log\left(\frac{[A^-]}{[HA]}\right) = 3.75 + \log\left(\frac{0.0020}{0.0105}\right) = 3.03$$

(Notice that mole ratios for HA and A^- were used instead of molarities. Why is this acceptable?)

c) To reach the equivalence point, we need to add enough OH^- to stoichiometrically react with the HA present. Initially, we had 0.0125 moles of HA, so we need 0.0125 moles of OH^-.

0.0125 moles of OH^- x 1 L/0.200 mol = 0.0625 L OH^- = 62.5 mL OH^- solution

d) At the equivalence point, a stoichiometric amount of OH^- has been added to the solution. All of the HA has been converted to A^- and there is no excess of OH^-. The pH of the solution is based on the K_b process for A^-. V_{total} = 62.5 mL + 50.00 mL = 112.5 mL.

$[A^-]_{initial}$ = moles A^- /volume solution = 0.0125 mol/0.1125 L = 0.111 M

	A^- (aq) +	H_2O (l) $\leftrightarrow$	HA (aq) +	OH^- (aq)
Concentrations:				
Initial	0.111		0	0
Change	-x		+x	+x
Equilibrium	0.111 – x		x	x

$$K_b = \frac{[HA][OH^-]}{[A^-]} = \frac{x^2}{0.111 - x} = 5.6 \times 10^{-11}$$

$$x = 2.49 \times 10^{-6} = [OH^-]; \quad pOH = 5.60; \quad pH = 8.40$$

e) pH after 64 mL have been added. At this point, we have added a 1.5 mL excess of OH^-. Determine moles of OH^- in the 1.5 mL excess, and divide by the total volume of solution:

(1.5 mL x 0.200 mol/L) / (64 mL+50 mL) = 2.63×10^{-3} M; pOH = 2.58; pH = 11.42

SOLUTION: a) initial pH = 2.18; b) pH after 10 mL OH^- = 3.03; c) volume needed to reach equivalence point = 62.5 mL; d) pH at equivalence point = 8.40; e) pH = 11.42. The answers to this problem seem reasonable: Originally, the solution contained a weak acid (pH<7) and at the equivalence point, it contained a weak base (pH>7).

Key Concept:

- *The amount of titrant needed to reach the equivalence point is based on the reaction's stoichiometry.* Compare the results from Exercises 14 and 15. Given that we started with the same amount of acid in each exercise, notice that it required the same amount of NaOH to neutralize the strong acid as it did the weak acid.

EXERCISE 16: Sketch rough titration curves to illustrate the titrations in Exercises 14 and 15.

STRATEGY: A titration curve is a plot of pH as a function of titrant volume. Use the data from Exercises 14 and 15 to construct the graph. Add extra data points if needed to round out the graph. (pH values at half-titration, 50.00 mL, 65.00 mL, and 75.00 mL were also calculated in order to construct more informative graphs. Do these calculations on your own.)

Exercise 14: HCl & NaOH		Exercise 15: Formic acid and NaOH	
Volume OH- (mL)	pH	Volume OH- (mL)	pH
0.00	0.60	0.00	2.18
10.00	0.76	10.00	3.03
31.25	1.11	31.25	3.75
50.00	1.60	50.00	4.34
62.50	7.00	62.50	8.40
64.00	11.42	64.00	11.42
65.00	11.64	65.00	11.64
75.00	12.30	75.00	12.30

SOLUTION:

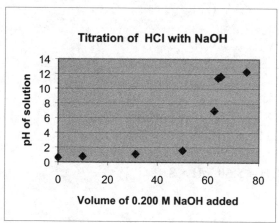

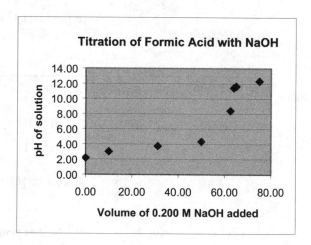

Titration of Polyprotic Acids

Key Concepts:

- Think of the titration of a polyprotic acid as a stepwise series of titrations of monoprotic acids.
- The number of acidic protons present in your polyprotic acid will dictate the number of equivalence points in your titration curve. (Example: a triprotic acid, H_3A, will show three equivalence points, with buffer regions between each equivalence point.)
- K_{a1} is much larger than K_{a2} for a polyprotic acid, so when a polyprotic acid is titrated, the first equilibrium reaction will occur completely before the second equilibrium reaction begins. For example, the acid H_2A will have this series of dissociations:

$$K_{a1}: \quad H_2A \text{ (aq)} + H_2O \text{ (l)} \leftrightarrow HA^- \text{ (aq)} + H_3O^+ \text{ (aq)}$$
$$K_{a2}: \quad HA^- \text{ (aq)} + H_2O \text{ (l)} \leftrightarrow A^{2-} \text{ (aq)} + H_3O^+ \text{ (aq)}$$

If we have a triprotic acid, then $K_{a1} \gg K_{a2} \gg K_{a3}$ and a similar series of reactions applies. Keep this concept clear in your mind and you should be able to visualize what components are present at any point along a titration curve for a polyprotic acid.

Helpful Hint
See Figure 18-7 in the text for a graphical breakdown of the major regions of a polyprotic acid's titration curve.

EXERCISE 17: Consider the titration of a weak polyprotic acid, H_2A, with strong base. Use the pictures shown to answer the questions. Color key: black = A^{2-}, white = H^+, grey = oxygen. Only major species are shown. Water molecules have been omitted for clarity. a) What is the pK_{a2} for H_2A? b) For which box will the pK_{a2} value be used in the Henderson-Hasselbalch equation to calculate pH? c) What is the pH at the first equivalence point? d) Which box illustrates the titration at the second equivalence point?

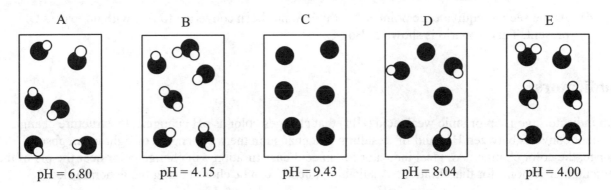

A	B	C	D	E
pH = 6.80	pH = 4.15	pH = 9.43	pH = 8.04	pH = 4.00

STRATEGY: This is the titration of a weak polyprotic acid with a strong base. Remember that any OH^- added to the container will react essentially 100% with the strongest acid available. The strongest acid remaining *after* the added OH^- has reacted will influence the pH of the solution: if H_2A is still present, then its acid dissociation process will dictate the pH; if HA^- is the only remaining acid, then its acid dissociation process will dictate the pH. The chemical equilibria taking place are:

$$K_{a1}: \ H_2A \ (aq) + \ H_2O \ (l) \ \leftrightarrow \ HA^- \ (aq) \ + \ H_3O^+ \ (aq)$$
$$K_{a2}: \ HA^- \ (aq) + \ H_2O \ (l) \ \leftrightarrow \ A^{2-} \ (aq) \ + \ H_3O^+ \ (aq)$$

It is a good idea to form a quick mental picture of what is happening in this titration: before any OH^- has been added, we will have only H_2A present as the dominant species. As we add each OH^-, one H_2A will get converted to one HA^- (and 1 water will also be formed). At some point, we will have added enough OH^- so that no H_2A remains, and only HA^- is present as the dominant species – this is the first equivalence point in the titration. Any additional OH^- added to the container will react with HA^- to form A^{2-}. This will continue until all of the HA^- has just been consumed. When we've added just enough OH^- to perfectly react with the HA^- so that the only dominant species present is A^{2-}, we'll be at the second equivalence point. Any more OH^- added to the pot will cause the pH to jump drastically. This visualization should help us answer the questions in this exercise.

SOLUTION:
a) The pK_{a2} for H_2A will be related to the process for which the second proton of H_2A is removed (i.e. going from HA^- to A^{2-}). We know that $pH = pK_a$ at half-titration, so pK_{a2} will equal the pH at the midpoint of the titration of HA^-. At this point, we have equal parts HA^- and A^{2-}. This scenario is illustrated in Box D, because Box D shows 3 HA^- and 3 A^{2-}.

b) Let's first fill in the Henderson-Hasselbalch equation as much as possible to help us answer this question. As in part "a", the pK_{a2} for H_2A will be related to the process for which the second proton of H_2A is removed (i.e. going from HA^- to A^{2-}), so the appropriate conjugate acid base pair will be: HA^-/A^{2-}.

$$pH = pK_{a2} + \log\left(\frac{[A^{2-}]}{[HA^-]}\right)$$

This equation is only used when we have a buffer, so we need to find the box that shows measurable amounts of both HA^- and A^{2-}. Again, Box D fits this description.

c)   The first equivalence point is the point at which all of the H_2A has been consumed, but no HA^- has undergone reaction with OH^- yet. This scenario is illustrated in Box A, so the pH at the first equivalence point will be 6.80.

d)   At the second equivalence point, all of the HA^- has been converted to A^{2-}, with no excess OH^- present. This scenario is shown in Box C.

Indicators

An indicator dye is an organic weak acid (HIn) that changes color as pH changes. Its structure changes when it converts between HIn and In⁻, resulting in a change in the wavelengths of light it can absorb, hence, the color change. An ideal indicator for an acid/base titration will change color near the pH of the stoichiometric point for that titration. A suitable indicator can be chosen using the general rule:

$$pH_{\text{at stoichiometric point}} = pK_{In} +/- 1$$

Helpful Hint
•   Table 18-2 lists several acid-base indicators.

EXERCISE 18: Would bromocresol green be a suitable indicator to use for titrating HCl with NaOH?

STRATEGY AND SOLUTION: A suitable indicator dye will change color near the stoichiometric point of the titration. The rule of thumb is $pH_{\text{at stoichiometric point}} = pK_{In} +/- 1$. The stoichiometric point of a strong acid-strong base titration is 7.00. According to Table 18.2 in the text, bromocresol green has a pK of 4.68, so it would change color from yellow to blue *before* the titration's equivalence point was reached. Bromocresol green is therefore not a good indicator to use for this titration.

18.4 SOLUBILITY EQUILBRIA

Key Terms:
•   **Solubility** – ability of a solid to dissolve in water. The technical definition is the number of grams of solid that can dissolve in 100 mL of water at 25°C.
•   **Saturated** – a solution that contains the maximum amount of solute that it can accommodate. The water is "saturated" with that particular solute. A saturated solution establishes a chemical equilibrium between the solid and aqueous phases.

Classification	K_{sp} range
Insoluble	<<<1
Slightly Soluble	10^{-2} to 10^{-5}
Soluble	$>10^{-2}$

Helpful Hints

- Remember to assess initial concentration of each reactant species based on V_{total}.
- If K_{eq} is large (i.e., a precipitation reaction), use the two-table method we established in Section 16.6: take the reaction to 100% completion in the first table, and then backtrack in the second table.
- Solubility changes greatly with temperature, so temperature is always specified in solubility problems.

EXERCISE 19: A salt called MX_3 was added to water until a saturated solution was made. The $[M^{3+}]$ in solution was 2.1×10^{-3} M. What is the K_{sp} of MX_3 at the temperature the solution was made?

STRATEGY: We can do a shortened version of the 7-step method. We need K_{sp}, so a K_{sp} expression is also needed:

$$MX_3 \text{ (s)} \leftrightarrow M^{3+} \text{ (aq)} + 3\, X^- \text{ (aq)}$$

According to the reaction stoichiometry, the $[X^-]$ at equilibrium will be three times that of M^{3+}. Using "x" to represent the $[M^{3+}]$ will illustrate how simple the math in this problem actually is:

$$K_{sp} = [M^{3+}][X^-]^3 = (x)(3x)^3 = 27x^4$$

$$K_{sp} = 27x^4 = 27(2.1 \times 10^{-3})^4 = 5.3 \times 10^{-10}$$

SOLUTION: K_{sp} at this temperature is 5.3×10^{-10}, which seems reasonable because the $[M^{3+}]$ is so small.

EXERCISE 20: Calculate the pH of a saturated solution of calcium hydroxide at 25°C. The pK_{sp} of calcium hydroxide in Appendix E of the text is 5.30.

STRATEGY AND SOLUTION: A saturated solution indicates that the solid and aqueous phases are in equilibrium. The K_{sp} expression is:

$$Ca(OH)_2 \text{ (s)} \leftrightarrow Ca^{2+} \text{ (aq)} + 2\, OH^- \text{ (aq)}; \quad K_{sp} = 10^{-5.30} = 5.0 \times 10^{-6}$$

We will have to find the equilibrium concentration of hydroxide ion to assess pH. (Recall that solids are not involved in the K_{eq} expression.)

	$Ca(OH)_2$ (s) $\leftrightarrow$	Ca^{2+} (aq) +	$2\, OH^-$ (aq)
Concentrations:			
Initial		0	0
Change		+x	+2x
At Equilibrium		x	2x

$$K_{sp} = 5.0 \times 10^{-6} = [Ca^{2+}][OH^-]^2 = (x)(2x)^2 = 4x^3$$

$x = 1.077 \times 10^{-2}$; $[OH^-] = 2x = 2.154 \times 10^{-2}$ M so pOH = 1.67 and pH = 12.33

SOLUTION: The pH of the solution will be 12.35. This value seems reasonable because it is greater than 7.

EXERCISE 21: How many grams of calcium hydroxide can be dissolved in 250.0 mL of water at 25°C?

STRATEGY AND SOLUTION: This question is asking us to prepare a saturated solution of calcium hydroxide. An equilibrium table must be constructed in order to determine the maximum amount of solid that will dissolve. We made such a table in Exercise 20, and it indicated that a saturated solution of calcium hydroxide is 1.08×10^{-2} M in Ca^{2+} ions.

$$\frac{1.08 \times 10^{-2} \, mol \, Ca^{2+}}{L} \; x \; 0.250 \, L \; x \; \frac{1 \, mol \, Ca(OH)_2}{1 \, mol \, Ca^{2+}} \; x \; \frac{74.0918 \, g}{mol} = 0.206 \, g \, Ca(OH)_2$$

Try It #5: How many grams of lead(II) iodide can dissolve in 500.0 mL of water at 25°C? The pK_{sp} of lead(II) iodide is 8.19.

Precipitation Equilibria

A precipitation equilibrium problem was done in Chapter 16, Exercise 18. The following exercise is a little more challenging.

EXERCISE 22: A student pours 50.00 mL of 0.500 M of NaI (aq) and 50.00 mL of 0.500 M NaCl (aq) into the same Erlenmeyer flask. She then adds 10.00 mL of 0.10 M aqueous silver nitrate to the flask. What are the concentrations of all solute ions in the flask? At 25°C, the K_{sp} of AgCl is 1.8×10^{-10} and the K_{sp} of AgI is 8.3×10^{-17}. Assume the process takes place at this temperature.

STRATEGY: We must inventory the initial concentrations of all ions before any reaction occurs. The total volume of the final solution is 110.00 mL (50.00 + 50.00 + 10.00). The initial concentration of each solute ion is therefore:

$$[Na^+] = \frac{M_1 V_1}{V_2} = \frac{(0.500 \, M)(100.00 \, mL)}{110.00 \, mL} = 0.455 \, M$$

* V_1 here is 100 mL because Na^+ came from the two 0.500 M solutions that were 50.00 mL each.

$$[Cl^-] = \frac{M_1 V_1}{V_2} = \frac{(0.500 \, M)(50.00 \, mL)}{110.00 \, mL} = 0.227 \, M \quad \text{This will also be the initial } [I^-].$$

$$[Ag^+] = \frac{M_1 V_1}{V_2} = \frac{(0.100 \, M)(10.00 \, mL)}{110.00 \, mL} = 0.00909 \, M \quad \text{This will also be the initial } [NO_3^-].$$

Now use the seven-step method to attack the problem.

1. **Determine what is asked for.** We need the concentrations of all solute ions in solution. Na^+ and NO_3^- are not involved in the equilibria because they are spectator ions, so their concentrations are already known. The real question is "how much AgI and AgCl precipitated?" – if we can answer that, we can determine the amount of solute ions remaining in solution.
2. **Identify the major chemical species.** The major reactants are the ones involved in equilibria: Ag^+, I^-, and Cl^-.

3. **Determine what chemical equilibria exist.** The equilibrium reactions will be:
$$Ag^+ (aq) + Cl^- (aq) \leftrightarrow AgCl (s); \; K_{eq} = 1/ K_{sp} = 1/ 1.8\times10^{-10} = 5.6\times10^9$$
$$Ag^+ (aq) + I^- (aq) \leftrightarrow AgI (s); \; K_{eq} = 1/ K_{sp} = 1/ 8.3\times10^{-17} = 1.2\times10^{16}$$

The equilibrium constant for the precipitation of AgI is several orders of magnitude larger than that of AgCl, so the AgI precipitation is the dominant equilibrium. The AgI will precipitate completely before any AgCl does.

4. **Write the K_{eq} expressions.**

$$K_{eq} = \frac{1}{[Ag^+][I^-]} = 1.2\times10^{16}$$

5. **Organize the data and unknowns.**
Since K_{eq} is so large, we should assume 100% reaction and backtrack. AgI (s) is not involved in the K_{eq} expression, but its theoretical values will be shown in the tables here to emphasize the direction of reaction in the second table.

	Ag^+ (aq) +	I^- (aq) $\leftrightarrow$	AgI (s)
Concentrations:			
Initial	0.00909	0.227	0
Change	-0.00909	-0.00909	0.00909
At 100% rxn:	0	0.218	0.00909

Concentrations:			
"Initial"	0	0.218	0.00909
Change	+x	+x	-x
Equilibrium	x	0.218 + x	0.00909 – x

6. **Substitute and calculate.**

$$K_{eq} = \frac{1}{[Ag^+][I^-]} = \frac{1}{x(0.218 + x)} = 1.2\times10^{16}$$

Since K_{eq} is so large, the amount of I^- that will reform is tiny, so "x" is tiny compared to "0.218+x" and can be dropped.

$x = 3.8\times10^{-16}$ (Our assumption that x is small was valid.)

This gives us $[I^-]$ at equilibrium: 0.218 M

Will any AgCl precipitate? We have concentrations of Ag^+ and Cl^-. Use them to compare Q to K_{eq} and assess first if a precipitate will form:

$$Ag^+ (aq) + Cl^- (aq) \leftrightarrow AgCl (s); \; K_{eq} = 5.6\times10^9$$

$$Q = \frac{1}{[Ag^+][Cl^-]} = \frac{1}{(3.8\times10^{-16})(0.227)} = 1.2\times10^{16}$$

$Q \gg K_{eq}$ so the reaction would proceed spontaneously in the reverse direction.
This means that no precipitate will form.

SOLUTION: To answer the question, the concentration of each solute ion is: $[Na^+] = 0.455$ M, $[NO_3^-] = 0.00909$ M, $[Ag^+] = 3.8 \times 10^{-16}$ M, $[I^-] = 0.218$ M, and $[Cl^-] = 0.227$ M.

7. **Does the result make sense?** The concentration of Ag^+ ions should be quite small due to the large equilibrium constant for AgI formation. Furthermore, the concentration of I^- should be lower than Cl^- (because some I^- precipitates).

Common Ion Effect

Key Concept:
- The solubility of a salt greatly decreases if a common ion is introduced.

EXERCISE 23: How many grams of lead(II) iodide can dissolve in 500.00 mL of 0.500 M NaI (aq) at 25°C? The pK_{sp} of lead(II) iodide at 25°C is 8.19.

$$PbI_2 \text{ (s)} \leftrightarrow Pb^{2+} \text{ (aq)} + 2I^- \text{ (aq)}; \quad K_{sp} = 10^{-8.19} = 6.5 \times 10^{-9}$$

STRATEGY: Set up an equilibrium table. 0.500 M I^- is present to start.

	PbI_2 (s) $\leftrightarrow$	Pb^{2+} (aq) +	$2I^-$ (aq)
Concentrations:			
Initial		0	.500
Change		+x	+2x
At Equilibrium		x	.500+2x

$$K_{sp} = 6.5 \times 10^{-9} = [Pb^{2+}][I^-]^2 = (x)(0.500 + 2x)^2$$

We can assume "2x" is small relative to "0.500+2x" since K_{sp} is so miniscule. Solving for x we get:
$$x = [Pb^{2+}] = 2.6 \times 10^{-8}$$

$$\frac{2.6 \times 10^{-8} \text{ mol } Pb^{2+}}{L} \times 0.500 \text{ L} \times \frac{1 \text{ mol } PbI_2}{1 \text{ mol } Pb^{2+}} \times \frac{461.0012 \text{ g}}{\text{mol}} = 6.0 \times 10^{-6} \text{ g } PbI_2$$

SOLUTION: 6.0×10^{-6} g of PbI_2 can dissolve in this solution.

In Try It #5, you determined the mass of lead(II) iodide that could be dissolved in 500 mL of water. Compare its value to the answer obtained in Exercise 23. It is easy to see that the addition of a common ion will greatly reduce the solubility of an insoluble salt. This is quite a useful concept/technique to understand.

Effects of pH

In Chapter 16, Exercise 12, we saw that Le Chatelier's Principle could be used to qualitatively illustrate the effect of an acidic environment on the solubility of magnesium hydroxide. What about a quantitative relationship?

EXERCISE 24: Show quantitatively that the concentration of calcium hydroxide will increase in an acidic environment.

STRATEGY AND SOLUTION: At 25°C, the pK_{sp} for calcium hydroxide is 5.30, so $K_{sp} = 5.0 \times 10^{-6}$. We can write an overall equilibrium reaction and compare its equilibrium constant to the K_{sp} for calcium hydroxide. Two H_3O^+ ions will be needed to react with the 2 OH^- ions generated by each $Ca(OH)_2$.

$$Ca(OH)_2(s) \leftrightarrow Ca^{2+}(aq) + 2\,OH^-(aq)$$
$$+ \quad 2\,[\,H_3O^+(aq) + OH^-(aq) \leftrightarrow 2\,H_2O\,(l)\,]$$

$$\text{Overall: } Ca(OH)_2(s) + 2\,H_3O^+(aq) \leftrightarrow Ca^{2+}(aq) + 4\,H_2O\,(l)$$

$$K_{overall} = K_{sp}\left(\frac{1}{K_w}\right)^2 = (5.0 \times 10^{-6})\left(\frac{1}{1 \times 10^{-14}}\right)^2 = 5.0 \times 10^{22}$$

Comparison of calcium hydroxide's tiny K_{sp} to the enormous overall equilibrium constant quantitatively illustrates that calcium hydroxide's solubility is greatly enhanced in acidic environments.

18.5 COMPLEXATION EQUILIBRIA

Stoichiometry of Complexes

Key Terms:
- **Ligand** – a species that bonds to a metal cation to form a complex. (To be a ligand, a chemical must have a lone pair that it can donate to the metal cation. NH_3 is an example.)
- **Bidentate ligand** – a species that bonds to a metal cation in two locations.
- **Coordination number** – total number of bonds between a metal center and its ligands.
- **Complex ion** – an ion composed of a metal ion center with one or more ligands attached to it.

Key Concepts:
- Coordination numbers must be determined by experiment.
- The formula for a complex ion is written in brackets to emphasize that the ligands are attached to the metal ion center. For example: $[Ni(H_2O)_6]^{2+}$ shows that 6 water molecules are attached to a Ni^{2+} cation. The coordination number of this ion is therefore 6. The overall charge on the complex ion is shown outside of the brackets. Water has no charge, so we know the charge on the nickel ion must be equal to the overall charge of the ion, 2+.

EXERCISE 25: Write the formula for the complex ion that forms between Pt^{4+} and Cl^- ions, given that the coordination number of this complex is 6.

> *STRATEGY:* The coordination number indicates that 6 chloride ions are present. Each has a 1- charge, so 6 Cl^- and one Pt^{4+} yields an overall charge of 2-.
>
> *SOLUTION:* The formula for the complex ion is $[PtCl_6]^{2-}$.

Complexation Calculations

Key Concepts:

- Ligands bond sequentially to the metal cation center, so a separate equilibrium reaction exists for the addition of each ligand.
- In this chapter, we will deal with calculations involving excess ligand, which means we can focus on the *overall* process of individual metal ions and ligands forming a final complex.

Practical Tip: Metal ion complexes tend to have very large complex formation equilibrium constants. This can be put to several good uses, such as removing heavy metal ions from a person's bloodstream (by adding a ligand that preferentially binds to the metal) or to increase the solubility of some insoluble metal by forming a metal complex.

The Chelate Effect

Key Terms:

- **Chelating agent** – a ligand that has two or more donor atoms in it. A bidentate ligand is a type of chelating agent.
- **Chelate effect** – the stabilization of a metal complex by a multidentate ligand.

Key Concept:

- If a chelating agent binds to a metal ion in one location, the other ligands on the chelating agent will be able to attach to the metal more easily than a nearby monodentate ligand.

Metal near ligand	Lone pair on ligand bonds to metal	Ligand rotates	Second lone pair bond to metal (bidentate)

Complex Formation and Solubility

EXERCISE 26: The pK_{sp} of cadmium sulfide is 28.85 at 25°C. How will the solubility of cadmium sulfide be affected by the addition of ammonia? The K_f for the complexation of Cd^{2+} to NH_3 (coordination number of 4) at this temperature is 1.0×10^7.

STRATEGY AND SOLUTION: Write equilibrium reactions for the processes.

$$CdS \text{ (s)} \leftrightarrow Cd^{2+} \text{ (aq)} + S^{2-} \text{ (aq)}; \ K_{sp} = 10^{-28.85} = 1.4 \times 10^{-29}$$

$$Cd^{2+} \text{ (aq)} + 4 NH_3 \text{ (aq)} \leftrightarrow [Cd(NH_3)_4]^{2+}; \ K_f = 1.0 \times 10^7$$

We can see from writing the two equilibria that the minute amount of cadmium ion present in solution from the K_{sp} process will be consumed in the formation of the complex ion. This is an application of Le Chatelier's principle. The added ammonia will remove Cd^{2+} in the solubility reaction, thereby driving the solubility reaction to the right to make more Cd^{2+}. The overall result is that the solubility of CdS will increase.

Chapter 18 Self-Test

You may use a periodic table, a calculator, Appendix E of the text, and Table 18.2 of the text for this test.

1. Which mixtures are buffers?

    a) 0.0100 moles of HCl and 0.0500 moles of NaOH in 100.0 mL of water

    b) 5.0 g NaH_2PO_4 and 2.0 g Na_2HPO_4 in 100.0 mL of water

    c) 5.00 g of NH_4Cl and 100.0 mL of water

2. Determine the pH of each solution in problem #1.

3. Describe how to prepare 500.0 mL of a pH 3.30 buffer whose total buffer concentration is 0.0500 M.

4. Morphine, $C_{17}H_{19}NO_3$, is an alkaloid, a naturally occurring base. Morphine has a K_b of 1.6×10^{-6} at 25°C. Determine the pH of 30.00 mL of 2.00×10^{-4} M morphine (aq).

5. The 30.00 mL of 2.00×10^{-4} M morphine in problem #4 is titrated with 1.00×10^{-4} M HCl.

    a) Determine the pH of the solution after a total of 10.00 mL HCl has been added to the original morphine solution.

    b) Determine the pH of the solution after a total of 30.00 mL HCl has been added to the original morphine solution.

    c) Determine the volume of 1.00×10^{-4} M HCl that would need to be added to the original morphine solution to result in a pH of 8.05.

    d) What is the pH of the solution at the equivalence point of the titration?

    e) What would be a suitable indicator dye to use for this titration?

6. Four cyanide ions bind to one Ni^{2+} ion to form a complex. The equilibrium constant for this process at some temperature is 1.0×10^{31}.

    a) What is the formula of the complex ion?

    b) 10.0 L of a solution that is 1.5×10^{-3} M in CN^- is treated with 5.0 g of nickel(II) sulfate. What is the concentration of CN^- in solution after the treatment?

Answers to Try Its:

1. 0.00 moles A⁻, 0.95 mol HA, 0.10 mol H_3O^+. The mixture is no longer a buffer.
2. A is a buffer
   B is a buffer (1 mol CN⁻ and 1 mol HCN are present)
   C is a buffer (6.25 mmol HS⁻ and 8.75 mmol H_2S)
   D is not a buffer because excess OH⁻ is present. A weak acid/base conjugate pair is not present as a major component.
3. We still have a buffer: 0.022 mol HOCl and 0.023 mol OCl⁻ are present after the NaOH is added. (Notice from this problem that a larger buffer capacity can accommodate larger quantities of H_3O^+ and OH⁻.
4. Mix 0.53 g of NaOH with 1.7 g $HONH_3Cl$ with water in a 250 mL volumetric flask. Dilute to the mark with water.
5. 0.27 g

Answers to Self-Test:

1. a) This is not a buffer (no weak acid/base pair).
   b) Yes, this is a buffer. It contains a mixture of $H_2PO_4^-$ and HPO_4^{2-} as major components.
   c) No, this is not a buffer. It contains NH_4^+ as a major component, but NH_3 is a minor component.

2. a) 13.60   b) 6.74   c) 4.64
3. There is more than one correct way to answer this question. Here is one:
   Nitrous acid and nitrite ion would be a good conjugate acid/base pair to use because the pK_a of HNO_2 at 25°C is 3.25. There are three general ways to make this buffer. For each, mix together in a 500.00 mL volumetric flask and dilute to mark with water: a) mix 0.0250 mol of HNO_2 and 0.0132 mol OH⁻ (ex. NaOH); b) mix 0.0250 mol of NO_2^- (ex. $NaNO_2$) and 0.0118 mol of H_3O^+ (ex. HCl); c) mix 0.0118 mol of HNO_2 and 0.0132 mol of NO_2^- (ex. $NaNO_2$).
4. 9.23
5. a) 8.90   b) 8.20   c) 35.1 mL   d) 6.19   e) bromocresol purple
6. a) $[Ni(CN)_4]^{2-}$; b) 1.07×10^{-8} M

Chapter 19: Electron Transfer Reactions

Learning Objectives

In this chapter, you will learn how to:
- Assign oxidation numbers for atoms.
- Balance redox reactions.
- Determine whether or not a given redox reaction will occur spontaneously.
- Determine the standard cell potential for an electrochemical cell.
- Calculate the standard free energy ($\Delta G°_{rxn}$) and the equilibrium constant (K_{eq}) for a reaction from standard potentials.
- Use the relationship between free energy, cell potential, and concentration to determine potentials and concentrations at non-standard conditions.
- Make a battery.
- Protect a metal from corrosion.
- Calculate the amount of a substance that will form in an electrolysis reaction.

Practical Aspects

Recall that most chemical reactions can be classified as either acid-base or redox. In Chapters 17 and 18, you learned about the thermodynamics of acid/base reactions. Here, we will focus on redox reactions. In this chapter, you will learn the chemistry concepts behind how to make a battery, how to prevent a metal from corroding, and how to electroplate an object such as silver-plated jewelry or a chrome-plated car bumper.

We will use simple molecules and ions in the examples in this chapter, but it is worth mentioning that the entire electron transport chain within our bodies also operates on basic redox principles. This chapter should therefore prove useful to chemists, biochemists, and biologists alike. This chapter also provides a direct link to physics, with the study of current flow and potentials. Finally, an environmental scientist will find this chapter useful, because many of the sensitive instruments used to measure low concentrations of contaminants in water samples are constructed from redox principles.

19.1 RECOGNIZING REDOX REACTIONS

Key Terms:
- **Oxidation** – loss of electrons.
- **Reduction** – gain of electrons.
- **Redox** – nickname for an oxidation-reduction reaction.
- **Oxidizing agent** – the substance that is gaining electrons in a redox reaction. The oxidizing agent helps another substance to undergo oxidation by taking electrons from it.
- **Reducing agent** – the substance that is losing electrons in a redox reaction. The reducing agent helps another substance to undergo reduction by supplying it with electrons.
- **Oxidation number (aka oxidation state)**– charge that an atom would have if 100% ionic bonding were assumed.

Key Concepts:
- Oxidation and reduction always occur together; one process cannot occur without the other.
- If any atoms within a reaction undergo a change in oxidation number, then the reaction is classified as a redox reaction.
- The sum of the oxidation numbers within a given substance must equal the overall charge on the substance.
- Remember that 100% ionic bonding is assumed when assigning oxidation numbers. This means the electron pair within a bond will be assigned to the more electronegative element.

Helpful Hints
- Section 19.1 in the text provides rules for assigning oxidation numbers.
- Oxygen within a compound typically has an oxidation number of –2.
- Hydrogen within a compound typically has an oxidation number of +1.

EXERCISE 1: Assign oxidation numbers to each: a) $CaCl_2$; b) $HClO_3$; c) NaH; d) Cr_2O_3; e) $Cr_2O_7^{2-}$.

STRATEGY: Use the guidelines established in the text and above for assigning oxidation numbers.

SOLUTION:
a) $CaCl_2$: This is an ionic compound, so the oxidation numbers are: Ca = +2 and Cl = –1.
b) $HClO_3$: The sum of all oxidation numbers must be zero because the overall compound has no charge. O is more electronegative than H or Cl, so it will take the electrons: the oxidation number of each O is –2, the oxidation number of H is +1, and the oxidation number of Cl is found by comparison:

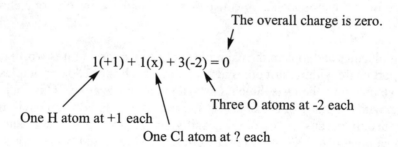

The overall charge is zero.

$$1(+1) + 1(x) + 3(-2) = 0$$

One H atom at +1 each

One Cl atom at ? each

Three O atoms at -2 each

Solving for x gives x = +5, so the oxidation number of Cl is +5.

c) NaH: Assuming 100% ionic bonding, H is more electronegative than Na, so Na = +1 and H = –1. Notice that here is one of the rare cases when H will have a –1 charge to it.
d) Cr_2O_3: This is an ionic compound so the charges are the oxidation numbers: O = –2 and Cr = +3.
e) $Cr_2O_7^{2-}$: O = –2. Use the method we used in part "b" to find the oxidation number of Cr:

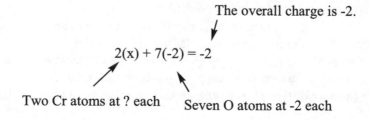

The overall charge is -2.

$$2(x) + 7(-2) = -2$$

Two Cr atoms at ? each

Seven O atoms at -2 each

Solving for x gives x = +6. The oxidation state of Cr in this ion is +6.

EXERCISE 2: Answer these questions for each reaction: Is this a redox reaction? What is being oxidized? What is being reduced? What is the oxidizing agent?

a) $2 CO (g) + O_2 (g) \rightarrow 2 CO_2 (g)$
b) $AgCl (s) \rightarrow Ag^+ (aq) + Cl^- (aq)$
c) $2 KCl (s) \rightarrow 2 K (s) + Cl_2 (g)$

STRATEGY: Assign oxidation numbers to every atom and see if changes in oxidation numbers occur. If they do, then it is a redox reaction.

The substance that is oxidized is the substance that is losing electrons in the process. The substance that is reduced is the substance that is gaining electrons in the process. By definition, the substance reduced is also called the oxidizing agent.

SOLUTION:

	Assign oxidation numbers:	Is this a redox rxn?	This is oxidized:	This is reduced. It is also the oxidizing agent:
a)	CO: $C = +2$, $O = -2$ O_2: $O = 0$ CO_2: $C = +4$, $O = -2$	**Yes** C lost electrons (+2 to +4) & O gained electrons (0 to –2)	C (in CO)	O_2
b)	AgCl: $Ag = +1$, $Cl = -1$ Ag^+: $Ag = +1$ Cl^-: $Cl = -1$	**No** Nothing gained or lost electrons	-	-
c)	KCl: $K = +1$, $Cl = -1$ K: $K = 0$ Cl_2: $Cl = 0$	**Yes** K gained electrons (+1 to 0) & Cl lost electrons (–1 to 0)	Cl^-	K^+

Notice that the oxidizing agent was listed specifically in part "a" to differentiate between which oxygen was being reduced.

Try It #1: Determine what is undergoing oxidation and what is undergoing reduction in this reaction:
$$2 C_2H_2 (g) + 5 O_2 (g) \rightarrow 4 CO_2 (g) + 2 H_2O (g)$$

19.2 BALANCING REDOX REACTIONS

Key Term:
- **Half-reaction** – one-half of a redox reaction. There is an oxidation half-reaction and a reduction half-reaction. When combined, these two half-reactions form the overall redox reaction.

Key Concept:
- All electrons must be accounted for in a balanced redox reaction.

Helpful Hint

- Many redox reactions are quite complex, so several methods have been established to balance them in a stepwise fashion. The text describes one method. Here, we will show a slight variation to the text's method, which emphasizes the importance of accounting for the electrons involved in the redox process.

EXERCISE 3: Separate each unbalanced redox reaction into two half-reactions. For each atom undergoing a change in oxidation number, determine the number of electrons gained or lost.

a) HNO_3 (aq) + Cu_2O (s) $\rightarrow$ $Cu(NO_3)_2$ (aq) + 2 NO (g)

b) SO_2 (aq) + $Cr_2O_7^{2-}$ (aq) $\rightarrow$ Cr^{3+} (aq) + SO_4^{2-} (aq)

STRATEGY: We're asked first to find the half-reactions. One half-reaction will show oxidation and the other will show reduction. Analyze the oxidation numbers of each atom within the reaction and determine which atoms are undergoing changes in oxidation number. We've already learned how to assign oxidation numbers, so this part will be shortcut.

SOLUTION:

a) N starts out with an oxidation number of +5 and ends with an oxidation number of +2 (in NO). The N in the $Cu(NO_3)_2$ didn't undergo a change, so it is not included here. In going from a +5 to a +2 oxidation number, the N atom gains 3 electrons. Electrons are gained, so this is reduction.

Reduction half-reaction: HNO_3 (aq) $\rightarrow$ 2 NO (g)

Cu starts out as Cu^{1+} and ends up as Cu^{2+}, so each Cu loses 1 electron.

Oxidation half-reaction: Cu_2O (s) $\rightarrow$ $Cu(NO_3)_2$ (aq)

b) S starts out with an oxidation number of +4 and ends up with a +6 oxidation number. S loses two electrons, so it is undergoing oxidation.

Oxidation half-reaction: SO_2 (aq) $\rightarrow$ SO_4^{2-} (aq)

Cr is originally Cr^{6+} and goes to Cr^{3+}, so each Cr gains 3 electrons. It is being reduced.

Reduction half-reaction: $Cr_2O_7^{2-}$ (aq) $\rightarrow$ Cr^{3+} (aq)

EXERCISE 4: Balance this reaction, which occurs under acidic conditions:

$$I_2 \text{ (aq)} + NO_3^- \text{ (aq)} \rightarrow IO_3^- \text{ (aq)} + NO_2 \text{ (g)}$$

STRATEGY: We're asked to balance the equation. This is a redox reaction because the oxidation numbers are changing on N and I. Follow the stepwise procedure to balance the reaction. (We'll leave out the phases until we get to the final step, just to save time.)

1. Break the unbalanced equation into half-reactions.

- $I_2 \rightarrow IO_3^-$
- $NO_3^- \rightarrow NO_2$

2. Balance each half reaction:

2a. Balance the element undergoing the redox change by adding coefficients.
The iodines need balancing, but the nitrogens are already balanced:

- $I_2 \rightarrow 2\ IO_3^-$
- $NO_3^- \rightarrow NO_2$

2b. Determine the number of electrons lost or gained by each substance undergoing a redox change. Account for those electrons in the half-reaction equation.

Iodine starts out with a 0 oxidation number and ends up with a +5 oxidation number. We have two iodines, and *each* lost 5 electrons, so 10 electrons total are lost by the Is.

Nitrogen starts out with a +5 oxidation number and ends up with a +4 oxidation number, so the nitrogen gains 1 e-.

- $I_2 \rightarrow 2\ IO_3^- + 10\ e^-$

- $1\ e^- + NO_3^- \rightarrow NO_2$

2c. Balance oxygen by adding water to the oxygen-deficient side of the reaction.

- $6\ H_2O + I_2 \rightarrow 2\ IO_3^- + 10\ e^-$

- $1\ e^- + NO_3^- \rightarrow NO_2 + H_2O$

2d. Balance hydrogens in this manner:

- *If the solution is acidic*, add H_3O^+ to the hydrogen-deficient side and the same number of H_2O molecules to the opposite side.
- *If the solution is basic*, add H_2O molecules to the hydrogen deficient side and the same number of OH^- ions to the opposite side.

This solution is acidic, so we can add H_3O^+ and H_2O molecules.

The iodine half-reaction needs twelve H atoms on the product side, so we need to add 12 H_3O^+ to that side and 12 H_2O to the reactant side. (Notice that this will give us a total of 18 H_2O on the reactant side.)

The nitrogen half-reaction needs two H atoms on the reactant side, so we need to add 2 H_3O^+ to that side and 2 H_2O molecules to the product side. (Notice that this gives us a total of 3 H_2O on the product side.)

- $18\ H_2O + I_2 \rightarrow 2\ IO_3^- + 10\ e^- + 12\ H_3O^+$

- $2\ H_3O^+ + 1\ e^- + NO_3^- \rightarrow NO_2 + 3\ H_2O$

3. The total electrons gained must equal the total electrons lost in the final equation. Account for this by multiplying each half-reaction by a factor so that each half-reaction shows the same number of electrons in the reaction.

The nitrogen half-reaction will have to be multiplied through by 10 to make the total electrons within each half-reaction the same.

- $18\ H_2O + I_2 \rightarrow 2\ IO_3^- + 10\ e^- + 12\ H_3O^+$

- $20\ H_3O^+ + 10\ e^- + 10\ NO_3^- \rightarrow 10\ NO_2 + 30\ H_2O$

4. Add the two half-reactions together, and cancel out like terms.

$20\ H_3O^+ + 10\ e^- + 10\ NO_3^- + 18\ H_2O + I_2 \rightarrow 2\ IO_3^- + 10\ e^- + 12\ H_3O^+ + 10\ NO_2 + 30\ H_2O$

Canceling like terms yields:

$$8\ H_3O^+ + 10\ NO_3^- + I_2 \rightarrow 2\ IO_3^- + 10\ NO_2 + 12\ H_2O$$

Do a quick inventory of all atoms and charges as a double-check:

Item	Reactants	Products
I	2	2
N	10	10
O	38	38
H	24	24
Charges	-2	-2

SOLUTION: The balanced chemical equation for the reaction (including phases) is:

$$8\ H_3O^+\ (aq) + 10\ NO_3^-\ (aq) + I_2\ (aq) \rightarrow 2\ IO_3^-\ (aq) + 10\ NO_2\ (g) + 12\ H_2O\ (l)$$

The answer seems reasonable because all atoms and charges were accounted for in the inventory.

EXERCISE 5: Balance this reaction, which occurs under basic conditions:
$$ClO^- (aq) + Cr(OH)_3 (s) \rightarrow Cl^- (aq) + CrO_4^{2-} (aq)$$

STRATEGY: We're asked to balance the equation. This is a redox reaction because the oxidation numbers are changing on Cl and Cr. Follow the stepwise procedure to balance the reaction.

1. Break the unbalanced equation into half-reactions.
- $ClO^- \rightarrow Cl^-$
- $Cr(OH)_3 \rightarrow CrO_4^{2-}$

2. Balance each half-reaction.

2a. Balance the element undergoing the redox change by adding coefficients. They're both already balanced.

2b. Determine the number of electrons lost or gained by each substance undergoing a redox change. Account for those electrons in the half-reaction equation.

Cl starts out with a +1 oxdiation number and ends up with a -1 oxidation number, so Cl gained two electrons.
- $2 e^- + ClO^- \rightarrow Cl^-$

Cr starts out with a +3 oxidation number and ends up with a +6 oxidation number, so Cr lost three electrons.
- $Cr(OH)_3 \rightarrow CrO_4^{2-} + 3 e^-$

2c. Balance oxygen by adding water to the oxygen-deficient side of the reaction.
- $2 e^- + ClO^- \rightarrow Cl^- + H_2O$
- $H_2O + Cr(OH)_3 \rightarrow CrO_4^{2-} + 3 e^-$

2d. Balance hydrogens in this manner:
- *If the solution is acidic*, add H_3O^+ to the hydrogen-deficient side and the same number of waters to the opposite side.
- *If the solution is basic*, add waters to the hydrogen-deficient side and the same number of OH^- ions to the opposite side.

This solution is basic.

The Cl half-reaction needs 2 H atoms on the reactant side, so add 2 H_2O molecules there and 2 OH^- ions on the product side:
- $2 H_2O + 2 e^- + ClO^- \rightarrow Cl^- + H_2O + 2 OH^-$

The Cr half-reaction needs 5 H atoms on the product side, so add $5H_2O$ to that side and 5 OH^- ions to the reactant side:
- $5 OH^- + H_2O + Cr(OH)_3 \rightarrow CrO_4^{2-} + 3 e^- + 5 H_2O$

Simplify the reaction, if desired, by canceling like terms (can wait till the end too).
- $H_2O + 2 e^- + ClO^- \rightarrow Cl^- + 2 OH^-$
- $5 OH^- + Cr(OH)_3 \rightarrow CrO_4^{2-} + 3 e^- + 4 H_2O$

3. The total electrons gained must equal the total electrons lost in the final equation. Account for this by multiplying each half-reaction by a factor so that they'll both have the same number of electrons in the reaction.

Multiply the Cl half-reaction by three. Multiply the Cr half-reaction by two.
- $3 H_2O + 6 e^- + 3 ClO^- \rightarrow 3 Cl^- + 6 OH^-$
- $10 OH^- + 2 Cr(OH)_3 \rightarrow 2 CrO_4^{2-} + 6 e^- + 8 H_2O$

4. Add the two half-reactions together, and cancel out like terms.

$10 OH^- + 2 Cr(OH)_3 + 3 H_2O + 6 e^- + 3 ClO^- \rightarrow 3 Cl^- + 6 OH^- + 2 CrO_4^{2-} + 6 e^- + 8 H_2O$

Canceling like terms yields:
$$4 OH^- + 2 Cr(OH)_3 + 3 ClO^- \rightarrow 3 Cl^- + 2 CrO_4^{2-} + 5 H_2O$$

Do a quick inventory of all atoms and charges as a double-check:

Item	Reactants	Products
O	13	13
H	10	10
Cr	2	2
Cl	3	3
Charges	-7	-7

SOLUTION: The balanced chemical equation for the reaction (including phases) is:

$$4\ OH^-\ (aq) + 2\ Cr(OH)_3\ (s) + 3\ ClO^-\ (aq) \rightarrow 3\ Cl^-\ (aq) + 2\ CrO_4^{2-}\ (aq) + 5\ H_2O\ (l)$$

The answer seems reasonable because all atoms and charges were accounted for in the inventory.

Try It #2: Balance the reaction, which occurs under basic conditions:

$$Al\ (s) + NO_3^-\ (aq) \rightarrow Al(OH)_4^-\ (aq) + NH_3\ (aq)$$

19.3 GALVANIC CELLS

Key Terms:

- **Direct electron transfer** – situation in which the two participants of the redox reaction come in contact with each other.
- **Indirect electron transfer** – situation in which the two participants of the redox reaction are kept separated from one another, but electron transfer can occur via a metal wire (or other similar medium) that separates them.
- **Electrochemical cell** – device in which electron flow associated with a redox reaction is routed through an external circuit. Two main types of electrochemical cells include:
  - **Galvanic cell** – contains components of a spontaneous reaction, whose electron flow can be used as an energy source (e.g., a battery). Figure 19-9 in the text illustrates the features of a galvanic cell.
  - **Electrolytic cell** – contains components of a non-spontaneous reaction. Electrical energy must be supplied to the cell to force the non-spontaneous reaction to occur (e.g., electroplating).
- **Electrode** – a conducting substance that allows electrons to be transferred between an aqueous phase and an external circuit. An electrode is classified in two ways:
  - **Active** – an electrode that participates in the redox reaction.
  - **Passive** – an electrode that doesn't participate in the redox reaction. Platinum and graphite are often used as passive electrodes, because they are unreactive.
- **Anode** – electrode within an electrochemical cell where oxidation occurs.
- **Cathode** – electrode within an electrochemical cell where reduction occurs.
- **Standard hydrogen electrode** – electrode made up of a hydrogen gas (at 1 atm), H_3O^+ (at 1 M), and a platinum electrode.

Key Concepts:

- A redox reaction will occur spontaneously if its ΔG is less than zero (recall Chapter 14).
- Indirect electron transfer will quickly stop working (because of charge build-up) unless some type of ion channeler is established between the two redox reactants in order to re-equilibrate the charge. Figure 19-7 in the text shows the ion transport channel within an electrochemical cell.
- A few types of ion transport channels are: porous plates, salt bridges, and cell membranes.

Helpful Hint
- *O*xidation occurs at the *a*node (both vowels). *R*eduction occurs at the *c*athode (both consonants).

EXERCISE 6: Use thermodynamic data to determine whether or not this redox reaction is spontaneous under standard conditions. Given: $\Delta G°_f$ for Zn^{2+} (aq) is -147.06 kJ/mol.

$$Zn\ (s)\ +\ 2\ Ag^+\ (aq)\ \rightarrow\ 2\ Ag\ (s)\ +\ Zn^{2+}\ (aq)$$

STRATEGY: A reaction is spontaneous if its ΔG is negative. Use Appendix D to determine $\Delta G°_{rxn}$:

$$\Delta G°_{rxn} = \Sigma\ (coeff)\ \Delta G°_f (products)\ -\ \Sigma\ (coeff)\ \Delta G°_f (reactants)$$

$$= [(2\ mol\ Ag)(0\ kJ/mol) + (1\ mol\ Zn^{2+})(-147.06\ kJ/mol)]$$
$$- [(1mol\ Zn)(0\ kJ/mol) + (2\ mol\ Ag^+)(77.107\ kJ/mol)]$$
$$= -301.27\ kJ/mol$$

SOLUTION: The reaction is spontaneous under standard conditions because its $\Delta G°_{rxn}$ is negative. The answer seems reasonable because Ag is a precious metal, so it should not be as easily oxidized as Zn.

EXERCISE 7: If the reaction in Exercise 6 were to be used in an electrochemical cell,
a) what half-reaction would take place at the anode? Write a balanced half-reaction.
b) what half-reaction would take place at the cathode? Write a balanced half-reaction.
c) draw a sketch of the electrochemical cell, given these specifications: 1) the Zn half-reaction contains an active electrode; 2) the Ag half-reaction contains a passive electrode; 3) the half-reactions are separated by a porous plate.

STRATEGY: Oxidation occurs at the anode and reduction occurs at the cathode. We can draw a sketch of the cell using Figure 19-9 in the text as a guide.

SOLUTION:
a) The reaction that occurs at the anode is the oxidation half-reaction: $Zn\ (s) \rightarrow 2\ e^- + Zn^{2+}$ (aq)
b) The reaction that occurs at the cathode is the reduction half-reaction: $2\ e^- + 2\ Ag^+$ (aq) $\rightarrow 2\ Ag$ (s)
c) Let's consider the required features of the cell before constructing it:
   1) The Zn half-reaction contains an active electrode – this means that the electrode is involved in the redox reaction, so the electrode *must* be made of Zn. Theoretically, the Zn can be immersed in any non-reactive salt solution, because this reaction will be *generating* Zn^{2+} ions, rather than using them. However, you will see in the next section that since an equilibrium process is established between Zn (s) and Zn^{2+} ions, it is customary to include Zn^{2+} in the original container. 1 M $ZnSO_4$ (aq) is a standard choice for supplying Zn^{2+} ions.
   2) The Ag half-reaction contains a passive electrode – this means the electrode is not involved in the redox reaction, which means Ag cannot be the electrode. A commonly-used passive (inert) electrode is platinum. We need a source of Ag^+ ions, so we'll immerse the Pt wire in an aqueous solution of $AgNO_3$ (or some other soluble silver salt).
   3) The half-reactions are separated by a porous plate. Draw a plate between the two solutions that can allow ions to pass through it, so that charge is maintained.

As mentioned previously, we want Ag^+ ions in the reduction half-cell and Zn^{2+} ions in the oxidation half-cell. The counter ions in the containers can be any anions that won't react in the process. For simplicity, we'll just write these as "-" charges.

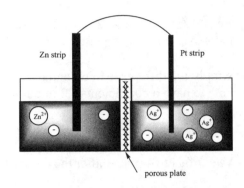

porous plate

Try It #3: Is the electrochemical cell in Exercise 7 an example of direct or indirect electron transfer?

19.4   CELL POTENTIALS

Key Term:
- **Standard reduction potential** – measure of the tendency that a substance has to undergo reduction under standard conditions (1 M, 1 atm, 25°C).

Key Concepts:
- The difference in electrical potential between two electrodes is termed "E" and is measured in volts.
- Standard reduction potentials are *relative* to this reference reaction:
$$2\ e^- +\ 2\ H_3O^+ \text{(aq)} \leftrightarrow\ H_2 \text{(g)} + 2\ H_2O \text{(l)},\ E° = 0.00\ V$$
- The greater the reduction potential, the more the substance will want to undergo reduction.
- In comparing two half-cells, the half-cell with the greater reduction potential will undergo reduction, and the half-cell with the lower reduction potential will undergo oxidation.
- Electrons always flow "downhill" (i.e., they'll leave the substance undergoing oxidation and flow towards the substance undergoing reduction. Electrons flow towards the substance that wants them).
- When a reaction is multiplied by an integer, the cell potential remains unchanged.

Useful Relationship:
- $E°_{cell} = E°_{cathode} - E°_{anode}$ Use this equation to determine the overall cell potential, $E°_{cell}$. The terms are subtracted to account for the oxidation portion of the rxn.

Helpful Hints
- Appendix F lists standard reduction potentials for many common half-reactions.
- When comparing two half-cell potentials, think of the values on a number line. The value farther to the right on the number line will be the reduction half-cell, and the other value will be the oxidation half-cell.

EXERCISE 8: Determine $E°_{cell}$ for the reaction shown in the electrochemical cell from Exercise 6.

STRATEGY: We're asked to find $E°_{cell}$, which is based on $E°_{cathode}$ and $E°_{anode}$. First, use Appendix F to determine $E°_{cathode}$ and $E°_{anode}$, then use the relationship $E°_{cell} = E°_{cathode} - E°_{anode}$ to determine $E°_{cell}$. We had already established that the two half-reactions are:
$$Zn \text{(s)} \leftrightarrow\ 2\ e^- +\ Zn^{2+} \text{(aq)}\ \text{and}\ 2\ e^- + 2\ Ag^+ \text{(aq)} \leftrightarrow 2\ Ag \text{(s)}$$
Remember, E° values are always written in the direction of reduction. Data from Appendix F provides:

$$2 \text{ e}^- + \text{Zn}^{2+} \text{ (aq)} \leftrightarrow \text{Zn (s)} \quad E° = -0.7618 \text{ V}$$
$$1 \text{ e}^- + \text{Ag}^+ \text{ (aq)} \leftrightarrow \text{Ag (s)} \quad E° = +0.7996 \text{ V}$$

Remember, too, that $E°_{cell}$ is simply the difference in potential between the two half-cells, so we *don't* have to account for the fact that we have a coefficient of 2 in our silver half-cell:

$$E°_{cell} = E°_{cathode} - E°_{anode} = E°_{\text{Ag half rxn}} - E°_{\text{Zn half rxn}} = 0.7996 \text{ V} - (-0.7618 \text{ V}) = 1.5614 \text{ V}$$

SOLUTION: $E°_{cell} = +1.5614 \text{ V}$ (sig figs to 4 places after the decimal, just like the data).

Key Concept:

- As you can see from the answer to Exercise 8, $E°_{cell}$ for a spontaneous process is positive.

Try It #4: Use standard reduction potentials to predict whether or not this reaction will occur spontaneously in the direction written: $\text{MgCl}_2 \text{ (aq)} + \text{Zn (s)} \leftrightarrow \text{ZnCl}_2 \text{ (aq)} + \text{Mg (s)}$

EXERCISE 9: Use relative standard reduction potentials to determine the direction of electron flow in this electrochemical cell:

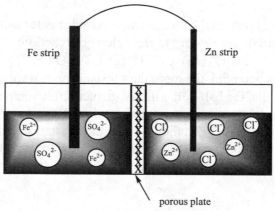

porous plate

STRATEGY: We're asked to determine the direction of electron flow in the apparatus. The electrons will leave the substance undergoing oxidation and flow towards the substance undergoing reduction. We must first therefore distinguish which substance is being oxidized and which is being reduced.

The half-reaction with the greater reduction potential is the reduction half-reaction, and the other reaction is the oxidation half-reaction. Determine the possible half-reactions that can occur. Write these half-reactions in the reduction direction, so that it will be easy to find them in Appendix F:

$$2 \text{ e}^- + \text{Fe}^{2+} \text{ (aq)} \leftrightarrow \text{Fe (s)} \quad E° = -0.447 \text{ V}$$
$$2 \text{ e}^- + \text{Zn}^{2+} \text{ (aq)} \leftrightarrow \text{Zn (s)} \quad E° = -0.7618 \text{ V}$$

The iron process has a greater reduction potential than the Zn process. (Remember to think of the number line: "-0.477" is to the right of "-0.7618" on the number line, so it represents the greater reduction potential.) The iron process will therefore be the reduction half-reaction (cathode) and the zinc process will be the oxidation half-reaction (anode). Electrons will flow from the zinc over the wire to the iron. Fe^{2+} ions in solution will pick up two electrons each and the resulting Fe atoms will deposit on the surface of the Fe cathode. Here is a picture to show electron flow:

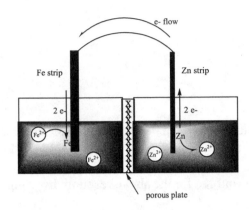

Try It #5: Indicate the direction of electron flow in the electrochemical cell from Exercise 7.

19.5   FREE ENERGY AND ELECTROCHEMISTRY

Key Terms:
- **pH meter** – instrument that uses differences in standard cell potentials to determine the concentration of hydronium ion in a given solution.
- **Current** – charge flow per unit time.

Key Concepts:
- It is possible to use a reaction's cell potential to calculate the reaction's free energy change and its equilibrium constant. *This is quite useful.*
- The moles of electrons transferred in an electrochemical process depend upon the magnitude of the electrical current and the amount of time the current is applied.

Useful Relationships:

- $\Delta G = -nFE$ This provides the relationship between spontaneity and cell potential. "n" is the number of moles of electrons being exchanged in the redox reaction and F = Faraday's constant. The convention is to use "n" as a unitless ratio, in order to obtain units of "J/mol" (or "kJ/mol") for ΔG.

- $E^{\circ} = \left(\dfrac{0.0592\ V}{n}\right) \log K_{eq}$ This equation is derived from $\Delta G^{\circ} = -RT \ln K_{eq}$ (from Chapter 14). It relates standard cell potentials to equilibrium constants. The derivation of this equation is in the text. Here, too, "n" is unitless.

- $E = E^{\circ} - \left(\dfrac{0.0592\ V}{n}\right) \log Q$ This is the Nernst equation and is used for calculating cell potentials under non-standard conditions. It is derived from $\Delta G = \Delta G^{\circ} + RT \ln Q$. The derivation of this equation is in the text. Again, "n" is unitless.

- $n = It/F$ This equation relates the moles of electrons that can be transferred within a given redox reaction when a current, I, is applied for a given length of time, t. "n" is *not* unitless here; keep units of moles in the equation.

Units and Conversions:
- $1 V = 1 J/C$
- Faraday's constant (F) = 96,485.34 C/mol
- Ampere (A) = unit for measuring current
- $1 A = 1 C/sec$

EXERCISE 10: Use standard reduction potentials to determine K_{eq} for this reaction at 25°C:
$$Sn^{4+} (aq) + 2 Fe^{2+} (aq) \leftrightarrow Sn^{2+} (aq) + 2 Fe^{3+} (aq)$$

STRATEGY: We're asked to find K_{eq} for the above reaction, by using E° values. The relationship to use is:

$$E° = \left(\frac{0.0592 V}{n} \right) \log K_{eq}$$

We will need to find E° for the reaction first. First, determine the two half-reactions and look up their standard reduction potentials in Appendix F. Remember that the reduction potential for the oxidation half-reaction will be written in the reverse direction in Appendix F.

Reduction half-reaction (cathode): $Sn^{4+} (aq) + 2 e^- \leftrightarrow Sn^{2+} (aq)$
Oxidation half-reaction (anode): $Fe^{2+} (aq) \leftrightarrow Fe^{3+} (aq) + 1 e^-$
Reduction potentials from Appendix F: $E°_{Sn\ rxn} = +0.151$ V and $E°_{Fe\ rxn} = +0.771$ V

$$E°_{cell} = E°_{cathode} - E°_{anode} = 0.151 V - 0.771 V = -0.620 V$$
(Notice that $E°_{cell}$ is a negative number, so this reaction is not spontaneous at standard conditions.)

The equation shows that there are two electrons transferred every time this reaction takes place, so n = 2. Plugging numbers into the above equation, we obtain:

$$-0.620 V = \left(\frac{0.0592 V}{2} \right) \log K_{eq} \qquad \text{Therefore, } K_{eq} = 1.13 \times 10^{-21}$$

SOLUTION: $K_{eq} = 1.13 \times 10^{-21}$ at 25°C. The answer seems reasonable because the reaction is non-spontaneous under standard conditions, which indicates that the equilibrium constant should be significantly less than 1 and it is.

EXERCISE 11: Use standard reduction potentials to determine $\Delta G°$ for this reaction:
$$Cu^{2+} (aq) + Ni (s) \leftrightarrow Ni^{2+} (aq) + Cu (s)$$

STRATEGY: We're asked to find the standard free energy for the reaction ($\Delta G°$) by using standard reduction potentials. The relationship to use is: $\Delta G° = -nFE°$. We need to find $E°_{cell}$ from Appendix F and we need n. This process shows an exchange of two electrons (a copper ion gains two electrons and a nickel atom loses two electrons, so two electrons are exchanged), so n = 2.

Reduction half-reaction (cathode): $2 e^- + Cu^{2+} (aq) \leftrightarrow Cu (s)$
Oxidation half-reaction (anode): $Ni (s) \leftrightarrow Ni^{2+} (aq) + 2 e^-$
Reduction potentials from Appendix F: $E°_{Cu\ rxn} = +0.3419$ V and $E°_{Ni\ rxn} = -0.257$ V

$$E°_{cell} = E°_{cathode} - E°_{anode} = 0.3419 V - (-0.257 V) = +0.599 V$$

$$\Delta G^\circ = -nFE^\circ = -2 \times \frac{96485.34\,C}{mol} \times \frac{0.599\,J}{C} \times \frac{1\,kJ}{1000\,J} = -115.57\,kJ/mol$$

SOLUTION: ΔG°_{rxn} = -116 kJ/mol (3 sig figs). This value seems reasonable because it indicates the reaction is spontaneous under standard conditions, as does the positive value for E°_{cell}.

Try It #6: Use standard cell potentials to determine ΔG°_{rxn} for the reaction below. Then, compare your answer to that obtained in Exercise 6.

$$Zn\,(s) + 2\,Ag^+\,(aq) \rightarrow 2\,Ag\,(s) + Zn^{2+}\,(aq)$$

EXERCISE 12: An electrochemical cell is constructed using copper metal immersed in 0.00100 M CuSO$_4$ (aq) for one half-reaction and nickel metal immersed in 1.00 M NiSO$_4$ (aq) for the other half-reaction.
a) What is the cell potential at 25°C?
b) What would be the copper ion concentration if the cell potential is 0.458 V and the concentration of NiSO$_4$ (aq) is 1.00 M?

STRATEGY: a) We're asked to find the cell potential when [Ni^{2+}] = 1.00 M and [Cu^{2+}] = 0.00100 M. These are non-standard conditions, because the concentration of copper ion is not 1 M. We'll need to use the Nernst equation, which is used for non-standard conditions:

$$E = E^\circ - \left(\frac{0.0592\,V}{n}\right)\log Q$$

In order to use this equation, we need to determine E°, n and Q first. We can easily find E° by determining what is undergoing oxidation and what is undergoing reduction. The substance with the greater reduction potential will undergo reduction. According to Appendix F, E° for each possible half-reaction is:

$2\,e^- + Ni^{2+}\,(aq) \leftrightarrow Ni\,(s)$ E° = -0.257 V ← this will undergo oxidation. This will be the anode.

$2\,e^- + Cu^{2+}\,(aq) \leftrightarrow 2\,Cu\,(s)$ E° = +0.3419 V ← this will undergo reduction. This will be the cathode.

$$E^\circ_{cell} = E^\circ_{cathode} - E^\circ_{anode} = +0.3419\,V - (-0.257\,V) = +0.599\,V$$

Here is the overall reaction: Ni (s) + Cu^{2+} (aq) $\leftrightarrow$ 2 Cu (s) + Ni^{2+} (aq). The reaction shows that two electrons are transferred in the process, so n = 2.

As we had learned in Chapters 14 and 16, Q is the reaction quotient, which can be calculated from concentrations. Recall that Q is the ratio of products over reactants, raised to the powers of their respective coefficients in the balanced equation, and that only solute concentrations and gaseous partial pressures are included in the ratio:

$$Q = \frac{[Ni^{2+}]^1}{[Cu^{2+}]^1} = \frac{1.00}{0.00100} = 1.00 \times 10^3$$

When we bring this all together, we get:

$$E = E^\circ - \left(\frac{0.0592\ V}{n}\right)\log Q = 0.599\ V - \left(\frac{0.0592\ V}{2}\right)\log 1.00 \times 10^3 = 0.51020\ V$$

b) We're asked to find the $[Cu^{2+}]$, given a cell potential of 0.458 V and $[Ni^{2+}] = 1.00$ M. These are non-standard conditions, because if the $[Cu^{2+}]$ was 1.00 M, then the cell potential would be 0.599 V. We need to use the Nernst equation again, but this time we're looking for $[Cu^{2+}]$, a component of Q:

$$E = E^\circ - \left(\frac{0.0592\ V}{n}\right)\log Q, \text{ so } \log Q = -n\left(\frac{E - E^\circ}{0.0592\ V}\right)$$

$$\text{Therefore, } Q = 10^{-n\left(\frac{E - E^\circ}{0.0592\ V}\right)} = 10^{-2\left(\frac{0.458\ V - 0.599\ V}{0.0592\ V}\right)} = 5.8011 \times 10^4$$

Now, we solve for the $[Cu^{2+}]$:

$$[Cu^{2+}] = \frac{[Ni^{2+}]^1}{Q} = \frac{1.00}{5.8011 \times 10^4} = 1.7238 \times 10^{-5}\ M$$

SOLUTION:

a) The potential for the cell under these conditions is 0.510 V (3 sig figs from data). This number seems reasonable, because the ratio of concentrations of products to reactants has increased tremendously, which means that the reaction should have less potential to be spontaneous in the forward direction than it would if the relative concentrations were equal.

b) The concentration of $[Cu^{2+}]$ is 1.72×10^{-5} M. (3 sig figs from data). This number seems reasonable because the potential is smaller than in part "a", and the concentration of the reactant Cu^{2+} is lower than in part "a". This cell has less potential to be spontaneous in the forward direction than the cell described in part "a".

Try It #7:   We saw earlier that the standard cell potential for the reaction:
$$Zn\ (s) + 2\ Ag^+\ (aq) \leftrightarrow 2\ Ag\ (s) + Zn^{2+}\ (aq)$$
is 1.56 V. Calculate the cell potential for the reaction when the concentration of silver ion is 0.100 M and the concentration of zinc ion is 0.750 M.

19.6 REDOX IN ACTION

Corrosion

Key Terms:
- **Galvanization** – the process of coating a metal with another, more easily-oxidized metal.
- **Passivation** – the process of coating a metal with an impervious metal oxide.

Key Concepts:
- Iron oxide is porous and does not adhere well to metal, so once iron begins corroding, the oxidation process can continue until all of the iron has corroded.

- Stainless steel is an alloy of iron with small amounts of Ni and Cr. Both Ni and Cr form strong oxide coatings, which help prevent Fe from corroding.

Batteries

Type	Used In	Redox Components	E° cell	Pros	Cons
Lead Storage	Car batteries	Pb, Pb^{2+}, Pb^{4+}	2 V	Rechargeable Long life	Corrosive (acid) Toxic (Pb)
Alkaline Dry Cell	Flashlights	MnO_2, Zn	1.5 V	Inexpensive	Not rechargeable Voltage drops over time
Nickel-Cadmium	Flashlights	Ni and Cd	1.35 V	Rechargeable	Toxic
Zinc-air	Watches, calculators, hearing aids, cameras	Zn and O_2	1.4 V	High capacity Constant, non-variable voltage	Expensive Deteriorates with air exposure

EXERCISE 13: Iron is the most abundant metal at the earth's surface, making it an obvious choice for construction materials. An underground iron storage tank, such as one used to house gasoline at a gas station, needs to be protected from corrosion. This can be done by linking the Fe tank to a reference metal that is above the ground's surface. The decomposition of the reference metal can be monitored, and the Fe tank is preserved. Here is a schematic picture:

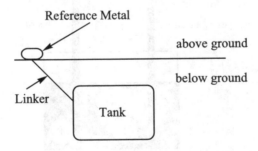

Given the options of using magnesium metal and copper metal, which metal should be used as the linker and which should be used as the reference? Use standard reduction potentials to determine your answer.

STRATEGY: According to Appendix F, the standard reduction potentials of the three metals are:
$$2\ e^- + Mg^{2+}\ (aq) \leftrightarrow Mg\ (s)\ \ E° = -2.37\ V$$
$$3\ e^- + Fe^{3+}\ (aq) \leftrightarrow Fe\ (s)\ \ E° = -0.037\ V$$
$$2\ e^- + Cu^{2+}\ (aq) \leftrightarrow Cu\ (s)\ \ E° = +0.3419\ V$$

Mg has the lowest reduction potential, so it will be the one most easily oxidized. Cu has the highest reduction potential, so it will be the least easily oxidized. As long as the three metals are connected to each other, *only the metal that is most easily oxidized will undergo oxidation.* The best metal to use as the reference at the ground's surface, then, would be Mg. The linker should be Cu because the linker should not corrode; it must maintain the connection between the tank and the reference metal. This method is actually used to prevent underground tanks from corroding. The Mg piece is monitored and when it gets small, it is replaced with a new piece.

19.7 ELECTROLYSIS

Key Terms:

- **Electrolysis** – the process of using electrical current to drive non-spontaneous redox reactions.
- **Electroplating** – the process of depositing one metal on top of another.
- **Competitive electrolysis** – situation in which the reaction vessel contains more than one possible substance that can undergo oxidation and/or reduction. As always, the most easily reduced substance will undergo reduction and the most easily oxidized substance will undergo oxidation.

Key Concepts:

- The amount of material electrolyzed depends upon the current flow and the amount of time the electrolysis is performed. ($n = It/F$)
- The moles of electrons that are exchanged in a redox reaction can be used in stoichiometric conversion factors for that reaction.
- If two half-reactions have similar potentials, the half-reaction that involves fewer electrons exchanged will most likely occur. (Kinetics wins out over thermodynamics in these cases.)

EXERCISE 14: Consider the cell shown in the picture. What is the minimum voltage that must be supplied in order to get a reaction to occur in this scenario? Write a balanced chemical equation for the reaction that will occur.

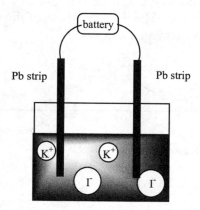

STRATEGY: First, determine which half-reactions could occur. The possibilities are based on what is initially present in the container: H_2O (l), Pb (s), I^-(aq), and K^+(aq). Pb is a metal element, so it may be oxidized. K^+ is in its oxidized form, so its only possible reaction is to be reduced to K. I^- can be oxidized to elemental I_2. Water can be oxidized or reduced. Use Appendix F to find the half-cell potentials.

Possible reduction half-rxns	E° (V)	Possible oxidation half-rxns	E° (V) (written as _reduction_ potentials)
K^+ (aq) + e^- $\leftrightarrow$ K (s)	-2.931	Pb (s) $\leftrightarrow$ Pb^{2+} (aq) + 2 e^-	-0.1262
2 H_2O (l) + 2 e^- $\leftrightarrow$ H_2 (g) + OH^-(aq)	-0.828	2 H_2O (l) $\leftrightarrow$ 4 e^- + 4 H^+ (aq) + O_2 (g)	+0.401
		2 I^- (aq) $\leftrightarrow$ I_2 (s) + 2 e^-	+0.5355

- Which reduction half-reaction will occur? The half-reaction with the greatest reduction potential will undergo reduction, so water will be reduced. This reaction will occur at the cathode. (Reduction occurs at the cathode.)
- Which oxidation half-reaction will occur? The half-reaction with the lowest reduction potential (i.e., the *greatest* oxidizing potential) will undergo oxidation, so Pb will be oxidized. This reaction will occur at the anode.
- Each half-reaction undergoes a two-electron change, so we don't need to multiply either reaction by a factor. The overall reaction will be:

$$Pb \text{ (s)} + 2 \text{ H}_2\text{O (l)} + 2 \text{ e}^- \leftrightarrow \text{H}_2 \text{ (g)} + \text{OH}^-\text{(aq)} + Pb^{2+} \text{ (aq)} + 2 \text{ e}^-$$

which simplifies to:

$$Pb \text{ (s)} + 2 \text{ H}_2\text{O (l)} \leftrightarrow \text{H}_2 \text{ (g)} + \text{OH}^-\text{(aq)} + Pb^{2+} \text{ (aq)}$$

$E°_{cell}$ for this process is:

$$E°_{cell} = E°_{cathode} - E°_{anode} = -0.828 \text{ V} - (-0.1262 \text{ V}) = -0.702 \text{ V}$$

SOLUTION: Just over 0.702 V will need to be supplied to this mixture in order for a reaction to occur. The reaction that will take place is: $Pb \text{ (s)} + 2 \text{ H}_2\text{O (l)} \leftrightarrow \text{H}_2 \text{ (g)} + \text{OH}^- \text{ (aq)} + Pb^{2+} \text{ (aq)}$.

EXERCISE 15: A car restorationist needs to electroplate a steel car bumper with chrome. To obtain an especially nice chrome finish, he performs a triple-plating process, in which copper is first plated, then nickel, then chromium. He intends to do this last step by reducing CrO_4^{2-}. Based on the size of the bumper, he knows he'll need to plate 86.2 g of chromium onto it in order to obtain the desired thickness. He uses a current of 4000. A for the plating process. After what amount of time should he remove the bumper from the CrO_4^{2-} (aq) solution vat?

STRATEGY: We're asked to determine the required time for the electroplating process. We've been given the current, the mass of chromium, and information on the chromium half-reaction taking place.

We should use the relationship: $n = It/F$ to solve this problem. We have the current, and F is a constant, so we just need to find n, the moles of electrons transferred in the process.

The conversion of chromate ion to metallic chromium involves a gain of six electrons:

$$CrO_4^{2-} \text{ (aq)} + 6 \text{ e}^- \leftrightarrow Cr \text{ (s)}$$

We can use this half-reaction information in a stoichiometric conversion factor:

$$86.2 \text{ g Cr} \times \frac{1 \text{ mol}}{51.996 \text{ g}} \times \frac{6 \text{ mol e}^-}{1 \text{ mol Cr}} = 9.94692 \text{ mol e}^-$$

$$n = \frac{It}{F}; \text{ so } t = \frac{nF}{I} = 9.94692 \text{ mol e}^- \times \frac{96485.34 \text{ C}}{\text{mol}} \times \frac{\text{sec}}{4000. \text{ C}} = 239.93 \text{ sec} = 3.999 \text{ min}$$

SOLUTION: The amount of time required for this process is 4.00 minutes. (3 sig figs based on mass data). The answer seems reasonable because the current is so huge.

Chapter 19 Self-Test

You may use a calculator, a periodic table, and the appendices from the text for this test.
Faraday's constant = 96,485.34 C/mol

1.   Balance this reaction, which occurs under acidic conditions:
$$MnO_4^- \text{ (aq)} + SO_2 \text{ (g)} \rightarrow SO_4^{2-} \text{ (aq)} + Mn^{2+} \text{ (aq)}$$

2.   The black tarnish on silver is composed of Ag_2S. This tarnish can be removed by placing the tarnished silver item on a piece of aluminum foil in a warm, aqueous salt solution.

   a) Is this an example of direct or indirect electron transfer?
   b) Write half-reactions for this redox process.

3.   a) Use standard reduction potentials to determine if this reaction is spontaneous under standard conditions.
$$4 H_3O^+ \text{ (aq)} + 10 NO_3^- \text{ (aq)} + 3 I_2 \text{ (aq)} \leftrightarrow 6 IO_3^- \text{ (aq)} + 10 NO \text{ (g)} + 6 H_2O \text{ (l)}$$

   b) What is the value of ΔG°_{rxn} for this reaction?

4.   Ag coated with AgCl can be used to detect chloride ions in solution because of this half-reaction that occurs:
$$AgCl \text{ (s)} + e^- \leftrightarrow Ag \text{ (s)} + Cl^- \text{ (aq)}$$

   An environmental chemist wanted to determine the concentration of chloride ions in a given water sample. She set up an electrochemical cell using these two half-reaction components: 1) a silver wire coated with AgCl immersed in the water sample, and 2) a solid copper electrode immersed in aqueous 1.00 M copper (II) sulfate. The resulting cell's potential measured 0.0437 V.

   a) Sketch the electrochemical cell.
   b) Indicate the direction of electron flow within this cell.
   c) Use standard reduction potentials to calculate the equilibrium constant for this reaction.
   d) What was the concentration of the Cl⁻ ion in the water sample?

5.   An electrolysis reaction was set up using 325 mL of 1.0 M KI (aq) and Pb electrodes. The overall reaction that occurred is shown below. What is the pH of the solution after a current of 9.45 mA has been applied to the reaction mixture for 12.0 min?

$$Pb \text{ (s)} + 2 H_2O \text{ (l)} \leftrightarrow H_2 \text{ (g)} + 2 OH^- \text{(aq)} + Pb^{2+} \text{ (aq)}$$

Answers to Try Its:

1. C is undergoing oxidation – its oxidation number goes from -1 in C_2H_2 to +4 in CO_2. O is undergoing reduction – its oxidation number goes from 0 in O_2 to -2 in the products.
2. $8\ Al\ (s)\ +\ 3\ NO_3^-\ (aq) +\ 5\ OH^-\ (aq)\ +\ 18\ H_2O\ (l) \rightarrow\ 8\ Al(OH)_4^-\ (aq)\ +\ 3\ NH_3\ (aq)$
3. Indirect. The two components undergoing a redox change aren't touching each other. The Ag^+ being reduced and the Zn being oxidized are in separate containers.
4. $E°_{cell}$ for this process is -1.61 V. A negative $E°_{cell}$ indicates the process is not spontaneous under standard conditions.
5. Zn is losing electrons, so electrons will flow from Zn over the wire to Ag. Each Ag^+ in solution undergoing reduction will pick up one electron to form Ag, which will deposit on the inert Pt cathode. This process will continue until one substance is extinguished, the charge balance is disrupted, or the connection between the containers is severed.
6. -301.30 kJ/mol; Exercise 6 obtained a value of -301.27 kJ/mol.
7. 1.50 V

Answers to Self-Test:

1. $6\ H_2O\ (l) + 2\ MnO_4^-\ (aq)\ +\ 5\ SO_2\ (g) \rightarrow\ 5\ SO_4^{2-}\ (aq)\ +\ 2\ Mn^{2+}\ (aq) + 4\ H_3O^+\ (aq)$
2. a) direct; the two substances undergoing the redox reaction are actually touching each other.
   b) $2e^- +\ Ag_2S\ (s) \rightarrow\ 2\ Ag\ (s) + S^{2-}\ (aq)$ and $Al\ (s) \rightarrow\ 3\ e^- + Al^{3+}\ (aq)$ (Silver sulfide is not water-soluble.)
3. a) No, it is not. $E°_{cell}$ for this process is -0.238 V. A negative value for $E°_{cell}$ indicates the reaction is not spontaneous under standard conditions. b) $\Delta G°_{rxn} = 689$ kJ/mol. (n = 30 electrons)
4. a) This sketch is shown with a porous plate, but you could use a salt bridge instead.

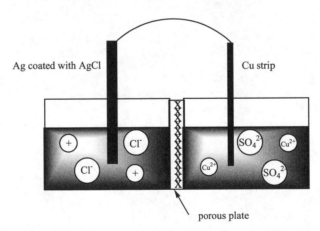

porous plate

   b) In this case, copper will be reduced and Ag will be oxidized, so the electrons leave the Ag, flow over the top metal wire, and onto the Cu. A Cu^{2+} ion in solution will pick up two electrons and deposit as Cu(s) on the copper cathode.
   c) 1.1×10^4; d) 0.0523 M
5. pH = 10.336

Chapter 20: The Transition Metals

Learning Objectives

In this chapter, you will learn how to:

- Predict properties of transition metal elements.
- Write names, formulas and structures for transition metal compounds.
- Use crystal field theory to predict properties (such as color and magnetic susceptibility) of transition metal compounds.
- Calculate the crystal field splitting energy for a transition metal complex, based on its color.
- Obtain a pure metal from its ore.

Practical Aspects

This chapter uses concepts from Chapter 7 (relation of light energy to wavelength and color), Chapter 8, (electron configurations and energy level diagrams), and Chapter 9 (shapes of molecules). In this chapter, you will learn why transition metal compounds are usually vividly colored. You will also learn about the many important societal uses of transition metals as well as the essential roles they play within biological systems.

20.1 OVERVIEW OF THE TRANSITION METALS

Key Concepts:
- Transition metals are good conductors of heat and electricity.
- Most transition metals can exist in several different oxidation states.
- Most transition metals are found in nature as oxides or sulfides, rather than in their elemental form.
- In transition metal solids, electrons fill in bonding orbital bands up through Group 6. After Group 6, electrons are added to delocalized antibonding orbital bands (a destabilizing factor).
- The physical properties of a transition element depend upon the element's electron configuration. For example, some periodic trends are:
  - **Melting point** – increases across a row as electrons are added to bonding orbital bands, but decreases sharply as soon as electrons are added to antibonding orbital bands.
  - **Density** – increases within a given column (as elements become more massive), and increases along a given row, while electrons are added to bonding orbital bands. When electrons start adding to antibonding orbital bands, density decreases.
  - **Ionization energy** – increases across a given row because Z_{eff} increases. (Recall that the definition of ionization energy is the energy required to remove a valence electron from an atom in its gaseous state, so delocalized orbitals are not an option here.) Ionization energy decreases down a column as electrons are held less tightly.
    * Remember that there are exceptions to every trend.

EXERCISE 1: Predict and explain. Which will have the greater: a) melting point, Mo or Tc? b) density, Zr or Hf? c) ionization energy, Sc or V?

STRATEGY: Use periodic trends to compare the elements.

SOLUTION:
a) Tc is to the immediate right of Mo on the periodic table, so it contains one more proton and electron than Mo. However, that extra electron is the first added to a destabilizing antibonding orbital band. The melting point of Mo is therefore greater than Tc, because it will require more thermal energy to disrupt Mo's structure than Tc's.
b) Hf is below Zr on the periodic table, making Hf more massive. Hf will therefore be more dense than Zr.
c) Sc and V are in the same row, but V has two more protons than Sc. The ionization energy depends upon the attraction of the electrons to the nucleus within a gaseous atom. V will therefore have a higher ionization energy than Sc.

EXERCISE 2: Ag is an excellent electrical conductor. Use the electron configuration of Ag to explain why it is such a good conductor.

STRATEGY: Electrical conductivity in a metal depends upon the ability of its electrons to delocalize.

SOLUTION: Ag is a d^9 transition element. A transition element with over 6 d electrons has electrons in delocalized antibonding orbitals. The electrons in Ag are highly delocalized, making Ag an excellent electrical conductor.

Try It #1: Use periodic trends to explain why Hg is a liquid at room temperature.

20.2 COORDINATION COMPLEXES

Key Terms:
- **Geometric isomers** – two complexes that have the same formula, but different 3-D orientations of the ligands. Two main categories:

  - **Cis vs. trans** – applies to complexes that contain two ligands of one type, ML_4X_2:
    "*cis*" indicates "same side"   "*trans*" indicates "across"

  - **Facial vs. meridianal** – applies to complexes that contain three ligands of one type, ML_3X_3:
    "*fac*" indicates facial and means   "*mer*" indicates meridianal and means that
    that the three ligands lie on one   the three ligands lie in one plane that goes
    face of the octahedron.   through the center of the complex.

- **Linkage isomers** – two complexes that have the same formula, but show different connectivities of the ligands to the metal center. This occurs only if a given ligand can attach to the metal using either of two donor atoms.

Key Concepts:
- Ligands are classified by the number of coordination sites they can bind to: monodentate (1), bidentate (2), tridentate (3), etc.
- In aqueous solution, water molecules will bind to any open coordination sites on a transition metal. For example, Co^{2+} (aq) is really $[Co(H_2O)_6]^{2+}$ (aq).
- The geometry of a coordination complex depends upon its coordination number.
- The most common coordination number is 6, which typically corresponds to an octahedral geometry.
- A coordination number of 4 will result in either a tetrahedral or square planar geometry.

Rules for Naming Transition Metal Complexes

Detailed rules for naming coordination compounds are provided in Section 20.2 of the text. Here is a summary:

All coordination compounds follow this pattern for the root of the name:
1. Name the cation first and the anion second.
2. Write the complex as one long word: name the ligands in alphabetical order, then the metal, then indicate the metal's charge with Roman numerals. Note that these neutral ligands have special names: H_2O = aqua, NH_3 = ammine (two "m"s), CO = carbonyl.

These suffixes and prefixes must be added accordingly:
- If the ligand is an anion, add the suffix "–o" to it.
- Use Greek prefixes to indicate how many ligands of a given type are present. If the ligand name contains a Greek prefix, then use these alternate prefixes to describe the number of ligands: "bis-" for two, "tris-"for three, "tetrakis-" for four. When using this alternate format, enclose the ligand name in parentheses.
- If the complex is an anion, add the suffix "–ate" to the metal name.
- Some complex anions take on the metal's Greek name. For example, Ag → argentate.

EXERCISE 3: Determine the oxidation state, coordination number, and number of d valence electrons for the metal in each: a) $[Co(NH_3)_3(NO_2)_3]$; b) $Na_2[MnCl_4]$; c) $[Ni(en)(H_2O)_4]SO_4$.

STRATEGY: Use guidelines for assigning oxidation states (numbers). The coordination number equals the steric number. The number of d valence electrons can be determined from the electron configuration.

SOLUTION:
a) $[Co(NH_3)_3(NO_2)_3]$: The overall charge is zero, each ammonia is neutral, each of the three nitrites is –1, so the oxidation state of cobalt is +3. There are six monodentate ligands present, so the coordination number is six. The Co^{3+} ion has the valence electron configuration d^6.

b) $Na_2[MnCl_4]$: The overall charge is zero, each sodium is +1, each chloride is –1, so the oxidation state of manganese is +2. There are four monodentate ligands present, so the coordination number is four. The Mn^{2+} ion has the valence electron configuration d^5.

c) $[Ni(en)(H_2O)_4]SO_4$: The overall charge is zero, each ligand is neutral, and sulfate is –2, so the oxidation state of nickel is +2. There is one bidentate ligand (en) and there are four monodentate ligands, so the coordination number is six. The Ni^{2+} ion has an electron configuration of d^8.

EXERCISE 4: Name each: a) $K_4[Fe(CN)_6]$; b) $[Ni(en)(H_2O)_4]SO_4$; c) $[Ni(en)_3]Cl_2$

STRATEGY: Apply the rules for naming coordination complexes.

SOLUTION:
a) $K_4[Fe(CN)_6]$
Cation and anion: The cation is potassium and the anion is the complex ion. The suffix "-ate" will need to be added.
Ligands: Six CN^- ligands = "hexacyano."
Metal charge/name: $4 K^+$ and $6 CN^- = -2$, so the charge on iron must be +2. (Fe^{2+}) = ferrate(II).
Name: Potassium hexacyanoferrate(II)

b) $[Ni(en)(H_2O)_4]SO_4$
Cation and anion: The cation is the complex ion and the anion is sulfate.
Ligands: One en and four waters. "aqua" is before "ethylenediamine" in the alphabet, so "tetraaquaethylenediamine" is the collective name for the ligands.
Metal charge/name: All of the ligands are neutral, so SO_4^{2-} is the only counter ion: nickel(II).
Name: Tetraaquaethylenediaminenickel(II) sulfate.

c) $[Ni(en)_3]Cl_2$
Cation and anion: The cation is the complex ion and the anion is chloride.
Ligands: Three ethylenediamine ligands. The Greek prefix "di-" is in the ligand name, so we must denote "three" with the alternate prefix, tris. The ligand name is tris(ethylenediamine).
Metal charge/name: en is neutral, and there are 2 Cl^- ions, so Ni^{2+} = nickel(II)
Name: Tris(ethylenediamine)nickel(II) chloride.

Try It #2: Name each compound: a) $[Co(NH_3)_3(NO_2)_3]$; b) $Na_2[MnCl_4]$

EXERCISE 5: Write the formula for each: a) Hexamminemolybdenum(III) chloride; b) Triaquatrifluorocobalt(III); c) potassium dicarbonyltetracyanochromate(II).

STRATEGY: Follow the rules for naming complex ions.

SOLUTION:
a) Hexamminemolybdenum(III) chloride: This contains 6 ammonia ligands attached to Mo^{3+}, and the counter ion is Cl^-. Ammonia is neutral, so 3 Cl^- ions are needed for the compound to be electrically neutral: $[Mo(NH_3)_6]Cl_3$.

b) Triaquatrifluorocobalt (III): Three waters and three fluorides are attached to a Co^{3+} ion. The three F^- ions counterbalance the Co^{3+}, making the complex neutral overall: $[Co(H_2O)_3F_3]$.

c) Potassium dicarbonyltetracyanochromate(II): two CO ligands and four CN^- ligands are attached to a Cr^{2+} metal ion. The charge on the complex is –2, so two K^+ ions are needed to counterbalance the charge: $K_2[Cr(CO)_2(CN)_4]$.

Try It#3: Write the formulas for: a) Triaquatrichlororuthenium(III); b) Tetraamminetitanium(II) chloride

EXERCISE 6: Name the two isomers of $[Pt(NH_3)_4(Cl)_2]Cl_2$, then draw structures for their complex ions.

STRATEGY: This complex contains two chloride ligands and four ammonia ligands (ML_4X_2 pattern), so the overall shape is octahedral. The two chloride ligands can be arranged next to each other (cis) or across from each other (trans). The complex has an overall charge of +2.

SOLUTION:

cis-tetraamminedichloroplatinum (IV) chloride   *trans*-tetraamminedichloroplatinum (IV) chloride

EXERCISE 7: Name and draw structures for the two isomers that fit this formula: $[Co(NH_3)_3(NO_2)_3]$.

STRATEGY: There are six ligands, so the geometry will be octahedral. When two sets of three ligands are attached to a metal center (ML_3X_3 pattern), all ligands of the same type can be clustered together on one face of the octahedron, or they can be arranged so that two of the same type are opposite each other. These are termed "facial" and "meridianal," respectively. The cobalt must have a +3 charge to counter the three –1 charges of the nitrites. Notice that within the complex "NO_2^-" is named "nitro."

SOLUTION:

fac-triamminetrinitrocobalt(III)   *mer*-triamminetrinitrocobalt(III)

Try It #4: Draw the structure for the dicarbonyltetracyanochromate(II) ion. If more than one isomer exists, draw both isomers.

20.3   BONDING IN COORDINATION COMPLEXES

Key Terms:
- **Crystal field splitting energy (Δ)** – difference in energy between non-degenerate d orbitals.
- **Pairing energy (P)** – destabilizing energy resulting from repulsive forces between two electrons that occupy the same orbital.
- **High-spin** – term used to describe electron arrangement in which the maximum unpaired spins are present. This occurs when $P > \Delta$, and it is easier for the electrons to occupy the higher energy d orbitals than to pair up in the lower energy orbitals.
- **Low-spin** – term used to describe electron arrangement in which the maximum paired spins are present. This occurs when $P < \Delta$.

- **Spectrochemical series** – list of ligands in order of increasing energy level splitting ability:

$$I^- < Br^- < Cl^- < F^- < OH^- < H_2O < NH_3 < en < NO_2^- < CN^- < CO$$

Key Concepts:
- All d orbitals are not degenerate. The set of d orbitals is split into two or more degenerate groups, based on slight differences in energy arising from electron-electron repulsions.
- The *magnitude* of the crystal field splitting energy depends upon these factors:
  - the charge on the metal ion – the greater the charge, the greater the splitting.
  - the size of the metal ion – the greater the *n*, the greater the splitting.
  - the coordination number – the higher the coordination number, the greater the splitting.
  - the identity of the ligand – the more tightly the ligand binds, the greater the splitting.
- The *pattern* of orbital splitting depends upon the geometry of the metal cation:

Geometry:	Octahedral	Tetrahedral	Square Planar
Crystal Field Energy Level Diagram:	(Energy) ——— ——— e_g ——— ——— ——— t_{2g}	(Energy) ——— ——— ——— t ——— ——— e	(Energy) ——— x^2-y^2 ——— xy ——— z^2 yz ——— ——— xz
Notes:	The t_{2g} set includes the d_{xy}, d_{xz}, and d_{yz} orbitals. The e_g set includes the d_{z^2} and $d_{x^2-y^2}$ orbitals.	The two sets are the reverse of octahedral. Splitting is smaller than in octahedral, so tetrahedral complexes are almost always high-spin.	The $d_{x^2-y^2}$ orbital is greatly destabilized relative to the other orbitals because of its orientation relative to the ligands in this geometry.

- The same rules apply for drawing energy level diagrams for transition metal cations as did for atoms - Hund's Rule, Aufbau Principle, Pauli Exclusion Principle.

Helpful Hints
- One way to remember the spectrochemical series is to notice that it is broken down into groups of what's attaching to the metal: "halogens < oxygen < nitrogen < carbon."
- The names "t_{2g}" and "e_g" are used to describe the two categories of d orbitals in an octahedral geometry. The names "e" and "t" are used to describe d orbitals in a tetrahedral geometry.
- An energy-level diagram for a transition metal ion is sometimes called a crystal field energy diagram.

EXERCISE 8: Draw a crystal field energy diagram for each complex ion. Then, write the electron configuration for each. a) $[Co(NH_3)_3(NO_2)_3]$; b) $Na_2[MnCl_4]$, tetrahedral; c) $K_4[CrF_6]$.

STRATEGY: Determine the splitting *pattern* from the orbital geometry. Then determine the relative magnitude of the Δ by assessing the factors that affect it: the size and charge of the metal ion, the coordination number, and the identity of the ligand.

SOLUTION:
a) $[Co(NH_3)_3(NO_2)_3]$: Co^{3+}, $3d^6$ arrangement, octahedral geometry. Relatively high oxidation state, fairly strong ligands. These factors indicate a high splitting energy. The complex will be low-spin.

b) $Na_2[MnCl_4]$, tetrahedral: Mn^{2+}, $3d^5$ arrangement. Tetrahedral complexes are almost always high spin, because their splitting energies are small. The weak Cl^- ligands shouldn't increase the splitting energy.

c) $K_4[CrF_6]$: Cr^{2+}, $3d^4$ arrangement, octahedral geometry. Lower oxidation state, weak ligands. These factors favor a low splitting energy. This complex will be high-spin.

Complex:	$[Co(NH_3)_3(NO_2)_3]$	$Na_2[MnCl_4]$	c) $K_4[CrF_6]$

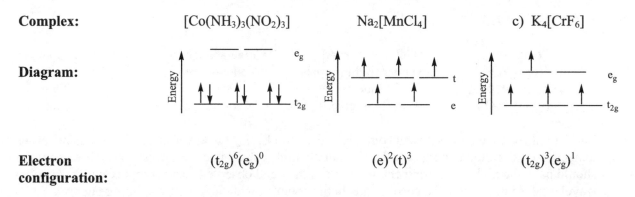

Electron configuration:	$(t_{2g})^6(e_g)^0$	$(e)^2(t)^3$	$(t_{2g})^3(e_g)^1$

Try It #5: Draw the crystal field splitting diagram and electron configuration for $[Pt(NH_3)_6]Cl_2$.

Magnetism and Color

Key Term:
- **Complementary colors** – colors related to each other by their absorption and transmission characteristics. Table 20-5 in the text lists complementary colors.

Key Concepts:
- Recall that paramagnetic substances can behave like magnets because they contain unpaired electron spins, while diamagnetic substances don't behave like magnets because all of their spins are paired.
- The crystal field splitting energy of a metal complex is related to its color.
- We can see reflected light (light that bounces off a substance) and transmitted light (light that passes through a substance). We see the complement of the color that the substance absorbs.

Useful Relationship:
- $\Delta E_{molecule} = \Delta = h\nu = hc/\lambda$ This relationship can be used to relate Δ to the wavelength of light that the complex absorbs.

EXERCISE 9: Which of the complexes listed in Exercise 8 will respond most strongly to a magnetic field? Which will not respond at all?

STRATEGY: A paramagnetic substance will respond to a magnetic field, and a diamagnetic substance will not. The substance with the highest overall spin will respond the most intensely.

SOLUTION: $[Co(NH_3)_3(NO_2)_3]$ has all spins paired, so it is diamagnetic. It will not respond to a magnetic field. Both $Na_2[MnCl_4]$ and $K_4[CrF_6]$ have unpaired spins, so they are both paramagnetic. $Na_2[MnCl_4]$, with five unpaired spins, will respond slightly more to a magnetic field than $K_4[CrF_6]$, with four unpaired spins.

EXERCISE 10: [Ni(en)₃]SO₄ is lavender, [Ni(en)(H₂O)₄]SO₄ is blue, and [Ni(H₂O)₆]SO₄ is teal. Explain the color differences.

STRATEGY: The color we can see is the complement of the color that the substance absorbs. Use Table 20-5 in the text to determine the color that each complex absorbed, then compare the relative energies of the absorbed colors.

SOLUTION: Table 20-5 shows that:

Complex	Color observed	Color absorbed	Wavelength absorbed
[Ni(en)₃]SO₄	Lavender (light purple)	Yellow-green	560 nm
[Ni(en)(H₂O)₄]SO₄	blue	Orange	610 nm
[Ni(H₂O)₆]SO₄	Teal (blue-green)	Red-orange	680 nm

The only difference between each complex is the ligands. H_2O is lower in the spectrochemical series than en, so the splitting energy for the all-water complex should be the lowest. The all-en complex should have the highest splitting energy. This is the trend observed. The complex with the shortest wavelength of light absorbed corresponds to the complex with the greatest splitting energy.

EXERCISE 11: The splitting energy of $[Rh(Cl)_6]^{3-}$ is 243 kJ/mol. a) Determine the wavelength of maximum absorption for this complex. b) What color is it?

STRATEGY: Use the relationship $E = hc/\lambda$ to determine the wavelength. Use Table 20-5 in the text to determine the color.

$$\frac{243 \times 10^3 \text{ J}}{\text{mol}} \times \frac{1 \text{ mol}}{6.022 \times 10^{23} \text{ photons}} = 4.035 \times 10^{-19} \text{ J/photon}$$

$$E = h\nu = \frac{hc}{\lambda}; \text{ so } \lambda = \frac{hc}{E} = \frac{6.626 \times 10^{-34} \text{ J sec}}{4.035 \times 10^{-19} \text{ J}} \times \frac{2.9979 \times 10^8 \text{ m}}{\text{sec}} \times \frac{10^9 \text{ nm}}{1 \text{ m}} = 492 \text{ nm}$$

SOLUTION: a) The wavelength of maximum absorption will be 492 nm. b) The color of this complex will be red-orange, which is the complement of the blue to blue-green light that is absorbed. The answers seem reasonable because Rh's d electrons are in $n=4$, and these large orbitals have larger splitting energies than smaller orbitals. Metals with $n=4$ and 5 d electrons tend to be low-spin because of their high splitting energies.

Try It #6: Approximate the crystal field splitting energy for a substance that appears green.

20.4 METALLURGY

Key Terms:
- **Metallurgy** – the process of purifying a metal from its ore. Metallurgy is a four-step process:
  1. **Separation** of the desired ore from other metal ores.
  2. **Conversion** of the pure ore to a form that is easily reduced.
  3. **Reduction** of the metal compound to the pure metal with a reducing agent or electrolysis.
  4. **Refining** the purity of the metal.
- **Roasting** – heating in the presence of air (oxygen), typically to convert sulfides to oxides.

- **Coke** – a form of carbon (charcoal) that has been heated to remove its impurities. Coke is used as the reducing agent for many reduction processes.

Key Concepts:
- Electrolysis requires enormous amounts of electrical energy, so it is only used in the reduction step if the metal is too reactive to use a reducing agent.

Helpful Hints
- Figure 20-20 in the text shows a general schematic diagram of the metallurgical process.
- Table 20-6 in the text summarizes the specific chemical processes used to separate a transition metal from its ore.
- Section 20.4 of the text provides detailed descriptions of separation methods for several metals.

Fast Facts – Iron:
- **Main natural source**: hematite, Fe_2O_3, and magnetite, Fe_3O_4.
- **Isolation process**: See Figure 20-22 in the text for an illustration of the blast furnace.
- **Main uses**: steel (700 million tons per year, worldwide).
- **Other:**
  - Iron is the most-used metal.

EXERCISE 12: Why are most transition metals found in an oxide or sulfide form, rather than in their elemental form? To answer this question, compare the standard reduction potentials of several transition metals to the positive standard reduction potentials for oxygen.

STRATEGY: The greater a substance's reduction potential, the more it wants to be in its reduced form.

SOLUTION:

Half-reaction	$E°$ (V)		Half-reaction	$E°$ (V)
$Cr^{3+} + 3e- \leftrightarrow Cr$	-0.744		$Co^{2+} + 2e- \leftrightarrow Co$	-0.28
$Cr^{2+} + 2e- \leftrightarrow Cr$	-0.913		$Mn^{2+} + 2e- \leftrightarrow Mn$	-1.185
$Fe^{2+} + 2e- \leftrightarrow Fe$	-0.447		$Ni^{2+} + 2e- \leftrightarrow Ni$	-0.257

All of these metals have negative standard reduction potentials. Reduction half-reactions for oxygen all show positive standard reduction potentials, which means that oxygen in contact with these metals will be the substance reduced, and the metal will be oxidized. In our oxygen-containing environment, then, these metals will exist in their oxidized form.

20.5 APPLICATIONS OF TRANSITION METALS

Fast Facts – Titanium:
- **Main natural source**: rutile (TiO_2) or ilmenite ($FeTiO_3$).
- **Properties**: strong, low density, stable at high temperatures.
- **Main uses**: metal - aircraft frames, jet engines, supplies for chemical industry (pipes, pumps, etc).
- **Other:**
  - 9th most abundant element in earth's crust.

- Alloys of titanium with tin or aluminum have the highest strength to weight ratio of all the engineering metals.
- TiO_2 – white pigment used for almost every white-colored commercial product on the market.

Fast Facts – Chromium:
- **Main natural source:** chromite ($FeCr_2O_4$).
- **Main uses:** Metal alloys (Nichrome for heat radiating wires; stainless steel), chrome plating.
- **Other:**
  - "Chromium" comes from the Greek "chroma" meaning color. Chromium compounds are often vibrantly colored.
  - Chromium(VI) is highly toxic.

Fast Facts – Copper:
- **Main natural source:** chalcopyrite ($FeCuS_2$) and other ores.
- **Main uses:** electrical wiring (50% of all Cu produced), pipes, alloys (bronze, brass).
- **Other:** The concentration of copper in the ore is often less than 1%, which makes copper refining quite expensive.

Fast Facts – Silver:
- **Main natural source:** usually mixed in with copper or zinc ores.
- **Main uses:** photography (#1 use), silverware, jewelry, mirrors, batteries.

Fast Facts – Zinc:
- **Main natural source:** sphalerite (ZnS).
- **Main uses:** prevention of steel corrosion, batteries, and alloys (brass and bronze).
- **Other:** Zinc oxide is used as a catalyst in vulcanizing rubber and as a white pigment.

Fast Facts – Mercury:
- **Main natural source:** cinnabar (HgS)
- **Main uses:** fluorescent lighting, thermometers, electrical switches, electrodes.
- **Other:** highly toxic; one of the two elements that is a liquid at room temperature.

Fast Facts – The Platinum Metals (Ru, Os, Rh, Ir, Pd, and Pt):
- **Main natural source:** contaminants in copper and nickel ores.
- **Main uses:** Catalysts – for example, automobile catalytic converters (Pt and some Rh and Pd) and catalytic hydrogenation of vegetable oils (Pd).

EXERCISE 13: TiO_2 is called both "titanium(IV) oxide" and "titanium dioxide." Why does it have two different names?

STRATEGY: Decipher the conventions for the names.

SOLUTION: "Titanium(IV) oxide" is the proper ionic compound name for TiO_2. "Titanium dioxide" is the proper covalent compound name for TiO_2. The fact that there are two acceptable names must be a result of TiO_2's ionic *and* covalent character.

EXERCISE 14: Why are catalytic converters so expensive?

STRATEGY AND SOLUTION: The platinum metals are used in making catalytic converters. They are found in nature in small quantities as contaminants in copper (mostly) and nickel ores. Copper is very expensive to refine, and the low abundance and large amount of work that goes into refining platinum makes it an even more expensive material.

20.6 TRANSITION METALS IN BIOLOGY

Key Terms:
- **Metalloprotein** – protein molecule that contains a metal center. The three main roles of metalloproteins are:
  - **Transport and storage** – For example, hemoglobin's function is oxygen transport; myoglobin's function is oxygen storage.
  - **Enzymes** – For example, carboxypeptidase catalyzes the breakdown of proteins by removing amino acids one at a time from a protein chain.
  - **Redox reagents** – For example, the cytochromes, copper blue proteins, and iron-sulfur proteins are all electron transport molecules that involve a change in oxidation state at their metal centers.

Key Concepts:
- A metalloprotein's function depends upon its ability to bind and release ligands.
- The physical and chemical properties of the metalloproteins depend upon the environment of the transition metal center.

Helpful Hint
- Section 20.6 of the text describes structures of several metalloproteins.

EXERCISE 15: Hemoglobin in its deoxygenated form contains five nitrogen ligands (4 from the porphyrin ring around it and one from a histidine). This form of hemoglobin is blue. When oxygen binds to the sixth binding site in hemoglobin, the structure of the compound changes, and it turns red. In terms of crystal field theory, what is causing this shift to red?

STRATEGY: Color depends upon the environment of the transition metal.

SOLUTION: The color we see is the complement of the color absorbed. Red light is lower in energy than blue light. If we *see* red light, then a higher energy wavelength was *absorbed* than if we see blue. When the color changes from blue to red, the structure of hemoglobin must change so that the crystal field splitting energy increases. For an electronic transition to occur, then, the molecule must absorb higher energy light, resulting in red light being reflected or transmitted to us.

Chapter 20 Self-Test

You may use a periodic table and a calculator for this test.

1. The anticancer drug, cisplatin, has the formula: $[Pt(NH_3)_2Cl_2]$. The molecule has a square planar geometry. Draw the structure of cisplatin and name it according to IUPAC rules.

2. a) What is the highest possible spin a metal ion can have? Draw a crystal field splitting diagram to illustrate this.
    b) Propose a complex ion that would have this spin.

3. Write the formula, draw the structure, and construct a crystal field energy diagram for:
    fac-triamminetriiodoplatinum(IV) bromide.

4. Why is it that zinc oxide (ZnO) and titanium(IV) oxide are white?

5. A typical blue copper protein contains a copper in a distorted tetrahedral environment. How does this geometry affect its color?

6. What commonly-used items would potentially increase in price if a copper mine were shut down?

7. Predict the properties:

    a) Rank these in order of increasing melting point: Ta, W, Pt
    b) Rank these in order of increasing density: Cr, Mo, W

Answers to Try Its:

1. Hg(l) consists of large atoms, each having ten electrons in its outermost d orbital. Many of these electrons are in delocalized antibonding orbitals, which destabilizes the bonding structure of Hg enough to make it a liquid at room temperature.

2. a) triamminetrinitrocobalt(III); b) sodium tetrachloromanganate(II)

3. a) $[Ru(H_2O)_3Cl_3]$; b) $[Ti(NH_3)_4]Cl_2$.

4.

5. $[Pt(NH_3)_6]Cl_2$, $(t_{2g})^6(e_g)^2$ (don't need to assign high/low spin because it's d^8)

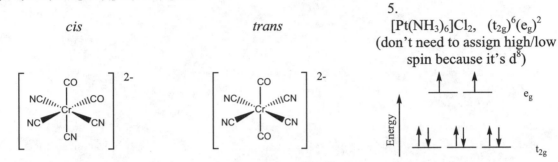

cis   *trans*

6. From Table 20-5 in the text, green will be observed if the crystal field splitting energy is roughly 166 kJ/mol. (Could have calculated it too from $E = hc/\lambda$.)

Answers to Self-Test:

1. *cis*-diamminedichloroplatinum(II)

2. a) If all five d orbitals contain one electron, the overall spin will be 2.5. This is the highest possible spin.

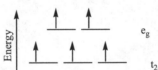

b) A complex that would have this electron arrangement would contain a d^5 metal cation in an environment where $P > \Delta$. Factors that would favor this would include: a small metal (3d valence electrons) with a low oxidation state, and relatively weak ligands, such as halides. A good choice would be: $[MnCl_6]^{4-}$.

3.
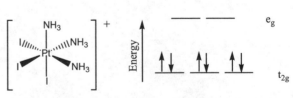

$[Pt(NH_3)_3I_3]Br$
Pt^{4+} has six 5d electrons in an octahedral environment. A large metal cation with a high charge in an octahedral environment favors low-spin.

4. Zn^{2+} is d^{10} and Ti^{4+} is d^0. Neither of these have electrons that can undergo a transition between non-degenerate d orbitals, so neither can absorb light. All light is reflected, resulting in white.

5. Tetrahedral geometry is almost always high-spin because of the small crystal field splitting energy. If we see blue, then the compound absorbs orange light (complementary color), which is relatively low in energy and indicates a small crystal field splitting energy.

6. Cu: electrical wiring and pipes. The Pt metals: catalytic converters, processed vegetable oils. Ag: photographs, mirrors, batteries, silverware, jewelry.

7. a) Pt < Ta < W: All are in same row, Pt has occupied anti-bonding orbital bands (lowest); W, from Group 6, has the maximum occupied bonding orbitals (highest).
   b) Cr < Mo < W. All are in the same column. Elements increase in density going down a column.

Chapter 21: The Main Group Elements

Learning Objectives

In this chapter, you will learn how to:

- Identify Lewis acids and bases.
- Classify a given Lewis acid or base as hard or soft.
- Apply the hard/soft acid-base concept to predict reaction products.
- Assess chemical and physical properties of the main group elements by applying what you've learned in Chapters 1-20 of the text.

Practical Aspects

The theme of this chapter is structure determines function. Many everyday uses for chemicals are explained based on their structures. This chapter introduces only two new themes: the concept of Lewis acids and bases and the concept of hard and soft acids and bases. This chapter then uses concepts from all previous chapters that emphasized structure, to explain the chemistry of the main group elements. Some examples are: drawing structures of covalent molecules, VSEPR theory, 3-D lattice arrays, band theory of solids, polymers, and redox and acid/base reactions.

21.1 LEWIS ACIDS AND BASES

Key Terms:
- **Lewis acid** – an electron pair acceptor.
- **Lewis base** – an electron pair donor.
- **Adduct** – product of a Lewis acid/base reaction: a Lewis base donates an electron pair to a Lewis acid, resulting in a new covalent bond formed between the two substances. In general, the formation of a Lewis acid-base adduct can be shown like this:

$$A + :B \rightarrow A–B$$

Key Concepts:
- A Lewis base must contain a pair of non-bonding electrons that are available for bond formation.
- A Lewis acid must be able to accept an electron pair, which means it either has vacant valence orbitals or is able to accommodate greater than 8 valence electrons. Most Lewis acids can be placed in one of these categories:
  - A molecule with vacant valence orbitals (For example, BH_3);
  - A molecule with delocalized π bonds involving oxygen (For example, SO_3);
  - A metal cation (For example, Fe^{3+}).
- The atoms involved in forming an acid-base adduct may need to re-hybridize in order for the adduct to be able to form.

EXERCISE 1: Identify each substance as a Lewis acid, a Lewis base, or neither. a) Co^{2+}; b) CO; c) CH_4; d) ethylenediamine; e) $BeCl_2$; f) $SnCl_4$.

STRATEGY: If the substance has an available (i.e., donatable) electron pair, then it is a Lewis base. If it can accommodate another electron pair, then it is a Lewis acid.

SOLUTION:
a) Co^{2+}: This is a metal cation, so it has space to accept an extra electron pair. It is a Lewis acid.
b) CO: Carbon monoxide has a triple bond between the C and O and a lone pair on each atom. It can act as a Lewis base.
c) CH_4: Carbon is an $n=2$ element, so it cannot accommodate more than the 8 valence electrons it already has in this structure. It cannot act as a Lewis acid. There is no available lone pair, so it cannot act as a Lewis base. CH_4 is neither.
d) ethylenediamine: recall that the structure of ethylendiamine contains two carbons singly bonded to each other, with an amine group attached to each carbon. Each nitrogen contains a lone pair, so ethylenediamine can act as a Lewis base.
e) $BeCl_2$: Be has 2 valence electrons, each of which has formed a single bond with a Cl. Be is an $n=2$ element; it has space to accommodate 4 more electrons. This is a Lewis acid.
f) $SnCl_4$: Sn in $SnCl_4$ has used up all of its valence s and p orbitals, but since it is an $n=5$ element, it has available d orbitals, which can accept an electron pair, so it is a Lewis acid.

Try It #1: Which lone pair on carbon monoxide is a better Lewis base?

EXERCISE 2: Identify the Lewis acid and base in each reaction, indicate the formation of the adduct with curved arrows, and draw the 3-D structures of the products.

a) $6\ H_2O\ (l)\ +\ Ni^{2+}\ (aq)\ \rightarrow\ [Ni(H_2O)_6]^{2+}\ (aq);$ b) $BF_3\ +\ F^-\ \rightarrow\ BF_4^-;$
c) $SO_2\ (g)\ +\ CaO\ (s)\ \rightarrow\ CaSO_3\ (s)$

STRATEGY AND SOLUTION: The Lewis acid will accept the lone pair that the Lewis base donates.

a) Ni^{2+} is the Lewis acid; H_2O is the Lewis base.

Bond formation

Final Adduct

b) BF_3 is the Lewis acid; F^- is the Lewis base.

Bond formation

Final Adduct

c) SO_2 is the Lewis acid; O^{2-} in CaO is the Lewis base.

Bond formation

Final Adduct

21.2   HARD AND SOFT LEWIS ACIDS AND BASES

Key Terms:
- **Polarizability** – the ease with which an atom's electron cloud can be distorted by an electrical field (introduced in Chapter 11).
- **Hard** – term used to describe a substance whose electron cloud is not very polarizable.
- **Soft** – term used to describe a substance with a highly polarizable electron cloud.
- **Hard-Soft Acid-Base (HSAB) Principle** – hard Lewis acids tend to combine with hard Lewis bases; and soft Lewis acids tend to combine with soft Lewis bases.
- **Metathesis reaction** – reaction in which bonding partners are exchanged.

Key Concepts:
- Polarizability depends upon how strongly the valence electrons are attracted to the nucleus. Recall that an electron's attraction to the nucleus depends upon distance (n value) and Z_{eff} (amount of screening).
- The terms "hard" and "soft" are relative to each other. There is no absolute scale for assessing hard and soft characteristics.

EXERCISE 3: Rank each group from hardest to softest. a) Pd^{2+}, Pt^{2+}, Ni^{2+}; b) Rb, In, Te; c) Co, Co^{2+}, Co^{3+}; d) Xe within: elemental Xe, XeF_2, or XeF_4.

STRATEGY: The more polarizable the electron cloud, the softer the substance.
SOLUTION:
a) Pd^{2+}, Pt^{2+}, Ni^{2+}: All of these elements are in the same column of the periodic table. All ions have the same charge, so we just need to compare the size of the electron clouds. Ni^{2+}, with the smallest electron cloud, will be the hardest. The softest will be Pt^{2+}, with the largest electron cloud.

b) Rb, In, Te: all of these elements have a valence shell of $n=5$, but in going from left to right across the periodic table, the Z_{eff} increases. Te feels the pull of 15 more protons than Rb does, so Te will be the least polarizable, and the hardest. In will be the next hardest. Rb, with the lowest Z_{eff} of the group, will have the most polarizable electron cloud and will be the softest.

c) Co, Co^{2+}, Co^{3+}: When an atom loses electrons, the remaining electrons can feel the attraction to the nucleus a little more strongly. Co^{3+}, with the smallest radius, will be the hardest. Co^{2+}, with the intermediate radius will be the next hardest. Co, with the largest radius, will be the softest.

d) Xe within: elemental Xe, XeF_2, or XeF_4? F is more electronegative than Xe, so it will pull electron density from Xe. The hardest will therefore be XeF_4, then XeF_2, and the softest will be Xe.

Try It #2: Which is harder, a) Sn^{2+} or Sn^{4+}?

EXERCISE 4: Predict whether or not a metathesis reaction will occur. If a reaction does occur, predict the products.
a) $TiI_4 + 2\,HgCl_2 \rightarrow$?   b) $CaS + H_2O \rightarrow$?   c) $CsF + LiI \rightarrow$?

STRATEGY: Hard acids prefer hard bases and soft acids prefer soft bases.

SOLUTION:

a) Ti^{4+} is harder than Hg^{2+}. Cl^- is harder than I^-. A metathesis reaction will therefore occur, to recombine the ions. The products will be: $TiCl_4 + 2\ HgI_2$.

b) H^+ is harder than Ca^{2+}. O^{2-} is harder than S^{2-}. A reaction will not occur because the hard acid is already combined with the hard base, and the soft acid is combined with the soft base.

c) Li^+ is harder than Cs^+. F^- is harder than I^-. A reaction will occur to form: $CsI + LiF$.

Try It #3: Will BH_3 prefer to react with NH_3 or PH_3?

21.3   THE MAIN GROUP METALS

Fast Facts – Aluminum:
- **Main natural source**: bauxite, $Al(O)OH$, contaminated with SiO_2, Fe_2O_3, clay and other hydroxides.
- **Isolation process:** Convert bauxite to Al_2O_3, then use the Hall-Heroult process (an electrolysis reaction).
- **Properties**: lightweight, strong, and forms a single-layer oxide coating for protection.
- **Uses**: alloys are used for everything from aircraft bodies to beverage cans, $AlCl_3$ is an important industrial catalyst, $Al_2(SO_4)_3$ is used in water purification.
- **Other:**
  - Third most abundant element in earth's crust, and the most abundant metal.
  - Stable as Al^{3+}, so it's difficult to form $Al\ (s)$.
  - Hard Lewis acid.

Fast Facts – Lead:
- **Main natural source**: galena, PbS
- **Isolation process:** heat with oxygen to form PbO, then reduce with charcoal (C).
- **Uses**: lead storage batteries (for cars); alloys such as pewter.
- **Other:**
  - Stable in +2 and +4 oxidation states.
  - Highly toxic, so its present-day use is limited. For many years, it was used in paints, solder, and as an anti-knock additive in cars.

Fast Facts – Tin:
- **Main natural source**: cassiterite, SnO_2
- **Isolation process:** add charcoal (C, a good reducing agent) at high temperature.
- **Properties**: relatively low melting point, resists corrosion.
- **Uses**: alloys such as pewter, bronze, and solder; coating for aluminum cans.
- **Other:**
  - Stable in +2 and +4 oxidation states.

EXERCISE 5: Explain why, in nature, Sn and Al exist as oxides while Pb exists as a sulfide.

STRATEGY: Compare cations and anions to see if there is a trend.

SOLUTION: Oxygen and sulfur are both in the same group; both oxide and sulfide have the same charge (-2). The difference between the two is that S is larger than O, making S^{2-} a softer base than O^{2-}. Similarly, Pb^{2+} is larger and softer than Sn^{4+} or Al^{3+}. Cations with high charges like +3 or +4 are typically considered hard. It seems reasonable, then, that Pb^{2+} will be more attracted to S^{2-}, while Sn^{4+} and Al^{3+} will be more attracted to O^{2-}.

21.4 THE METALLOIDS

Fast Facts – Silicon:
- **Main natural source**: silicon dioxide, SiO_2, and related silicate anions.
- **Isolation process:** to obtain up to 98% purity, heat with coke (charcoal) at extremely high temperature. A multi-step process is required for high-purity silicon for semiconductors.
- **Properties**: semiconductor.
- **Uses**: computer chips, glass, and silicone polymers.
- **Other:**
  - Second-most abundant element in earth's crust.

Fast Facts – Other Metalloids:
- **B** – unique properties due to its small size: high ionization energy, has no available d orbitals.
- **Sb** – alloyed with lead in a lead storage battery to minimize chance of water undergoing electrolysis during battery recharging.
- **Ge** – semiconductor.
- **As** – pesticide.
- Binary compounds of some metalloids, such as GaAs or InSb, have the same number of valence electrons as Si or Ge, and exhibit semiconductor properties.

EXERCISE 6: Compare the reduction potentials of Sb and water to determine how the presence of Sb in a lead storage battery can minimize the chance of water electrolysis during battery recharging. Given:

$$Sb + 3\ H^+ (aq) + 3\ e\text{-} \leftrightarrow SbH_3,\ E° = -0.150\ V.$$

STRATEGY: Look up the standard reduction potential for the reduction of water in Appendix F of the text and compare its value to that of the antimony half-reaction's reduction potential. The substance with the greatest reduction potential will get reduced.

SOLUTION:

Half-Reaction	Standard Reduction Potential $E°$ (V)
$Sb + 3\ H^+ (aq) + 3\ e\text{-} \leftrightarrow SbH_3$	-0.510
$2\ H_2O\ (l) + 2\ e\text{-} \leftrightarrow H_2\ (g) + OH^-(aq)$	-0.828

The reduction of Sb (in an acidic environment) and the reduction of water are both non-spontaneous processes. When the battery is recharged, electrical potential is put into the battery to drive the non-spontaneous processes in the forward direction. The Sb half-reaction has a higher reduction potential than the water half-reaction, so Sb will preferentially undergo reduction.

21.5 PHOSPHORUS

Fast Facts – Phosphorus:
- **Main natural source**: Apatite, $Ca_5(PO_4)_3X$, where $X = F$, OH, or Cl.
- **Isolation process**: heated with silica and coke (C, a good reducing agent) at high temperature, in the absence of oxygen.
- **Properties**: Phosphorus exists in three forms:
  - White phosphorus (P_4) – individual small molecules; highly reactive and highly toxic.
  - Red phosphorus – polymer made from heating white phosphorus. Less reactive and less toxic than white phosphorus.
  - Black phosphorus – contains long, crosslinked chains of P atoms. More stable than red phosphorus.
- **Uses**: phosphoric acid (one of the top 10 chemicals produced in the U.S.), phosphate fertilizers, polyphosphate detergents, herbicides, insecticides, and biological molecules like ATP.
- **Other:**
  - Almost 90% of phosphoric acid produced goes towards making phosphate fertilizers.
  - The most important phosphate fertilizer is $(NH_4)_2HPO_4$.
  - Elevated concentrations of phosphates in the environment (primarily from detergents), cause excess algae growth in lakes, which deplete the lakes of oxygen, killing off aquatic life.

EXERCISE 7: The three forms of elemental phosphorus all have phosphorus atoms bonded to each other with sp^3 hybridization. All bond angles in white phosphorus are 60°. Some of the bond angles in red phosphorus are close to 109.5°. Which form of phosphorus has a) weaker bonds?; b) longer bonds?; c) is more reactive?

STRATEGY: The ideal bond angle for sp^3 hybridized orbitals is 109.5°. Bond strength depends upon how well the two bonding orbitals can overlap with each other. The better the overlap, the shorter the bond, the stronger the bond, the more stable the substance.

SOLUTION: White phosphorus, with greater deviations from the ideal 109.5° bond angle for sp^3 hybridized orbitals, will form weaker, longer bonds than red phosphorus. This will make white phosphorus more reactive than red phosphorus.

EXERCISE 8: Water that contains relatively high concentrations of calcium and magnesium is classified as hard water. Polyphosphates such as sodium pyrophosphate, $Na_4P_2O_7$, can complex with the calcium or magnesium ion, making them useful in detergents. The pyrophosphate anion, $P_2O_7^{4-}$, is shown here:

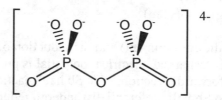

a) What is the hybridization on each phosphorus?
b) Describe the orbitals used to make the double bond between each P and O.
c) Can this ion act as a bidentate ligand to a metal cation?

STRATEGY AND SOLUTION:

a) An atom's hybridization depends upon its steric number within the molecule/ion. Each P has a steric number of four, so it will be sp^3 hybridized.

b) Each P=O is constructed from the same orbitals, so we only need to describe one of the double bonds. Bonds are always made with valence electrons: O has its valence electrons in $n=2$ (s and p orbitals are therefore available) and P has its valence electrons in $n=3$ (s, p and d orbitals are available). The first bond an atom forms is always a σ bond, the head-on overlap of two atomic orbitals. P will use sp^3 hybrid orbitals in making the σ bond, as mentioned in part "a." O will use its 2s orbital for the σ bond (recall only the central atoms hybridize). The second bond in the double bond is a π bond, made by the sideways overlap of two orbitals. P has already used its p orbitals, so it must use a d orbital for the π bond. O will use a p orbital. The two bonds can therefore be described as 2s-3sp^3 σ and 2p-3d π.

c) A bidentate ligand is one which binds to a metal center in two locations. This can only occur if the molecule can wrap around in such a way that lone pairs from two of its atoms can attach to the metal center. Typically, this will work if a five- or six-membered ring is formed. This molecule can do that, as indicated by numbering the atoms:

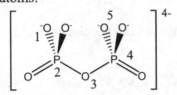

The metal center will act as atom #6 in a six-membered ring.

21.6   OTHER NON-METALS

Fast Facts – Sulfur:

- **Main natural sources:**
    - sulfide minerals, such as pyrite (FeS_2), molybdenite (MoS_2), chalcosite (Cu_2S), cinnabar (HgS), and galena (PbS);
    - H_2S in natural gas;
    - sulfur-compounds in crude oil and coal;
    - elemental form, S_8, near hot springs and volcanoes.
- **Isolation process:** melted S_8 can be extracted directly from earth; H_2S in oil refineries can be oxidized to S_8 by treatment with SO_2.
- **Properties:** yellow, crystalline solid, containing 8 atoms in a ring.
- **Uses:** Sulfuric acid.
    - **Uses of sulfuric acid:** fertilizer production, HF manufacturing, isolation of Ti from its ore, production of $Al_2(SO_4)_3$ for water treatment, dye industry.
- **Other:**
    - Approximately 90% of sulfur goes towards making sulfuric acid. Sulfuric acid is the #1 produced chemical in the U.S. It is used in every major chemical-related industry.

Fast Facts – Chlorine:

- **Main natural source:** NaCl (sea water).
- **Isolation process:** electrolytic oxidation by the chlor-alkali process.
- **Properties:** good oxidizing agent.
- **Uses:** water treatment, ethylene dichloride and vinyl chloride production (for manufacturing vinyl chloride and PVC), starting materials for organic syntheses, preparation of inorganic halides.

Fast Facts – Other Halogens:
- **Main natural source**: F: fluorite (CaF_2); Br, I: NaBr, NaI.
- **Isolation process:** F: treatment with sulfuric acid produces HF;
  Br, I: treatment of brines with Cl_2 (g).
- **Properties**: Br and F are highly toxic.
- **Uses**:
  - Fluorine: fluorinated hydrocarbons; toothpaste; production of UF_6, which is used to purify uranium for nuclear reactors.
  - Iodine – iodized salt (the hormone thyroxine contains I).
  - Bromine – AgBr for photography, reactant for many organic syntheses, insecticide.

EXERCISE 9: One of the reasons sulfuric acid is so corrosive is that it is a strong dehydrating agent, which means it can pull water out of molecules (like sugars, paper, and skin), leaving behind elemental carbon. Determine how many molecules of water could be removed from each sugar when sulfuric acid dehydrates it. a) Blood sugar (aka glucose) - $C_6H_{12}O_6$; and b) table sugar (sucrose) – $C_{12}H_{22}O_{11}$.

STRATEGY: Determine how many water molecules can be made from the hydrogen and oxygen atoms in the sugars.

SOLUTION: a) $C_6H_{12}O_6 = C_6(H_2O)_6$, so six waters could be removed from each glucose molecule. $C_{12}H_{22}O_{11} = C_{12}(H_2O)_{11}$, so eleven water molecules could be removed from each sucrose molecule.

Side note: The name of the chemical class for sugars, "carbohydrate," originated from "hydrated carbon."

EXERCISE 10: Sodium hypochlorite is a good bleach because it can oxidize large, colored compounds and convert them to smaller, colorless compounds that are soluble in detergents. Here is the structure of FD&C Blue #1, a colored compound that can be broken down by hypochlorite. What must the hypochlorite be able to do to this structure to convert it to a colorless substance?

STRATEGY AND SOLUTION: Carbon-based compounds may be colored if they contain a delocalized pi system. This structure does, as evidenced by its alternating single and double bonds, which all lie in one plane. The delocalized pi system would need to be destroyed by the bleach in order for the chemical to lose its color.

Chapter 21 Self-Test

You may use a periodic table for this test.

1. Identify the Lewis acid and base in this reaction: Ag^+ (aq) + Br^- (aq) → $AgBr$ (s)

2. Do you expect this reaction to occur? Why or why not?
$$AlBr_3 + 3 NaF \rightarrow AlF_3 + 3 NaBr$$

3. Heavy-metals act as poisons by binding to sulfur-containing amino acids within an enzyme, inactivating the enzyme. Which metal do you expect to better bind with S, Sn^{4+} or Pb^{2+}?

4. Will this reaction have a large or small equilibrium constant? Briefly explain.
$$CH_3HgOH + HSO_3^- \rightarrow CH_3HgSO_3^- + HOH$$

5. Explain the trend in pK_{sp}s for the silver halides:

Silver Halide	pK_{sp}
AgCl	9.74
AgBr	12.27
AgI	16.08

6. At room temperature, BH_3 exists as a dimer, diborane. The structure is shown here:

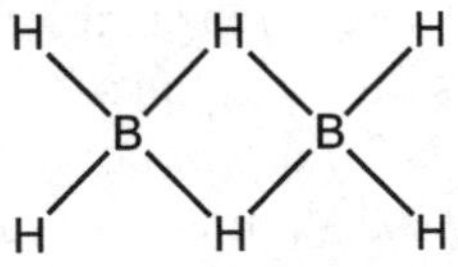

Re-draw the structure to better illustrate the bonding on the central hydrogen atoms.

7. One of the major uses for elemental chlorine is in the production of the polymer PVC. These reactions show the production of the monomer, vinyl chloride:

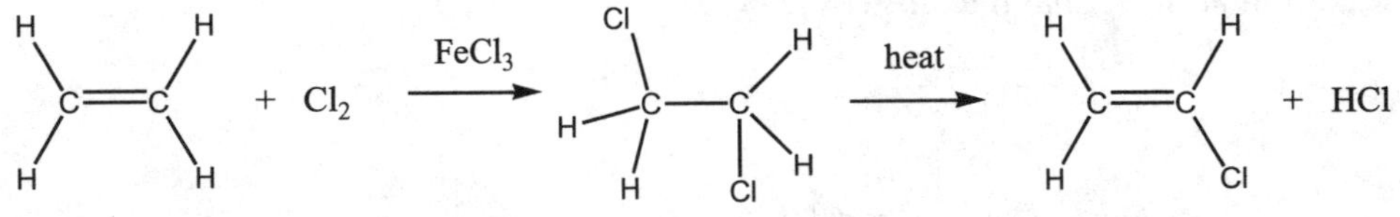

ethene   1,2-dichloroethane   vinyl chloride

a) Identify the oxidizing agent in the first step of the reaction.
b) What is the role of $FeCl_3$?
c) How do the hybridization and geometry around the carbon atoms change during step 2 of this process?
d) Show a portion of the PVC polymer that will form from vinyl chloride. Show four repeat units.
e) Categorize PVC as a plastic, a fiber, or an elastomer.

Answers to Try Its:

1. The lone pair on the carbon is held much less tightly than the lone pair on the oxygen (both are $n=2$, but O has a greater Z_{eff} with two more protons), so carbon's lone pair will be the better Lewis base.
2. Sn^{4+} will be harder than Sn^{2+} because it has fewer electrons. Its electron cloud is held more tightly to the nucleus, making it less polarizable than Sn^{2+}.
3. BH_3 is small and non-polarizable, so it is a hard acid. It will prefer to react with a harder base. N is smaller than P ($n=2$ vs $n=3$, respectively), so NH_3 should be harder than PH_3. BH_3 will prefer to react with NH_3.

Answers to Self-Test:

1. Ag^+ is the Lewis acid and Br^- is the Lewis base.
2. Hard and soft are relative terms, so even though Na^+ is small and therefore relatively hard, it is softer than the extremely hard Al^{3+}. Br^- is much more polarizable, and therefore softer than F^-. The reaction should occur, because in doing so, it would combine hard Al^{3+} with hard F^- and (relatively) soft Na^+ with soft Br^-.
3. S atoms are relatively soft bases, so S will prefer to bind with the softer Lewis acid. Pb^{2+} is softer than Sn^{4+} for two reasons: it is larger, and it has a smaller positive charge. Pb^{2+} will therefore bind more strongly to sulfur (in enzymes) than Sn^{4+}.
4. This reaction should have a large equilibrium constant because the reaction shows the harder H^+ combining with the harder OH^- and the softer CH_3Hg^+ combining with the softer SO_3^{2-}, as is predicted by HSAB theory.
5. The smaller the pK_{sp}, the larger the K_{sp}, the more solid will dissolve in a given amount of water. The most soluble salt is the one in which the ions are least attracted to each other. Ag^+ is a large cation with a low charge, so it is considered a soft Lewis acid. Of the three halide ions shown, I^- is the softest and Cl^- is the hardest. The trend seems reasonable: Ag^+ will be most attracted to I^- and least attracted to Cl^-, making AgCl the most soluble of the set.
6. Hydrogen is $n=1$, so it only has an s orbital to use in bonding. It cannot make two bonds. The central H atoms can better be shown as making half-bonds with the boron atoms:

7. a) Cl_2 is being reduced, so it is the oxidizing agent.
   b) $FeCl_3$ is a heterogeneous catalyst in this reaction.
   c) The carbons start out with sp^3 hybridization in a tetrahedral geometry, and end up with sp^2 hybridization and trigonal planar geometry.
   d)

   e) PVC is a plastic (made from free radical polymerization).

Chapter 22: Nuclear Chemistry and Radiochemistry

Learning Objectives

In this chapter, you will learn how to:

- Predict whether or not a given nuclide is stable.
- Calculate the binding energy of a given nuclide.
- Write a balanced nuclear equation.
- Use the half-life of a given radionuclide to predict the age of an object.
- Distinguish between the different types of nuclear decay processes and why they occur.
- Calculate energy released in a fission or fusion reaction.
- Assess for what useful applications a given radionuclide can be used.

Practical Aspects

In today's highly technological society, everyone can benefit from a basic understanding of nuclear chemistry. Use the basic knowledge you obtain from this chapter to further research nuclear issues, so that you may have an educated opinion about the nuclear reactions that affect your life.

22.1 NUCLEAR STABILITY

Key Terms:
- **Nuclear chemistry** – study of the changes in matter which occur within the nucleus of the atom.
- **Nucleon** – term used in nuclear chemistry to describe particles within the nucleus (protons and neutrons).
- **Nuclide** – term used to describe a given nucleus using its atomic number (Z) and its mass number (A).

Key Concept:
- The mass number (A) is the sum of the protons and neutrons within the atom: $A = Z+N$

EXERCISE 1: Write the nuclear symbol for each: a) the isotope of oxygen that contains 9 neutrons; b) the nuclide that contains 27 protons and 33 neutrons.

> *STRATEGY:* Each nuclide has a specific atomic number and mass number, which must be described in the nuclear symbol.
>
> $$^A_Z \text{Element}$$

> *SOLUTION:*
>
> a) Oxygen is Z=8, and 9 neutrons indicates that A = 17 (A = Z+N): $^{17}_8\text{O}$
>
> b) Z = 27 corresponds to Co. A = Z+N = 27 + 33 = 60: $^{60}_{27}\text{Co}$

Nuclear Binding Energy

Key Term:
- **Binding energy** – energy required to remove a proton or neutron from the nucleus of an atom.

Key Concepts:
- A nucleus is more stable than its separated nucleons.
- The mass of an atom's nucleus is less than the sum of the masses of the individual nucleons.
- Mass is converted into energy when a nucleus forms. A small change in m results in a huge change in E.
- Each nuclide has a unique binding energy. When performing energy calculations involving nuclides, be sure to use the molar mass for the particular *isotope*, and not the element's average molar mass.
- The most stable nuclide of all is ^{56}Fe because it has the highest binding energy per nucleon.
- Activation energy barriers for nuclear reactions are enormous. Coulomb's Law can be used to measure these activation energies.

Useful Relationships:
- $E = mc^2$ or $\Delta E = (\Delta m)c^2$ This is Einstein's equation for relating mass to energy.
- $\Delta E = (\Delta m)(8.988 \times 10^{10} \text{ kJ/g})$ This is a shortcut equation of $\Delta E = (\Delta m)c^2$.
- Proton mass = 1.007276 g/mol; neutron mass = 1.008665 g/mol; electron mass = 0.0005486 g/mol.

EXERCISE 2: Calculate: a) the total binding energy, and b) the binding energy per nucleon, for one mole of sodium-23 (isotopic mass = 22.989770 g/mol). Report answers in units of "kJ/mol."

STRATEGY: The binding energy depends upon the difference in mass between the individual components of the atom and the nuclide. We're given the nuclide mass (22.989770 g/mol). We can use $\Delta E = (\Delta m)c^2$ or the shortcut $\Delta E = (\Delta m)(8.988 \times 10^{10} \text{ kJ/g})$ to find the total binding energy.

Sodium-23 contains eleven protons, eleven electrons and twelve neutrons:

Protons:	11(1.007276 g/mol) =	11.080036 g/mol
Electrons:	11(0.0005486 g/mol) =	0.0060346 g/mol
Neutrons:	12(1.008665 g/mol) =	12.103980 g/mol
	Total molar mass =	23.190051 g/mol

The total binding energy can be found from:
$$\Delta E = (\Delta m)(8.988 \times 10^{10} \text{ kJ/g})$$
$$= (22.989770 \text{ g/mol} - 23.190051 \text{ g/mol})(8.988 \times 10^{10} \text{ kJ/g})$$
$$= -1.800 \times 10^{10} \text{ kJ/mol}$$
(The nuclide is more stable than its individual components, so energy is released.)

Divide the total binding energy by the number of nucleons to get the binding energy per nucleon:
$$(-1.800 \times 10^{10} \text{ kJ/mol})/23 = -7.827 \times 10^8 \text{ kJ/mol of nucleons}$$

SOLUTION: a) The total binding energy of sodium-23 is -1.800×10^{10} kJ/mol. b) The binding energy per nucleon is -7.827×10^8 kJ/mol of nucleons. These answers seem reasonable. The total binding energy of ^{23}Na is greater than that of ^{4}He (-2.731×10^9 kJ/mol), as expected. Also, the binding energy per nucleon is slightly less negative than that of iron-56, the most stable nuclide.

EXERCISE 3: In a neutron capture reaction, a free neutron collides with a nuclide, and the nuclide incorporates the neutron. What activation barrier does this reaction have?

STRATEGY: Use Coulomb's Law to predict the barrier: $E = kq_1q_2/r$.

SOLUTION: Coulomb's law shows that the energy barrier depends upon the charges of the two particles brought together. A neutron has no charge to it, so $q_{neutron} = 0$. E is therefore 0. A neutron capture reaction has no energy barrier. (Incidentally, this is the only type of nuclear reaction that has no energy barrier to it.)

Stable Nuclides

Key Term:
- **Belt of stability** – the region within a plot of neutrons vs. protons for all known nuclides that defines which nuclides are stable. See Figure 22-5 in the text for the plot and its belt of stability.

Key Concepts:
- Lighter stable nuclides have a N/Z ratio of 1:1. The ratio increases up to 1.54:1 for the heaviest stable nuclides.
- Nuclides with even numbers of protons and neutrons tend to be stable, while nuclides with odd numbers of protons and neutrons tend to be unstable.
- There are no stable nuclides above Z=83 (bismuth).

EXERCISE 4: Predict which nuclides are stable and which are unstable: a) $^{14}_{6}C$; b) $^{60}_{27}Co$; c) $^{81}_{35}Br$

STRATEGY: Stable nuclides have N/Z ratios of 1:1 up to 1.54:1. Use Figure 22-5 in the text as a guide for assessing borderline cases. Remember that odd-odd nuclides are typically unstable and any nuclide over Z=83 is unstable.

SOLUTION:

a) $^{14}_{6}C$: The N/Z ratio of $^{14}_{6}C$ is 8:6 = 1.33:1. The ratio of neutrons to protons is too high for such a small element. This won't be a stable nuclide.

b) $^{60}_{27}Co$: The N/Z ratio of $^{60}_{27}Co$ is 33:27 = 1.22:1. From Figure 22-5 in the text, the ratio looks as if it lies within the belt of stability. However, this nuclide contains an odd number of protons (27) and neutrons (33), so it is probably unstable.

c) $^{81}_{35}Br$: The N/Z ratio of $^{81}_{35}Br$ is 46:35 = 1.31:1. This appears to lie within the belt of stability, so it is probably a stable nuclide.

Try It #1: Do you predict this nuclide to be stable or unstable: $^{210}_{85}At$?

22.2 NUCLEAR DECAY

Key Terms:
- **Nuclear decay** – the spontaneous decomposition of one nuclide into a new, more stable, nuclide.
- **Radioactive** – term used to describe nuclides that undergo nuclear decay.

Key Concepts:

- The driving force for radioactive decay is to create a stable nucleus.
- Three things characterize a given nuclear decay (and will be covered in this section):
    1) the product of the nuclear decay reaction;
    2) the type of energy released during the nuclear decay; and
    3) the rate of the nuclide's decay.
- Matter is converted into energy during a nuclear decay process.
- For any given radioactive decay, the total mass number and total electrical charge are conserved. Therefore, a balanced nuclear equation exists when:
    - sum of subscripts in reactants equals sum of subscripts of products (accounts for charge).
    - sum of superscripts in reactants equals sum of superscripts of products (accounts for mass).

Decay Processes

Summary of the Five Types of Nuclear Decay Processes

Name	Symbol	Equivalent To:	Why it Occurs:	What is Happening During this Process:
Alpha Emission	$^4_2\alpha$	A helium nucleus (He^{2+})	The nuclide is trying to decrease its overall mass (it is too massive to be stable).	He^{2+} gets ejected from the nucleus.
Beta Emission	$^0_{-1}\beta$	An electron $^0_{-1}e^- = {}^0_{-1}\beta$	The N:Z ratio is too *high* and the nuclide is trying to get it back into balance.	A neutron converts to a proton and ejects an electron from nucleus: $$^1_0n \rightarrow {}^1_1p + {}^0_{-1}e^-$$
Positron Emission	$^0_{+1}\beta$	Particle with same mass as an electron, but with a positive charge.	The N:Z ratio is too *low* and the nuclide is trying to get it back into balance.	A proton converts to a neutron and a positron: $$^1_1p \rightarrow {}^1_0n + {}^0_{+1}\beta$$ The positron annihilates an electron. Two photons of γ ray energy are released: $$^0_{-1}e^- + {}^0_{+1}\beta \rightarrow 2\,{}^0_0\gamma$$
Electron Capture	$^0_{-1}e^-$	An electron	The N:Z ratio is too *low* and the nuclide is trying to get it back into balance.	A 1s electron is captured by the nucleus. The electron combines with the proton to make a neutron. $$^1_1p + {}^0_{-1}e^- \rightarrow {}^1_0n$$ A high-energy photon is released (X-ray region) when the 1s hole is filled.
Gamma Ray	$^0_0\gamma$	A high-energy photon (γ region)	Usually, another type of emission has occurred and the nuclide has extra "excitation" energy to release.	Excitation energy is released as γ rays.

Key Term:
- **Metastable** – term used to describe a highly unstable, excited nuclide that will undergo γ emission. A metastable nuclide is indicated with a superscript "m" after the nuclide symbol.

Key Concepts:
- Positron emission and electron capture are not observed directly. Rather, high energy photon emission indicates if these processes have occurred.
- In a balanced nuclear equation, always indicate electron *capture* as a *reactant* and an *emission* as a *product*. For example, Radon-222 undergoing alpha emission would be shown like this:

$$^{222}_{86}Rn \rightarrow \, ^{218}_{84}Po + \, ^4_2\alpha.$$

- The type of decay process is an indicator of the energy released during the process. This will be covered in Section 22.6.
- If the product of a nuclear decay is also unstable, then it, too, will undergo nuclear decay. The process continues until a non-radioactive substance is produced. This often occurs with particularly heavy elements.

EXERCISE 5: Recall that odd-odd nuclides tend to be unstable. What types of decay can an odd-odd nuclide undergo, in order to stabilize itself to an even-even nuclide?

STRATEGY: Let's deal with the protons first: If a nuclide has an odd number of protons, then to form a nuclide with an even number of protons would require that it gain or lose one proton. The decay processes that change the atomic number by one are: beta emission ($^0_{-1}\beta$), positron emission ($^0_{+1}\beta$), and electron capture ($^0_{-1}e^-$).

Now to address the number of neutrons. In each decay process mentioned, the mass number of each decay particle is zero. When the number of protons changes by 1, the number of neutrons does too – this maintains a zero change in mass. Therefore, all three emission types work.

SOLUTION: Beta emission, positron emission, or electron capture will convert an odd-odd nuclide into an even-even nuclide.

EXERCISE 6: What type of emission is predicted for each radioisotope? a) ^{239}Pu; b) ^{14}C; c) ^{99}Tc^m

STRATEGY: Determine why the nuclide is unstable, and predict the best type of decay process to eliminate that instability.

SOLUTION:
a) ^{239}Pu – the overall mass is too high. It will probably undergo α decay to decrease the overall mass.
b) ^{14}C – the N:Z ratio is too high. β emission simultaneously decreases the number of neutrons *and* increases the number of protons by one, so beta emission is the mode of decay that ^{14}C will undergo.
c) ^{99}Tc^m: m = metastable. This nuclide will undergo γ-ray decay to eliminate its excitation energy.

EXERCISE 7: Write a nuclear reaction to illustrate these nuclear decays:
   a) Plutonium-239 undergoes alpha decay;   b) Carbon-14 is a beta-emitter;
   c) ^{99}Tc^m undergoing decay;   d) ^{40}K undergoes electron capture.

STRATEGY: Determine the Z and A for the nuclide and write its symbol. This is the reactant. The products will be determined from the decay type and what remains.

a) $^{239}_{94}Pu \rightarrow ^{4}_{2}\alpha + ^{239-4}_{94-2}X$; so X has atomic #92. The product formed is $^{235}_{92}U$.

b) $^{14}_{6}C \rightarrow ^{0}_{-1}\beta + ^{14-0}_{6-(-1)}X$; so X has atomic #7. The product formed is: $^{14}_{7}N$

c) $^{99}_{43}Tc^{m} \rightarrow ^{0}_{0}\gamma + ^{99}_{43}Tc$; a metastable substance will release a γ - ray.

d) $^{40}_{19}K + ^{0}_{-1}e^{-} \rightarrow ^{40-0}_{19+(-1)}X$; so X has atomic #18. The product formed is: $^{40}_{18}Ar$

SOLUTION: a) $^{239}_{94}Pu \rightarrow ^{4}_{2}\alpha + ^{235}_{92}U$; b) $^{14}_{6}C \rightarrow ^{0}_{-1}\beta + ^{14}_{7}N$; c) $^{99}_{43}Tc^{m} \rightarrow ^{0}_{0}\gamma + ^{99}_{43}Tc$

d) $^{40}_{19}K + ^{0}_{-1}e^{-} \rightarrow ^{40}_{18}Ar$

Try It #2: Determine the identity of the nuclide formed in each process: a) Krypton-92 undergoes beta decay. It also releases a gamma ray in the process. b) ^{26}Si undergoes positron emission.

Rates of Nuclear Decay

Key Term:
- **Half-Life** – time it takes for half of the radioactive substance to undergo nuclear decay.

Key Concepts:
- Radioactive decay follows first-order kinetics; therefore the half-life of a radioisotope is constant and can be determined from the relationship: $t_{1/2} = \ln2/k$.
- The half-life for a radioisotope is constant. (Recall that half-lives for first-order processes do not change.)
- Half-lives of radioisotopes can range from a fraction of a second to over a billion years.
- Particles emitted during a nuclear decay are so energetic that it is possible to measure them individually. (See Useful Relationship below.)

Useful Relationship:
- $\ln\left(\dfrac{N_0}{N}\right) = \dfrac{t \ln 2}{t_{1/2}}$. This equation is derived from $t_{1/2} = \ln2/k$ and is used to relate the number of

emissions that occur at the present time (N) to the number of emissions that occurred at t=0 (N_0).

EXERCISE 8: The half-life of Thorium-234 is 24.10 days. What fraction of a sample of Thorium-234 remains after 472 hours?

STRATEGY: The fraction of ^{234}Th remaining after a given time will be N/N_0. First, convert the times into the same units: 472 hours = 19.67 days. Use the relationship between half-life and amount of nuclide present:

$$\ln\left(\frac{N_0}{N}\right) = \frac{t \ln 2}{t_{1/2}} = \frac{19.67 \text{ days} (0.693)}{24.10 \text{ days}} = 0.56561$$

$$\left(\frac{N_0}{N}\right) = e^{0.56561} = 1.7605; \text{ The fraction remaining} = \frac{N}{N_0} = \frac{1}{1.7605} = 0.568$$

SOLUTION: 0.568 (or 56.8%) of the original ^{234}Th remains after 472 hours. This seems reasonable, because less than one half-life has passed.

Try It #3: What is the half-life of Cobalt-60, given that 16.887% of it remains after 13.50 years?

22.3   INDUCED NUCLEAR REACTIONS

Key Terms:

- **Induced nuclear reaction** – nuclear reaction that occurs as a result of two particles colliding. Two types of induced reactions are:
  - **Neutron capture** –induced reaction in which a nuclide picks up a free neutron. A metastable nuclide is generated, which in turn emits either a γ-ray or proton.
  - **Bombardment** – induced reaction in which two nuclides collide and generate a new nuclide.
- **Compound nucleus** – the product of a reaction between two nuclei, which also emits energy by emitting a small particle such as a proton, neutron, or beta particle.

Key Concepts:

- Neutron-capture reactions are always exothermic because the incoming neutron feels such a strong attraction to the nucleus.
- Free neutrons are not readily available, so neutron capture reactions typically occur within laboratories (using equipment designed to generate free neutrons) or in the upper atmosphere.
- Two particles must be accelerated to tremendous speeds in order to successfully form a compound nucleus when they collide. (They must overcome the repulsive forces of bringing the two positively charged nuclei together.) Two types of accelerators used are:
  - **Cyclotron** (useful for generating elements with atomic numbers under 101).
  - **Linear accelerator** (useful for generating elements with atomic numbers over 101).

Helpful Hint

- Figures 22-7 and 22-8 in the text show how a cyclotron and a linear accelerator work.

EXERCISE 9:   Classify each type of reaction and determine the missing piece(s) in each equation:

a) $^{238}_{92}U + ^{1}_{0}n \rightarrow ^{239}_{94}Pu + X$;   b) $^{14}_{7}N + ^{4}_{2}\alpha \rightarrow [?] \rightarrow ^{1}_{1}p + X$;   c) $^{10}_{5}B + ^{1}_{0}n \rightarrow [?] \rightarrow ^{4}_{2}\alpha + X$

STRATEGY: In a balanced nuclear equation, charge and mass must be conserved. The identity of an unknown nuclide is determined by the atomic number.

SOLUTION:
a) This is a neutron-capture reaction. A compound nucleus is formed during this process (^{239}U), but it is not shown in the overall equation.

$^{238}_{92}U + ^{1}_{0}n \rightarrow ^{239}_{94}Pu + ^{238+1-239}_{92-94}X$;   X has no mass and a charge of - 2, which corresponds to 2 $^{0}_{-1}\beta$ particles.

b) This is a bombardment reaction of nitrogen-14 with an alpha particle. A compound nucleus is formed, with an atomic number of (7+2)=9 and an atomic mass of (14+4)=18. The compound nucleus is ^{18}F. A proton is then emitted as well as another nuclide, X.

$$^{14+4-1}_{7+2-1}X = \, ^{17}_{8}X; \text{ atomic } \#8 \text{ is oxygen, so the nuclide is } ^{17}_{8}O.$$

c) This is a neutron-capture reaction. $^{10}_{5}B + \, ^{1}_{0}n \rightarrow [?] \rightarrow \, ^{4}_{2}\alpha + X$. The compound nucleus is $^{11}_{5}B$ and the unknown final product is $^{7}_{3}Li$.

Try It #4: Dubnium-260 can be made by bombarding californium-249 with a high energy ^{15}N ion. a) Identify the compound nucleus formed. b) What other particles are produced in the process?

22.4 NUCLEAR FISSION

Key Terms:

- **Fission** – splitting of an atomic nucleus into two or more smaller nuclides. At least one neutron is also generated.
- **Chain reaction** – continuing series of nuclear fissions.
- **Critical mass** – minimum mass of fissionable material needed to sustain a chain reaction.

Key Concepts:

- Fission follows neutron capture by a large nuclide. Naturally occurring ^{235}U and artificially generated ^{239}Pu or ^{233}U are fissionable materials.
- Unlike natural nuclear decays that produce one specific nuclide, fission reactions yield a large range of product nuclides.
- Fission releases an enormous amount of energy, much more than in a natural nuclear decay, or a non-nuclear reaction.
- In order for a chain reaction to occur, at least one neutron generated in a fission process must be able to be absorbed by another fissionable nuclide.
- The natural isotopic abundance of ^{235}U is too low to sustain a chain reaction and therefore must be purified before it can be used for a nuclear fission process.
- ^{239}Pu is made from neutron capture by ^{238}U.

EXERCISE 10: What is the missing nuclide in this fission process?

$$^{235}_{92}U + \, ^{1}_{0}n \rightarrow \, ^{94}_{36}Kr + X + 3 \, ^{1}_{0}n$$

STRATEGY: Conserve mass and charge in the balanced nuclear equation. Notice that three neutrons are produced and must be accounted for in the reaction:

$$^{235}_{92}U + \, ^{1}_{0}n \rightarrow \, ^{94}_{36}Kr + \, ^{235+1-94-1-1-1}_{92-36-0}X + 3 \, ^{1}_{0}n; \text{ so X has atomic } \# = 56$$

SOLUTION: The missing nuclide is $^{139}_{56}Ba$.

Try It #5: Consider the reaction in Exercise 10. Indicate the features in the reaction that define it as a nuclear fission process.

EXERCISE 11: Calculate the amount of energy released during the fission of 1 g of ^{235}U via the process shown in Exercise 10. The isotopic masses are provided here: ^{235}U =235.043625 g/mol , ^{94}Kr = 93.9343620 g/mol, ^{139}Ba = 138.908836 g/mol.

STRATEGY: We're asked to find the energy released from the fission of 1 g of ^{235}U by this process:

$$^{235}_{92}U + ^{1}_{0}n \rightarrow ^{94}_{36}Kr + ^{139}_{56}Ba + 3 ^{1}_{0}n$$

This is a nuclear change, so mass is not conserved. We must find Δm for the process and determine how much energy it generates using $\Delta E = (\Delta m)c^2$.

$$\Delta m = \Sigma \text{ of isotopic masses of products} - \Sigma \text{ of isotopic masses of reactants}$$
$$= [93.9343620 + 138.908836 + 3(1.008665)] - [235.043625 + 1.008665]$$
$$= -0.183097 \text{ g/mol of reaction}$$

$$\Delta E = (\Delta m)c^2 = \frac{-0.183097 \text{ g}}{\text{mol of rxn}} \times \left(\frac{2.998 \times 10^8 \text{ m}}{\text{sec}}\right)^2 \times \frac{1 \text{ kg}}{1000 \text{ g}} = -1.64568 \times 10^{13} \text{ J / mol of rxn}$$

$$\frac{-1.64568 \times 10^{13} \text{ J}}{\text{mol of rxn}} \times \frac{1 \text{ mol rxn}}{1 \text{ mol } ^{235}U} \times \frac{1 \text{ mol}}{235.043625 \text{ g}} = -7.002 \times 10^{10} \text{ J/g } ^{235}U$$

SOLUTION: The reaction releases -7.002×10^{10} J/g. This number seems reasonable because it is so tremendously large.

Nuclear Weapons & Nuclear Power Plants

Key Concepts:
- A nuclear bomb uses an *uncontrolled* nuclear fission chain reaction. This is achieved by rapidly pushing together sub-critical portions of fissionable material to create a critical mass, and then forcing that critical mass to stay together until the chain reaction escalates exponentially.

- A nuclear power plant is a *controlled* nuclear fission chain reaction. The chain reaction is controlled by maintaining a slow, constant rate of nuclear fissions. This is achieved by raising and lowering "control rods" (neutron absorbers such as ^{10}B) into the fissionable material.

Helpful Hint
- Figures 22-12 and 22-13 in the text illustrate how the fission process is used in a nuclear power plant.

EXERCISE 12: Draw schematic pictures of nuclear fission processes to differentiate between the chain reactions that occur within: a) a nuclear bomb; and b) a nuclear power plant. Show two steps of the chain reaction, with three neutrons generated per step.

STRATEGY: A nuclear bomb is an uncontrolled nuclear fission, so every neutron generated in the first fission could induce another fission process. A nuclear power plant uses controlled nuclear fission; the number of free neutrons is regulated with neutron absorbers, so that fission occurs at a controlled rate.

SOLUTION: Using "Fis" to represent the fissionable material, "P" to represent a fission product; "n" to represent a neutron, and "A" to represent a neutron absorber:

a) nuclear bomb   b) nuclear power plant

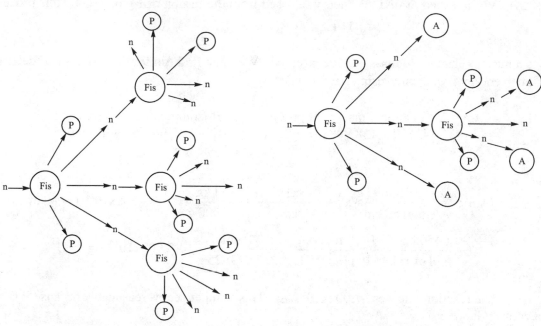

Notice in the figures that after two fission steps, the nuclear bomb has generated nine neutrons that can initiate nine more nuclear fissions, while the nuclear power plant shows one neutron available for the next nuclear fission.

Try It #6: Consider the uncontrolled nuclear fission process (nuclear bomb) in the drawing in Exercise 12. a) How many neutrons would be generated in the third step of the fission process? b) How many neutrons would be generated in the fourth step?

22.5 NUCLEAR FUSION

Key Terms:
- **Nuclear fusion** – the joining or fusing of two small nuclides to form a larger nuclide. Fusion reactions typically release even more energy than fission reactions.
- **Threshold temperature** – minimum temperature required for a given fusion reaction to occur.
- **First-generation star** – star that is formed from the fusion of hydrogen atoms. Further cycles of fusion reactions eventually occur to produce nuclides up to ^{56}Fe.
- **Second-generation star** – star that is formed from the products of a first-generation star that had exploded in a supernova. Here, hydrogen *and* heavier nuclei undergo the initial fusion reactions.
- **Third-generation star** – star formed from the products of a second-generation star that had exploded in a supernova. Here, much heavier nuclei are available in the initial fusion reactions than in a first- or second-generation star.

Key Concepts:
- In order for fusion to occur, the two colliding nuclides must have incredibly high kinetic energies (i.e., they must be at incredibly high temperatures). The energy released from a fission reaction is sometimes used to initiate a fusion reaction.
- Starting materials for fusion reactions are readily available and the products are stable, so nuclear waste contamination and disposal are not a concern.
- Fusion reactions occur at such high temperatures that they melt any container used to hold them. One way to contain a fusion reaction is to suspend the materials in a magnetic field, so that they don't actually come in contact with the container.

EXERCISE 13: Which would require higher temperatures for the fusion reaction to occur: fusion of two ^{4}He atoms or fusion of two ^{12}C atoms?

STRATEGY: We're asked to find which of two fusion processes has the higher threshold temperature. The question just asks for *relative* temperatures, which we can establish by determining the *relative* nuclear repulsive forces that must be overcome to get the two nuclei to fuse. The repulsive forces indicate the kinetic energy (i.e., temperature) needed by the nuclei for fusion to occur. We can approximate the energy requirements using Coulomb's law: $E = kq_1q_2/r$.

The "q"s are the nuclear charges of the atoms, 2+ for helium and 6+ for carbon. "r" is the distance between the centers of the two nuclei when they are brought together to the point of just touching. An atom's nucleus is roughly 10,000 times smaller than the entire atom (an estimation from Chapter 2). The nuclei are proportionally smaller than the atoms, so we can simply use atomic radii to establish *relative* energies. Helium's atomic radius is 32 pm and carbon's is 77 pm.

$$E_{He} = \frac{kq_1q_2}{r} = \frac{k(2+)(2+)}{2(32)} = 0.0625k/pm \qquad E_C = \frac{kq_1q_2}{r} = \frac{k(6+)(6+)}{2(77)} = 0.234k/pm$$

SOLUTION: Both energies are positive, which indicates repulsive forces that must be overcome. Carbon's energy is higher than helium's, indicating that the fusion of two ^{12}C atoms requires a higher threshold temperature than the fusion of two ^{4}He atoms. (Read about first-generation stars in this section of the text and you will see that these two processes do occur at different temperatures.)

22.6 EFFECTS OF RADIATION

Key Terms:
- **Ionizing radiation** – radiation that generates ions as it passes through matter. For example, one α particle can generate 150,000 cations. α particles, β particles, γ-rays, and X-rays are all forms of ionizing radiation.
- **rem** – unit of measurement to describe radiation exposure. This unit is based upon the combined exposure risks associated with the amount, the type, and the energy content of the specific radiation source.

Key Concepts:
- Cells that divide most rapidly tend to be most easily damaged by ionizing radiation. This includes hair cells, gastrointestinal tract cells, blood cells, and cancer cells.

- Long-term effects of radiation exposure include irreversible damage to DNA (that can be passed on to offspring) and increased risk of cancer.
- Different types of emitters require different levels of shielding for protection: α particles can be stopped by a piece of paper, β particles require a medium-thick piece of aluminum foil, and γ-rays require a thick piece of lead.

22.7 APPLICATIONS OF RADIOACTIVITY

Nuclear chemistry knowledge can be applied to a wide variety of applications. The radionuclide's mode of decay, its decay product, and its half-life will determine for what applications that particular nuclide should be used.

Dating Using Radioactivity

Key Concepts:
- The isotope or its decay product(s) must be present in the object to be dated.
- There must be a detectable amount of the isotope present within the object. (i.e., the radioisotope's half-life is important).
- Carbon-14 dating is used to date previously-living substances. A living organism has a steady concentration of ^{14}C in it because it is constantly exchanging carbon with the environment. When the organism dies, it no longer takes in ^{14}C, so the loss of ^{14}C can be monitored. The amount of ^{14}C in the sample is compared to the amount of ^{14}C in a present-day sample. (The half-life of ^{14}C is 5730 years.)
- Moon rocks, meteorites, and earth rocks can be dated using ^{238}U (half-life = 4.51×10^9 years) or ^{40}K (half-life = 1.28×10^9 years). The dating is done by taking a ratio of the original nuclide to one of its decay products. For example, the ratio of ^{40}K to its decay product ^{40}Ar, can identify the age of a rock.

EXERCISE 14: A case of 1940 Chateau M. is going to be auctioned tomorrow. It is considered to be one of the finest wines in the world. Radioactive dating has been performed on the wine to verify its authenticity. The dating was performed by measuring for tritium, ^3H, which has a half-life of 12.3 years. Given the data in the table below, determine whether or not the wine was made in 1940.

Sample	This sample contains:
1940 Chateau M.	1.91 nanograms of tritium
2000 wine, grown in the same vineyard	40.0 nanograms of tritium

STRATEGY: We're asked to determine the wine's authenticity. One way to assess this is to determine if 1.91 ng correlates to 60 years (2000-1940). Use the relationship for calculating amounts from half-lives:

$$\ln\left(\frac{N_0}{N}\right) = \frac{t \ln 2}{t_{1/2}} ; \ln\left(\frac{40.0\,\text{ng}}{1.91\,\text{ng}}\right) = \frac{0.693t}{12.3\,\text{years}} ; \text{solving for t, we get 54 years.}$$

SOLUTION: This wine was made in 1946, not 1940. This wine is a fake; don't buy it!

Note: We could also have determined the authenticity by calculating the amount of tritium expected to remain after 60 years has passed. That number would have been 1.36 ng, which does not correlate to the 1.91 ng present in the wine, indicating the wine is a fake.

Other Applications

Radioactive Tracers
- ^3H_2O can be used to detect the location of an underground pipe leak.
- ^{24}Na (a β and γ emitter with a half-life of 15.0 hrs) can be used to detect obstructions in the circulatory system.
- Melvin Calvin mapped out the steps of photosynthesis using ^{14}C as a tracer.

Food Irradiation
- Irradiation of food using ^{60}Co (a γ and β emitter) can kill molds, bacteria, and pathogens like salmonella and listeria. The ^{60}Co is set up in a similar fashion to ^{60}Co in radiation treatment.

EXERCISE 15: a) Why is ^{60}Co a good choice for food irradiation and radiation therapy? b) What two features do ^{24}Na and ^{11}C share that make them good nuclides to inject into a patient for medical diagnostic purposes?

STRATEGY: The type of emitter, the product formed, and the half-life may influence for what applications a given nuclide is used.

SOLUTION: a) ^{60}Co emits γ rays, which will pass right through a patient's body or a food sample. Thus, the radiation won't stay inside the food or patient. The half-life of ^{60}Co is 5.26 years, so a single source of ^{60}Co can be used for a long period of time.

b) ^{24}Na and ^{11}C both have short half-lives and are γ–ray emitters. (^{24}Na also emits β particles.) Short half-lives guarantee that the nuclides will decay to safe levels within a very short period of time (days). γ- or β-emitting nuclides are used for medical diagnoses because they have such strong penetrating power that, once injected, they can pass through tissue and exit the patient's body. α-emitters, which can be stopped by a piece of paper, are not used in situations where the nuclide is injected into or ingested by the patient.

Chapter 22 Self-Test

You may use a periodic table and calculator for the test.

Proton mass = 1.007276 g/mol; neutron mass = 1.008665 g/mol; electron mass = 0.0005486 g/mol.

1. Is ^{10}Be expected to be stable?

2. ^{30}P undergoes positron emission. What new nuclide is formed?

3. Write a nuclear reaction to illustrate the beta decay of tritium (^{3}H).

4. When uranium-238 is bombarded with deuterons (^{2}H), a compound nucleus is formed, which decomposes to a new nuclide, X, and a proton.

   a) Identify the compound nucleus.
   b) Identify the new nuclide.
   c) Is the new nuclide stable?

5. Calculate the energy released by the two γ-rays produced during positron emission when a positron annihilates an electron.

6. Americium-241 is used in smoke detectors in this manner: a steady stream of nuclear emissions from ^{241}Am hits a target detector. The alarm is triggered if the stream is disrupted by any particulates such as smoke or water vapor. What type of emitter is ^{241}Am?

7. Propose an experiment using ^{131}I which would indicate that the process shown below is dynamic:
$$AgI \text{ (s)} \leftrightarrow Ag^{+} \text{(aq)} + I^{-} \text{(aq)}$$

8. ^{11}C has a half-life of 20.3 min and is used in positron emission tomography. Given that it takes 54.2 minutes from the time the ^{11}C is generated until it is injected into the patient,

   a) what percentage of the ^{11}C generated is actually present at the time of injection?
   b) what percentage of the original ^{11}C remains inside the person after 8 hours?

Answers to Try Its:

1. unstable; Z > 83, and it contains odd numbers of protons and neutrons.
2. a) $^{92}_{37}Rb$; b) $^{26}_{13}Al$
3. 5.261 years
4. a) $^{264}_{105}Db$; b) 4 neutrons. (Notice we need 0 charge units and 4 mass units, so 4 $^{1}_{0}n$.)
5. features: neutron capture; formation of two, smaller nuclides; generation of at least one new neutron.
6. a) 27 neutrons; b) 81 neutrons

Answers to Self-Test:

1. No, ^{10}Be would be unstable because its N/Z = 1.50, which is too high for such a small element.
2. $^{30}_{14}Si$
3. $^{3}_{1}H \rightarrow {}^{0}_{-1}\beta + {}^{3}_{2}He$
4. a) $^{240}_{93}Np$; b) $^{239}_{92}U$; c) no, U has Z > 83, so it won't be stable.
5. 1.638×10^{-16} kJ (note: a positron and an electron have the same mass. Zero mass remains after the two particles annihilate each other.)
6. It is an alpha emitter: 1) the nuclide is too large to be stable, so it will try to decrease its overall mass with alpha emission; and 2) alpha particles are easily stopped by particulates like smoke, so the detector signal can be interrupted. Beta or γ-rays would go through the smoke or water, so they could not be used in this type of detector.
7. Prepare a saturated solution of AgI, leaving some solid in the bottom of the container. Prepare another saturated solution of AgI, but this time use AgI labeled with $^{131}I^-$. Gently pour some of this second solution into the first and let it sit for a couple of hours. Separate the solid from the liquid above it, and test the solid for radioactivity. The solid originally was not radioactive, so if dynamic equilibrium exists, then some $Ag^{131}I$ (s) would have formed during the waiting period and would be detected in the solid.
8. a) 15.7%; b) 0.00000765%